Energy Storage Options and Their Environmental Impact

ISSUES IN ENVIRONMENTAL SCIENCE AND TECHNOLOGY

TITLES IN THE SERIES:

How to obtain future titles on publication

A subscription is available for this series. This will bring delivery of each new volume immediately on publication and also provide you with online access to each title *via* the Internet. For further information visit http://www.rsc.org/issues or write to the address below.

For further information please contact:
Sales and Customer Care, Royal Society of Chemistry, Thomas Graham House, Science Park, Milton Road, Cambridge, CB4 0WF, UK
Telephone: +44 (0)1223 432360, Fax: +44 (0)1223 426017, Email: booksales@rsc.org
Visit our website at www.rsc.org/books

ISSUES IN ENVIRONMENTAL SCIENCE AND TECHNOLOGY

EDITORS: R. E. HESTER AND R. M. HARRISON

46
Energy Storage Options and Their Environmental Impact

ROYAL SOCIETY
OF CHEMISTRY

Issues in Environmental Science and Technology No. 46

Print ISBN: 978-1-78801-399-4
PDF ISBN: 978-1-78801-553-0
EPUB ISBN: 978-1-78801-627-8
Print ISSN: 1350-7583
Electronic ISSN: 1465-1874

A catalogue record for this book is available from the British Library

The Royal Society of Chemistry is a charity, registered in England and Wales, Number 207890, and a company incorporated in England by Royal Charter (Registered No. RC000524), registered office: Burlington House, Piccadilly, London W1J 0BA, UK, Telephone: +44 (0) 20 7437 8656.

For further information see our web site at www.rsc.org

Printed in the United Kingdom by CPI Group (UK) Ltd, Croydon, CR0 4YY, UK

Preface

The growth of renewable energy technologies, mainly wind and solar, demands the development of practical and economically viable energy storage technologies in order to balance the availability of supply with demand. This book explores the current state-of-the-art of energy storage and examines the likely environmental impacts of the main categories based on the types of energy involved.

The first chapter, written by Robert Lynch of the University of Limerick, Ireland, and his collaborators, provides an overview of energy sources and electricity supply grids, including fossil fuels, nuclear, renewables, biomass and geothermal sources. Relevant issues such as variable demand, smart grids, distributed generation, ramp times and the growth of electric vehicles are explored in relation to such issues as load levelling and stability of electricity grids. By far the largest providers of mechanical energy storage are pumped hydroelectric storage (PHS) and compressed air energy storage (CAES); these methods are examined in some detail in the second chapter by David Evans of the British Geological Survey (BGS) and his colleagues. Both storage technologies are well developed and offer the potential for better integration and penetration of renewable electricity sources and thus the reduction of greenhouse gas (GHG) emissions to the atmosphere. The chapter reviews the issues relating to the general operational parameters and also the legislative and environmental aspects of the two storage types, mainly in the context of the UK but also with reference to worldwide developments.

The third chapter, written by Noel Buckley of the University of Limerick and his collaborators, provides a wide-ranging overview of electrochemical storage systems with a focus on the three important types: rechargeable batteries, fuel cells and flow batteries. The relative strengths and weaknesses of these, from lead–acid batteries, through lithium and lithium-ion batteries, sodium–sulfur and nickel–metal hydride types, to the several kinds of fuel cell and flow battery are considered from the points of view of both the technologies and their applications and also their potential environmental impacts. This is followed in the fourth chapter, written by Fernando Rhen,

Issues in Environmental Science and Technology No. 46
Energy Storage Options and Their Environmental Impact
Edited by R.E. Hester and R.M. Harrison
© The Royal Society of Chemistry 2019
Published by the Royal Society of Chemistry, www.rsc.org

also of the University of Limerick, and his collaborators, by a review of electrical storage technologies involving devices such as supercapacitors and supercapatteries, flywheels, superconducting magnets and synchronous condensers. These devices, having fast response times, can serve to correct short-duration fluctuations of electricity supply and demand that can cause instability in electrical grids and utility systems.

In the fifth chapter, written by Alexander Cowan and his colleagues at the University of Liverpool, photochemical energy storage methods are reviewed. This chapter highlights energy storage strategies that utilise solar energy to drive the formation of chemicals, fuels and feedstocks. The production of solar fuels that can be stored and transported is an attractive way to address the intermittency of terrestrial solar and provide sustainable access to the fundamental feedstocks on which society has come to rely. The solar energy-driven reactions considered here are the splitting of water to produce hydrogen and oxygen, and the coupled oxidation of water and reduction of CO_2 to produce a variety of higher-value carbon products and oxygen. The chapter provides an introductory overview to both direct (photochemical) and indirect solar (photovoltaic-enabled electrolysis) routes to these fuels. The sixth chapter, the final chapter on storage technologies, deals with thermal (sensible heat and latent heat) and thermochemical energy storage and is written by Yukitaka Kato and Takahiro Nomura of the Tokyo Institute of Technology and Hokkaido University, respectively. This introduces the concepts of phase-change materials and chemical heat pumps and discusses encapsulation, composite materials, heat exchangers, application to concentrated solar power plants and the various types of chemical heat pump currently under development.

Smart energy systems are reviewed in the seventh chapter, by Rasmus Lund and his colleagues at Aalborg University, Denmark. A smart energy system is a combination of the currently isolated energy sectors, such as electricity, heating and transport, and it includes three smart energy grid infrastructures, namely the electric, thermal and gas grids. These grids connect the energy resources with the demands, energy production, energy storage and interconnection points. From the case studies examined in detail, it is concluded that hydroelectric storage, batteries in electric vehicles, thermal storage in district heating systems and storage of renewable electrofuels are important and provide a cost-efficient flexibility to the overall energy system, although large-scale batteries on the grid level and stationary batteries in buildings are not feasible from an energy system perspective. In the eighth chapter, Heidi Hottenroth of Pforzheim University, Germany, and her colleagues review the application of life-cycle assessment (LCA) for determining environmental impact in the context of stationary energy storage systems. The LCA technique is applied to three different case studies involving pumped hydroelectric storage, lithium-ion batteries and combined heat and power plants in order to determine which energy storage system is the best option in a specific setting. Finally, business opportunities and the regulatory framework are examined in the ninth chapter, by Reinhard

Madlener and Jan Martin Specht of Aachen University, Germany. These involve assessment of the economic viability and cost competitiveness of the different storage methods and the various flexibility options competing with each other to balance supply and demand. Market and regulatory conditions and also underlying uncertainties are considered in relation to business models and return on investment.

We are pleased to have engaged this international group of experts to produce wide-ranging overviews of the important area of energy storage. We are confident that this volume will provide a valuable resource for decision makers, scientists and engineers, and equally for practitioners and students involved with the globally ongoing sustainable energy transition.

Ronald E. Hester
Roy M. Harrison

Contents

Issues in Environmental Science and Technology No. 46
Energy Storage Options and Their Environmental Impact
Edited by R.E. Hester and R.M. Harrison
© The Royal Society of Chemistry 2019
Published by the Royal Society of Chemistry, www.rsc.org

Life-cycle Analysis for Assessing Environmental Impact   **261**
Heidi Hottenroth, Jens Peters, Manuel Baumann,
Tobias Viere and Ingela Tietze

Editors

Ronald E. Hester, BSc, DSc (London), PhD (Cornell), FRSC, CChem

Ronald E. Hester is now Emeritus Professor of Chemistry in the University of York. He was for short periods a research fellow in Cambridge and an assistant professor at Cornell before being appointed to a lectureship in chemistry in York in 1965. He was a full professor in York from 1983 to 2001. His more than 300 publications are mainly in the area of vibrational spectroscopy, latterly focusing on time-resolved studies of photoreaction intermediates and on biomolecular systems in solution. He is active in environmental chemistry and is a founder member and former chairman of the Environment Group of the Royal Society of Chemistry and editor of 'Industry and the Environment in Perspective' (RSC, 1983) and 'Understanding Our Environment' (RSC, 1986). As a member of the Council of the UK Science and Engineering Research Council and several of its sub-committees, panels and boards, he has been heavily involved in national science policy and administration. He was, from 1991 to 1993, a member of the UK Department of the Environment Advisory Committee on Hazardous Substances and from 1995 to 2000 was a member of the Publications and Information Board of the Royal Society of Chemistry.

Roy M. Harrison, BSc, PhD, DSc (Birmingham), FRSC, CChem, FRMetS, Hon MFPH, Hon FFOM, Hon MCIEH

Roy M. Harrison is Queen Elizabeth II Birmingham Centenary Professor of Environmental Health in the University of Birmingham. He was previously Lecturer in Environmental Sciences at the University of Lancaster and Reader and Director of the Institute of Aerosol Science at the University of Essex. His more than 400 publications are mainly in the field of environmental chemistry, although his current work includes studies of human health impacts of atmospheric pollutants as well as research into the chemistry of pollution phenomena. He is a past Chairman of the Environment Group of the Royal Society of Chemistry for whom he edited 'Pollution: Causes, Effects and Control' (RSC, 1983;

Fifth Edition 2014). He has also edited "An Introduction to Pollution Science", RSC, 2006 and "Principles of Environmental Chemistry", RSC, 2007. He has a close interest in scientific and policy aspects of air pollution, having been Chairman of the Department of Environment Quality of Urban Air Review Group and the DETR Atmospheric Particles Expert Group. He is currently a member of the DEFRA Air Quality Expert Group, the Department of Health Committee on the Medical Effects of Air Pollutants, and Committee on Toxicity.

List of Contributors

Manuel Baumann, Institute for Technology Assessment and Systems Analysis (ITAS), Karlsruhe Institute for Technology (KIT), P.O. Box 3640, 76021 Karlsruhe, Germany

D. Noel Buckley, Department of Physics, Bernal Institute, University of Limerick, Limerick, Ireland, and Department of Chemical and Biomolecular Engineering, Case Western Reserve University, Cleveland, Ohio, USA. Email: noel.buckley@ul.ie

Gideon Carpenter, Evidence, Policy and Permitting Directorate, Natural Resources Wales, Pembrokeshire, UK

Thomas Conway, Department of Electronic and Computer Engineering, University of Limerick, Limerick, Ireland

Alexander J. Cowan, Department of Chemistry and Stephenson Institute for Renewable Energy, The University of Liverpool, Liverpool L69 7ZD, UK. Email: acowan@liverpool.ac.uk

Peter Duffy, Enercomm International, Enercomm House, Lisduff, Longford, Ireland, and Schwungrad Energie, Parsons House, Clara Road, Tullamore, Offaly, Ireland

David J. Evans, British Geological Survey, Keyworth, Nottingham, UK. Email: dje@bgs.ac.uk

Gareth Farr, British Geological Survey, Cardiff, UK

Colin Fitzpatrick, Department of Electronic and Computer Engineering, University of Limerick, Limerick, Ireland

Mark Forster, Department of Chemistry and Stephenson Institute for Renewable Energy, The University of Liverpool, Liverpool L69 7ZD, UK

Heidi Hottenroth, Pforzheim University, Institute for Industrial Ecology, Tiefenbronner Strasse 65, 75175 Pforzheim, Germany. Email: heidi.hottenroth@hs-pforzheim.de

Yukitaka Kato, Laboratory for Advanced Nuclear Energy, Institute of Innovative Research, Tokyo Institute of Technology, 2-12-1-N1-22, O-okayama, Meguro-ku, Tokyo 152-8550, Japan. Email: yukitaka@lane.iir.titech.ac.jp

Henrik Lund, Department of Planning, Aalborg University, Rendsburggade 14, Aalborg 9000, Denmark

Rasmus Lund, Department of Planning, Aalborg University, A. C. Meyers Vænge 15, Copenhagen 2450, Denmark. Email: rlund@plan.aau.dk

Robert P. Lynch, Department of Physics, Bernal Institute, University of Limerick, Limerick, Ireland, and Department of Chemical and

Biomolecular Engineering, Case Western Reserve University, Cleveland, Ohio, USA. Email: robert.lynch@ul.ie

Reinhard Madlener, Institute for Future Energy Consumer Needs and Behavior (FCN), School of Business and Economics/E.ON Energy Research Center, RWTH Aachen University, Aachen, Germany. Email: rmadlener@eonerc.rwth-aachen.de

Padmanathan Narayanasamy, Advanced Energy Materials Group, Micro Nano System Centre, Tyndall National Institute, University College Cork, Cork, Ireland

Gaia Neri, Department of Chemistry and Stephenson Institute for Renewable Energy, The University of Liverpool, Liverpool L69 7ZD, UK

Takahiro Nomura, Center for Advanced Research of Energy and Materials, Faculty of Engineering, Hokkaido University, Kita 13 Nishi 8, Kita-ku, Sapporo 060-8628, Japan

Colm O'Dwyer, School of Chemistry and Tyndall National Institute, University College Cork, Cork, Ireland

Susana Paardekooper, Department of Planning, Aalborg University, A. C. Meyers Vænge 15, Copenhagen 2450, Denmark

Jens Peters, Helmholtz Institute Ulm (HIU), Karlsruhe Institute for Technology (KIT), Helmholtzstrasse 11, 89081 Ulm, Germany

Nathan Quill, Department of Physics, Bernal Institute, University of Limerick, Limerick, Ireland

Kafil M. Razeeb, Advanced Energy Materials Group, Micro Nano System Centre, Tyndall National Institute, University College Cork, Cork, Ireland

Fernando M. F. Rhen, Department of Physics, Bernal Institute, University of Limerick, Limerick, Ireland. E-mail: fernando.rhen@ul.ie

Han Shao, Advanced Energy Materials Group, Micro Nano System Centre, Tyndall National Institute, University College Cork, Cork, Ireland

Jan Martin Specht, Institute for Future Energy Consumer Needs and Behavior (FCN), School of Business and Economics/E.ON Energy Research Center, RWTH Aachen University, Aachen, Germany

Ingela Tietze, Pforzheim University, Institute for Industrial Ecology, Tiefenbronner Strasse 65, 75175 Pforzheim, Germany

Tobias Viere, Pforzheim University, Institute for Industrial Ecology, Tiefenbronner Strasse 65, 75175 Pforzheim, Germany

Energy Sources and Supply Grids – The Growing Need for Storage

PETER DUFFY, COLIN FITZPATRICK, THOMAS CONWAY AND
ROBERT P. LYNCH*

ABSTRACT

Efficiently exploiting renewable, sustainable and green energy resources is one of the most critical challenges facing our world today.
For example, as part of this challenge, Germany aims to generate 65%
of its electricity from renewable sources by 2020 and Ireland aims to
generate 40%. Renewable energy sources, *e.g.* solar and wind energy,
are plentiful and sufficient to power our ever-increasing demand for
more devices, technology and transportation. However, the increased
demand for electricity at peak times, the increased instantaneous
penetration of the grid by energy from non-conventional generation
systems (such as wind turbines and solar photovoltaic) and the intermittent and non-dispatchable nature of renewable energy sources are
threatening the stability of the electricity grid and limiting the ability of
the transmission system operator to respond to sudden changes in
generation or demand. This is particularly an issue in isolated grids
such as on the island of Ireland, where the failure of a single generator
results in the loss of a significant fraction of the overall grid capacity in
an instant. However, in mainland Europe, the electricity grid of each
nation is interconnected and synchronised, allowing the loss of a
single generator in one region to be compensated for by increasing the
output of the many other generators on the continent by a small
amount. In the future, there will be a need for significant grid-scale

*Corresponding author.

Issues in Environmental Science and Technology No. 46
Energy Storage Options and Their Environmental Impact
Edited by R.E. Hester and R.M. Harrison
© The Royal Society of Chemistry 2019
Published by the Royal Society of Chemistry, www.rsc.org

storage, load levelling and stabilisation of the grid. Electric vehicles will become more prevalent and the fraction of renewables on the grid will increase significantly. These technologies and the way in which they interact with the grid will greatly affect the stability of the electricity grid. Smart and innovative interaction of these technologies with the grid raises the possibility of optimising the level of energy storage required for stable and reliable grid operation. However, lack of planning in these areas could make future cost-effective, sustainable and reliable energy solutions hard to achieve.

1   Introduction

In the last 50 years, global anthropogenic greenhouse gas (GHG) emissions (see Figure 1a) have almost doubled as our farming, deforestation, industrialisation, transportation and population have expanded rapidly.[1] GHG emissions are composed primarily of CO_2, CH_4 and N_2O. Burning of fossil fuels, *e.g.* coal, natural gas and oil, is the largest single contributor to GHG emissions (see Figure 1b), accounting for 57% of all GHG emissions and over three-quarters of all anthropogenic CO_2 emissions.[2,3] Concerns regarding security of energy supply and the impact of humans on the sustainability of our planet have led to significant changes in policies that attempt to reduce our dependence on fossil fuels by increasing our harvesting of energy from renewable resources and increasing the use of electricity for transport and for heat supply *via* heat pumps.

Driven by our increased use of renewable energy, the demand for energy stability and electricity-balancing technology is growing rapidly.[4] Wind and

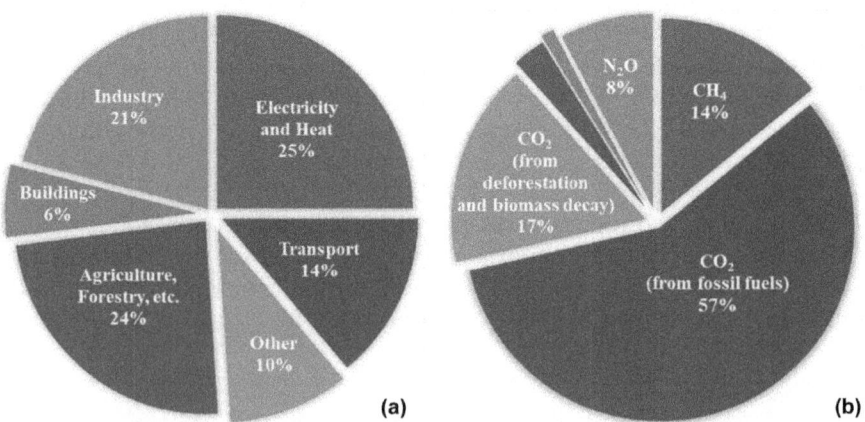

Figure 1   Pie charts of the breakdown of global anthropogenic greenhouse gas emissions (a) by economic sector source (in 2010)[1] and (b) by gas emission type (in 2004).[3] Note that percentages are of CO_2 equivalent mass and the two minor percentages not indicated are 3% other CO_2 and 1% other gases.

solar power are non-synchronous and volatile, requiring the transmission system operator (TSO) to limit the instantaneous system non-synchronous penetration (SNSP), *e.g.* energy from wind and solar. Currently, sufficient synchronous 'inertia' and system services, necessary for grid stability and power quality management, are provided by conventional generation. The stability of the electricity grid is determined over several time scales. The fast reaction speeds of electrochemical systems and long operational life spans of many of these technologies make them ideal candidates for grid stabilisation and load levelling. However, such technologies cannot provide the stability currently provided by synchronous 'inertia' and, if they were to provide a large enough buffer for the variations in energy supply and demand, the cost of electricity would increase significantly. Therefore, a combination of smart grids and additional system services for the stabilisation of supply in conjunction with significant additional energy storage are required to facilitate the reduction in burning of fossil fuels and development of renewables as our source of energy for electricity, heat and transport.

2 Energy Sources

The transition to a lower carbon fuel mix, largely driven by the need to combat climate change, continues, with renewables being the largest source of energy growth.[5,6] The energy company BP's Energy Outlook estimates that by 2040 oil, coal, natural gas and non-fossil fuels will each provide around 25% of the world's energy.[6] That will be the most diversified fuel mix the world will have ever seen by a considerable margin. The expected increase in the standard of living worldwide will continue to drive an increase in global energy demand and, in order to mitigate the effects of climate change, this demand must be met by innovative clean energy solutions. Currently, this trend is being observed in the European Union, where renewables account for 80% of new capacity and wind power is predicted to become the leading source of electricity shortly after 2030.[7]

Coal and natural gas are the most used energy fuels for generating electricity. In 2014, the share of world energy consumption for electricity generation by source was coal at 40.8%, natural gas at 21.6%, nuclear at 10.6%, hydro at 16.4% and other sources (solar, wind, geothermal, biomass, *etc.*) at 6.3%. Oil accounts for only 4.3% of electricity generation, even though oil (petroleum and other liquids) provides the largest quantity of energy according to the World Energy Resources Report.[5]

2.1 *Generation of Electricity from Combustion of Fossil Fuels*

In a thermal power plant, thermal energy from the combustion of fossil fuels, such as coal, oil and natural gas, is converted to electrical energy. Although there can be notable differences between different types of a given fuel in the level of emissions per unit energy (for example, differences in coal

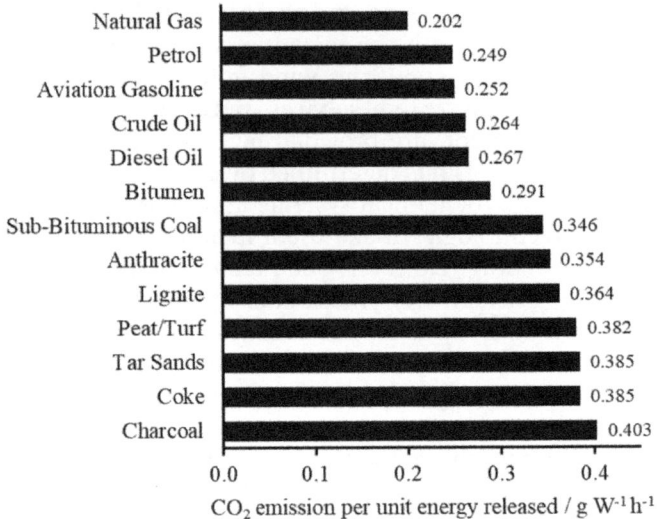

Figure 2   Bar chart of the mass of carbon dioxide emitted per unit of total energy
released by fully burning common fossil fuels.
Data from Intergovernmental Panel on Climate Change Report 2014.[1]

types; see Figure 2), there are some more general trends in the relative
emissions between coal, oil and gas. Compared with oil and natural gas, coal
produces the most CO_2 per unit of energy – CO_2 emissions from oil and
natural gas are about 77 and 58%, respectively, of those from coal, as shown
in Figure 2. This difference between the CO_2 emissions of different fossil
fuels derives primarily from their carbon content – since coal consists primarily of carbon–carbon bonds the most significant product of its combustion is CO_2, whereas natural gas consists primarily of carbon–hydrogen
bonds and therefore results in less CO_2 per kWh.

The combustion of any fossil fuel is damaging to the environment, but
coal generally contains a small percentage of sulfur that, if released into the
atmosphere as sulfur oxides, can lead to acid rain. However, oil is a small
player in the power sector and therefore the following sections will focus on
coal and gas, *i.e.* the major fossil fuels employed in the supply of electricity.

Coal-fired technology has a relatively low conversion efficiency of stored
chemical energy to electrical energy compared with modern gas-fired combined cycle plants (of the order of 40% *versus* 58%), exacerbating coal's
emissions problem. This gives gas-fired power plants a significant environmental advantage over both coal- and oil-fuelled generators. Nevertheless,
coal remains a large fuel source in many countries, particularly in countries
such as China and India where there are huge reserves.

2.1.1   Coal.   Conventional power stations turn heat energy from fossil
fuels into high-pressure and high-temperature steam that is then used to
generate electricity. For example, in a typical coal-fired power plant,

distillate oil is used to raise the temperature of a boiler's combustion chamber before admitting coal. When the chamber is at the correct temperature, coal ground to dust in ball mills is blown into the combustion chamber, where the pulverised fuel burns and generates heat. The design ensures good mixing of fuel and air to achieve complete combustion while ensuring that any bottom ash and fly ash produced as by-products of the process are captured. The gases from the combustion carry the heat to a boiler where water absorbs the heat and generates steam. The generated steam, when it reaches sufficient pressure and temperature, is admitted to the steam turbine, changing the internal energy of the steam into rotational kinetic energy (cooling the steam), which drives a synchronous electricity generator. (See the chapter Electrical Storage for more details on the operation of synchronous machines.) After this energy has been extracted, the steam is condensed and the water is circulated back to the boiler. As the plant output is increased by the plant operators at the request of a grid dispatch centre, the plant's control system increases the fuel feed to the combustion chamber, thereby delivering more steam to the steam turbine and generating more electricity. The overall thermal efficiency is typically between 35 and 40%, with some sophisticated systems achieving >40%.

In large-scale coal generation units (*ca.* 400 MW), the power is generated at around 15–20 kV, then enters a generator transformer and is stepped up to the local transmission voltage; this is typically in the hundreds of kilovolts. Coal plants generally have long start-up times, typically taking up to 6 h to go from cold to full load output. Of course, in situations where the plant is hot already, this delay can be significantly shortened to about 2 h.

Coal is currently still the most widely used fuel in the world for electricity generation, accounting for 37% of the total electricity generated.[8] In particular, some large economies, including China, the USA, India and Germany, use significant amounts of coal for electricity generation. With the emphasis in recent years on cleaner air, reduced emissions and combating climate change, there has been a shift away from coal, but progress has been slow. Data from the International Energy Agency (IEA) show that coal's share in electricity generation remained significant at 41% in 2014, but is estimated to have decreased since then.[7] Coal-fired power generation in the major developed countries is on a steep downward trajectory, in particular in the USA owing to competitive gas prices and the growth in renewables, whereas developing countries are still experiencing coal generation growth. In India, the third-largest coal consumer in the world, coal-fired power generation increased by 3.3% in 2015 as their economy continues to grow at a rapid pace. The widespread availability of coal in many countries makes it difficult for renewable energy technologies to compete.

In recent years, China has made policy decisions to reduce excess coal production. This has depressed global coal demand, particularly in the electricity sector, where it has typically been replaced by natural gas and renewable energy sources.[6] Britain's relationship with coal has almost come full circle, with the closure of Britain's last three underground coal mines and consumption decreasing to where it was roughly 200 years ago, around the time of the industrial revolution. Britain's electricity sector recorded its first-ever coal-free day in April 2017, thought to be the first time the nation had not used coal to generate electricity since the world's first centralised public coal-fired generator opened in 1882, at Holborn Viaduct in London.[9]

2.1.2   Natural Gas.   Although coal is currently the primary source of electricity, it is likely that natural gas will soon take its place, as it is a much more environmentally friendly fuel. This is clear from the CO_2 data in Figure 2, where natural gas results in significantly less CO_2 emissions per unit of energy released. However, gas turbines often consume significant quantities of water, which is used in the reduction of NO_x gas emissions.

As alluded to previously, gas, compared with coal or oil, has the additional advantage that a greater fraction of the chemical energy released during its combustion can be converted to electricity [in large-scale combined-cycle gas turbine (CCGT) plants]. Since products from the combustion of natural gas contain very few tars or particulates, it can be combusted in a gas turbine that is connected to a synchronous generator. The most common type of gas-fired plants are open-cycle gas turbines (OCGTs) that can be fuelled by gas or oil distillate. These plants are primarily designed for use during peaks in demand and as a backup to forecasting errors and rapid drops in generation from wind. Their design allows them to turn on quickly so as to take on a fraction of their maximum load and to ramp up to full load in less than half an hour. In addition, since the combustion of the fuel happens in the turbine, the response of such systems is much quicker than that of a conventional thermal power plant (where flow of steam can be controlled quickly but the ramp rate of heating of water in the boiler is much slower). However, flue gases that are exhausted from the gas turbine result in a huge loss of thermal energy to the atmosphere. The overall thermal efficiency of OCGT systems is typically less than 40%, *i.e.* similar to that of coal-fired systems.

CCGT plants capture a large quantity of this 'waste' heat from the flue-gases. These gases are passed through a waste-heat recovery boiler to generate steam, which is then used to drive a steam turbine and hence generate further electricity. The result is that the thermal efficiency of a CCGT plant can be close to 60%. However, a large decrease in efficiency occurs if the plant is operating at a fraction of its rated load or it switches over to operating as an open-cycle plant.

OCGT plants are normally used as peaking plants, *i.e.* to supply electricity for short durations when demand is high. Many system operators use OCGT plants during peak demand times or when some baseload plants trip out or fail to turn on when dispatched; in essence they frequently provide back-up power to the electricity grid and typically might run only a couple of hours daily, if at all. CCGT plants are extensively used as baseload plant, *i.e.* to meet the minimum load of the network. They typically run at full output for

16 h per day, so as to maximise the efficiency of the system, and at 80% load during the night so as to maximise stability.

Given the significantly greater thermal efficiency of CCGT plants, one would expect them to be the most common type of gas-fired power plant being constructed. However, CCGT plants, typically sized at 500 MW, are designed to run and deliver these high performances as baseload plant. It follows that, owing to the dramatic recent increase in intermittent generation – mainly wind turbine and solar photovoltaic – OCGT or other peaking plants with high part-load efficiency and flexibility are the most common type of new power plant being constructed. There is a growing need for and interest in 'wind-chasing' plants where there is a high penetration of intermittent generation. Wind chasers tend to be highly flexible, with good efficiency and high part-load performance. Indeed, large reciprocating gas engines with dual-fuel capability to burn back-up distillate fuel are becoming common. For example, a 160 MW power plant could be composed of 10 such gas engines and would therefore be able to turn on incrementally with relatively high efficiency.

2.1.3 Oil. Oil-fired generating plant lies in third place behind coal and gas, with only 4.3% according to world energy statistics.[6] Typically, oil is burned in conventional boilers with combustion chambers in the form of heavy fuel oil (HFO, termed 3000-second oil), light fuel oil (LFO, termed 200-second oil) and distillate or diesel oil. HFO and LFO must be heated prior to injection into the combustion chamber under pressure so that the atomised oil is completely and efficiently combusted. Distillate or diesel oil is normally used in conventional coal- and oil-fired plants during start-up, but these plants change over seamlessly to LFO or HFO (or coal in the case of coal-fired plant) when the HFO/LFO is sufficiently hot for use. The design of these oil-fired plants is similar to that of plants that burn coal, but oil-fired boilers are generally more efficient than similar-sized coal-fired plants as the latter have higher 'house loads' due to the additional power consumed in fuel handling, grinding and pulverising the coal prior to blowing it into the combustion chamber.

A more widespread use of distillate is as a fuel for OCGT plant where there is no gas transmission system. Distillate is also used widely as a back-up or secondary fuel for gas-fired power plants so that these plants are available for running if the gas supply is unavailable for any reason.

2.2 Nuclear Power

Nuclear power plants are normally steam-driven systems in which electrical energy is produced from splitting of atoms *via* nuclear fission.[10] More energy per mass of fuel can be produced in this way than by any other electricity generation system. The first large nuclear reactors were built in the 1940s as a method for producing plutonium for use in nuclear weapons. These plants used uranium as fuel, which had been shown to have fissile properties in

1938 by Otto Hahn, Fritz Strassman and Lise Meitner.[11] They showed that the mass of the products of the fission of uranium was less than the original mass – a verification of Einstein's theory of special relativity. In 1942, Enrico Fermi's team were the first to initiate a self-sustaining chain reaction as part of the Manhattan Project.[11]

For a nuclear reactor to sustain a chain reaction and produce atomic energy, atoms first have to be split, *i.e.* nuclear fission must occur. Uranium is the most commonly used fuel. In such reactors,[12] the fission of a small number of nuclei results in the release of neutrons capable of striking and splitting other nuclei within the fuel. The neutrons from this chain reaction are slowed by the moderator that encases the uranium fuel rods, optimising the probability of them splitting further atoms. Light water (H_2O), heavy water (deuterium oxide) and graphite are used as moderators.

The loss in mass, m, during these reactions corresponds to the release of energy, E, stored within the atom, where $E = mc^2$ and c is the speed of light in a vacuum. Therefore, since c^2 is a large multiplier, a small loss of mass results in a large release of energy. In a nuclear power station, the heat generated from the splitting of the atoms is discharged through a cooling system that heats water, or some other coolant, that then is used to heat the water in a boiler, creating steam that drives a steam turbine. Therefore, in order to achieve high thermal efficiency, the coolant must operate at high temperatures and therefore high pressures. Furthermore, the nuclear reactions that occur in the reactor/pile produce radioactive products. Therefore, the reactor has to be encased in steel and concrete safety shields needed to contain the pressure and the reactive products, and a range of safety measures and protocols have to be closely followed to ensure safety.

Nuclear power stations have even longer ramp-up times than coal-fired and CCGT plants and therefore they are ideally suited to meeting the grid's baseload. Furthermore, nuclear plants operate most efficiently when run continuously at close to their rated output. However, in France, where 70–80% of electricity is generated from nuclear fission,[13] nuclear power plants are also used to meet peak demand, requiring plants to be designed so as to accommodate this ability without resulting in a significant loss of efficiency. This is a unique situation, resulting in very low off-peak electricity prices and encouragement of demand-side management through off-peak space and water heating and, more recently, charging of electric cars. However, the reliance on one source of energy has resulted in security of supply issues and price spikes, especially during cold weather; *e.g.* France's electricity demand rises by 2.3 GW per 1 °C drop in temperature, resulting in France changing from being a net exporter to a net importer of electricity during cold weather.[14]

In 2015, nuclear power provided 4.9% of total primary energy supply and 10.6% of electricity generation worldwide, requiring 65 000 tonnes of uranium.[15,16] Since nuclear power plants operate in a similar manner to conventional power plants, by boiling water to produce steam to drive a steam turbine, they also have similar efficiencies, *i.e.* 33% for a boiling water reactor and 30% for a pressurised water reactor, the two main reactor types in the USA.[10]

In addition to the energy efficiency, another important consideration is the fuel efficiency. The fuel in a nuclear reactor is made up of fissile material, *e.g.* U-233, U-235 or Pu-239, and fertile material, *e.g.* U-238 or Th-232.[12] The fissile material, when hit by neutrons, releases, in addition to a large amount of energy, at least two neutrons that can split other fissile material, resulting in a continued chain reaction. The fertile material, on the other hand, cannot undergo fission directly but must instead capture neutrons produced in the fission process, leading to it becoming fissile. This process is called breeding and it can replenish the concentration of fissile material. All nuclear reactors breed some fuel. In a typical reactor, the most common isotope formed is Pu-239 through the capture of a neutron by U-238 (the fertile material), which then undergoes beta decay to yield the fissile isotope. This isotope can then take part in the chain reaction, producing similar energy to that in the fission of U-235 (the main fissile material in a typical reactor).[17]

The risks associated with the operation of nuclear reactors are significant and the radioactive wastes produced can have very long half-lives and therefore require careful disposal. These concerns result in substantial political pressure that limits the adoption of nuclear power in many countries and has resulted in the decommissioning of nuclear plants (without replacement) in countries such as Germany.[14] These concerns are primarily driven by major accidents, *e.g.* at Chernobyl in 1986 and, most recently, at Fukushima–Daiichi in 2011.[18] However, nuclear reactors do have many safeguards, including intrinsic negative feedbacks such as a negative temperature coefficient of reactivity, and they do not cause air pollution or GHG emissions, making them an important alternative to fossil fuel-fired power stations, especially for meeting the electricity grid's baseload. Furthermore, there has been a resurgence in research in this area, with significant research into alternative nuclear reactor designs such as the liquid fluoride thorium reactor (LFTR).[19,20] Such reactor types promise several operational advantages, including a strong negative temperature coefficient of reactivity, very reliable failsafe mechanisms (owing to the liquid nature of the fuel, all the fissile material can be rapidly removed from the reactor though a plug in its base, quenching the chain reaction), high thermal efficiency and long fuel cycles between charges.[19] Since, compared with uranium, thorium is more abundant and proliferation resistant, and has a high fuel burnup (*i.e.* most of the fuel in an LFTR is consumed per fuel cycle as opposed to <1% in a typical uranium reactor) and short half-life products, LFTRs would result in increased security of supply, decreased costs and reduced damage to the environment.[19,20]

2.3 Renewables: Solar, Wind, Wave, Tidal and Hydro

2.3.1 Solar. The Sun pours energy onto the Earth's surface at rates that dwarf even the current rates of use of non-renewable fuels, including fossil and nuclear fuels. Only one part in 20 million of the Sun's radiation strikes our atmosphere and about half of that energy reaches the Earth's surface. This energy originates from nuclear reactions in the Sun, *i.e.* conversion of energy stored as mass in the nuclei in the Sun to energy. We typically convert solar energy into electrical energy in one of two ways, using photovoltaics (PV) or concentrated solar power (CSP).

In PV, the energy is harvested by large arrays of small solar cells, typically made of silicon. These cells convert light (photons from the Sun) into electricity. These cells typically operate at low efficiency, with less than 20% of the photons that strike the device leading to an excited electron. Although this low efficiency is undesirable in terms of land usage, the ubiquitous nature of solar energy makes any method of harvesting it desirable. As advanced semiconductor devices, PV cell arrays are both expensive and energy intensive to produce, which can significantly offset their green credentials.

In CSP, the energy is harvested in a similar manner to that in conventional thermal generators. Heat from the Sun is concentrated using specialised optics and used to heat water, which can generate steam and turn a turbine. This system can also be adapted to incorporate some thermal energy storage using a fluid from which heat can later be harvested (see the chapter Thermal and Thermochemical Storage for further details).

Solar power provided just 1% of total global electricity generation in 2016 but this is rising rapidly. In 2014, the International Energy Agency predicted that solar penetration could be as high as 27% by 2050, making it the world's largest source of renewable electricity.[21]

2.3.2 Wind. Solar energy absorbed in our atmosphere can reappear as lightning, wind, *etc.* Winds arise from different degrees of heating of the Earth's atmosphere by radiation from the Sun. Although windmills have made innumerable contributions to our industrial progress, the need to site them in often remote locations seriously limits their applications. However, the use of similar wind turbines for the generation of electricity has decoupled the need for colocation of generation and load, greatly increasing the usefulness of this energy source.

The harvesting of energy from the wind has seen explosive growth in recent decades. This is due to the many advantages that wind generators have over other sources of electrical energy. Wind energy is, in principle, available almost anywhere, produces no GHGs or other harmful emissions during operation and consumes no water and little land. However, large wind turbines may present a risk to certain species of birds and are often lamented for their impact on rural scenery.

As with most renewable sources, the major drawback of using wind energy to supply the electricity grid is the intermittent and non-synchronous (see the section on synchronous machines in the chapter Electrical Storage) nature of its output. Modern weather forecasting techniques allow the wind, and thus the amount of wind energy generated, to be predicted with reasonable accuracy. However, there are still discrepancies between forecasted and actual wind levels. Furthermore, even if the wind output is stable and matches forecasts, the output of a wind generator cannot be increased, except for short durations, to accommodate a shortfall in generation or an increase in demand, *i.e.* it is non-dispatchable. As a result, when the fraction of wind power (or other renewables) on the grid is high, a number of backup generators must keep burning fuel at low output, ready to come online should demand suddenly increase or should the wind suddenly decline. This has a negative impact on the overall efficiency of a grid with a high level of renewable penetration.

As with more traditional systems (*e.g.* thermal power plants), electrical energy is generated by converting the kinetic energy of the wind into rotational energy by having it turn a large electricity-generating turbine. In traditional power stations, the speed of rotation of the generator is fixed by the grid frequency (*e.g.* 50 Hz in Europe and 60 Hz in the USA) and all of these generators operate in a synchronous manner. Wind turbines can be operated in either a synchronous or non-synchronous manner.[22] Although synchronous wind turbines, operating at constant speed, were at one time more common, these have largely been supplanted by non-synchronous units, operating at variable speed.[23] While non-synchronous designs do not provide a synchronous inertial response to the grid, such designs can be used to maximise the energy harvested from the wind by operating at the optimum rotating speed for the current wind speed. In addition, depending on design, the rotational kinetic energy stored in the angular momentum of the turbine can deliver reserve power for short durations.[22,24–26]

Wind turbines can be located either on land or at sea. Offshore wind turbines have a number of advantages, including access to the more frequent and powerful winds often available offshore and the lack of objections, since local people are non-existent. However, the installation and maintenance costs are much higher than for land-based systems.

Wind penetration has increased considerably and is now estimated to account for 4% of total global electricity generation.[7] This increase is expected to continue for the foreseeable future and is largely driven by the move away from fossil fuels and the ubiquitous nature of wind energy.

2.3.3 Wave. Renewable energy can be harvested from the motion of water waves and converted into electricity. This can be done in many ways, including harvesting the kinetic energy of a floating body as it continuously rises and falls with the waves or forcing the rising water into an enclosed space, compressing air and using it to turn a turbine. This technology is promising as it is typically not visible to humans and would harvest an abundant and currently untapped source of renewable energy. This technology has seen relatively little deployment owing to the corrosive environment of the sea, resulting in a requirement for expensive materials and regular maintenance. These systems may also pose a hazard to various types of marine life, may encroach on productive fishing grounds and could provide a navigation hazard to seagoing vessels.[27] Several small- to medium-scale (<10 MW) commercial systems have been commissioned, but most systems that have been installed are still in the developmental stage and are not yet commercially viable.[28]

2.3.4 Tidal.

Renewable electricity can be generated from the motion of the tides. This has the significant advantage that the tides (unlike wind or sunshine) can be accurately predicted, making tidal energy one of the few non-volatile forms of renewable electricity. Indeed, careful planning of the location and phase difference between tides across Europe or other similar regions could allow tidal energy to meet reliably some of the baseload of electricity grids.[29] However, only certain regions have sufficiently large tidal ranges to make this technology competitive.[30] Tidal generators can work in a number of ways but typically involve using the tide to turn hydroelectric turbines. One method of achieving this is by capturing the water at high tide in a reservoir or dam and slowly releasing this water through the turbines after the tide has fallen. In some regions where the flow of water is constricted (*e.g.* by natural straights or artificial bridges), the flow of water due to changing tides can reach a relatively high velocity and a strategically placed turbine can generate a significant amount of power.

Tidal generation has seen relatively little deployment. This is often due to the lack of suitable locations but these systems also potentially pose a threat to marine wildlife and suffer from the same corrosion issues as are experienced by wave generators. The nascent stage of the technology's development also contributes to the high up-front cost of these systems. A handful of plants have been commissioned worldwide, with the largest having outputs of a few hundred megawatts.[31]

2.3.5 Hydro.

Water evaporated by energy from the Sun can condense on high land, converting solar energy to potential energy. As this water makes its way back to lower ground, hydroelectric generators can convert this energy to electricity. The flow of water rotates turbines in these generators. While these turbines can be used to harvest tidal energy, hydroelectric generation typically refers to large systems of dams and turbines in which the water is kept at an elevated level by the dam and slowly released through a tunnel containing a turbine. The motion of the water causes the turbine to rotate, generating electricity. These systems can be operated in a continuous manner or can hold their reserves for times of increased demand, *e.g.* pumped hydro storage (see the next chapter on Mechanical Systems for Energy Storage for further details).

Although there are some locations that provide a natural water drop which can be exploited for hydroelectric generation, typically the construction of large dams, along with the canals used to redirect water during and after the dam's construction, is necessary. This can lead to a very large up-front capital cost for these systems. However, once built, these systems generate very little emissions. Hydroelectric generation is one of the oldest types of electricity generation and 70% of all electrical energy from renewable sources was generated from hydroelectric sources in 2017, amounting to almost 16% of total electricity generation globally.[15] However, there are a number of drawbacks in addition to the high capital costs. The redirection of rivers and creation of large new lakes can lead to the displacement of large numbers of people in addition to disrupting the habitats of a wide range of flora and fauna.[32]

2.4 Geothermal, Combined Heat and Power, Biomass Combustion and Waste Incineration

2.4.1 Geothermal.
Geothermal energy is harvested from the heat of the Earth's molten interior. This heat originates from radioactive decay within the Earth and, to a much lesser extent, near the surface, from stored solar energy.[33] On average, the temperature increases at about 30 $°C\,km^{-1}$ towards the centre of the Earth, resulting in a total rate of heat flow to the Earth's surface of 42 TW or 82 $mW\,m^{-2}$. This energy may be harnessed to provide energy for a variety of purposes, including direct heating to nearby communities or industries, or to generate electricity or provide heating using geothermal heat pumps.

Direct heating in the form of hot springs was used in ancient Roman and Chinese times for bathing and washing. Today, direct heating is employed in locations worldwide that have hot springs close to towns and communities in the form of district heating schemes. Additionally, the heat may be delivered as hot water or industrial steam to nearby process industries. The drive to combat climate change coupled with the availability of hot springs mean that district heating schemes are well developed in Iceland and are being looked at more closely in other countries.[34] These schemes can be augmented by fossil-fuel boilers and more recently by renewable sources such as wind and solar PV.[35]

Geothermal heat can be used for electricity generation. Energy can be extracted from the Earth by injecting water deep underground. The returning steam (or hot water, which is later converted to steam) drives turbines connected to generators. The depth of injection and location *inter alia* have a significant impact on both the return temperature of the water/steam and the quantity of energy extracted. If the steam is superheated and at sufficient temperature, it may be used to drive a steam turbine directly. In most cases, it is likely that the return geothermal steam or water will require routing through a boiler to bring the steam conditions up to those necessary for admission to the steam turbine. Although the capital cost can be enormous in extracting geothermal heat, the operating costs are generally low; these costs are largely the energy costs for continuous recirculation of water.

Geothermal energy contributes a significant share of electricity generation in several countries. There is 13.2 GW of geothermal power capacity worldwide, with 43% of this based on island nations and 72% of the installed capacity near tectonic plate boundaries or hotspots where the energy is more accessible.[5] The Philippines are the second-largest geothermal electricity producer after the USA, at about 10 TWh of electricity per annum, which equals approximately 14% of the Philippines' total electricity generation. Kenya is the seventh-largest producer of electricity from geothermal energy at about 5 TWh of electricity per annum, but it has the largest share of its total electricity generation from geothermal energy at about 44%.

In many areas of the world, particularly in Europe and the USA, buildings require heating during the winter and cooling during the summer, owing to fluctuations in ambient air temperatures. Although air temperatures above ground change throughout the day and with the seasons, temperatures 3 m below the Earth's surface are consistently between 10 and 15 °C. This consistent below-ground temperature can be used to heat and cool buildings by employing geothermal heat pumps. These transfer heat from the ground into buildings during the winter and reverse the process in the summer.

According to the US Environmental Protection Agency (EPA), geothermal heat pumps are the most energy-efficient, environmentally clean and cost-effective systems for heating and cooling buildings.[36] All types of buildings can use geothermal heat pumps but a limiting factor is the amount of heat transfer surface required to extract or dissipate the heat from/to the ground.

2.4.2 Combined Heat and Power.   Combined heat and power (CHP) is the simultaneous production of utilisable heat and electricity from an integrated thermodynamic process. Usually this involves capturing and utilising the waste heat from the production of electricity and thereby increases the overall process operating efficiency. Waste heat recovery from power plants can provide steam and/or hot water to serve local requirements. This can include steam or hot water for industrial use, adsorption chilling and district heating schemes providing heated water for residential developments.

CHP plants are frequently given priority dispatch so that the plant generates electricity whenever supplying heat to contracted customers. If the plant is dispatched only in accordance with merit order then there may be significant periods when the plant would not be in operation and those relying on heat will have to provide their own back-up heat by less environmentally friendly means. There is an electricity cost to be borne when the CHP plant is given priority as more efficient plants are displaced, often resulting in the market operator applying a Public Service Obligation levy on all electricity users so as to recover the additional out-of-merit-order costs arising from the plant being given priority.

Governments have recognised the importance of CHPs in helping to achieve national emissions targets and combat climate change. The migration towards large-scale business parks with all services on site, including power and heat, is providing benefits to industrial steam users and nearby residential communities using district heating and is therefore a major boost to the CHP sector. Government policies may seek to attract more investment into this sector by enabling investors to qualify for certain capital and depreciation allowances in addition to not being liable for carbon taxes.

Although these may not appear significant, they may be just sufficient to enable some projects to reach financial viability.

2.4.3 Combustion of Biomass. Biomass is fuel that is developed from organic materials such as wood waste, forest trimmings, bagasse (*i.e.* dried pulp from sugar cane), straw and animal manure. Some crops, such as willow and *Miscanthus* grass, are specifically grown for biomass use. These renewable and arguably sustainable sources of energy are used to generate renewable electricity or create other forms of renewable energy. The combustion of biomass can be considered carbon neutral because the carbon dioxide produced when it is burned is offset by the carbon dioxide captured during its production. When the biomass waste material is used in the production of electricity, it produces GHG emissions, but if the waste was openly burned, dumped or allowed to rot into the soil, these emissions would still occur.

A key element that is often missed is that the combustion of biomass should be as close as possible to the source of the biomass, as fuel costs and carbon emissions due to long distance transportation can significantly increase carbon emissions from non-renewable sources, undermining the concept of biomass carbon neutrality, *e.g.* forest trimmings transported from the mid-western USA to Europe by ship and then transported further by road to a biomass-fired power plant. This has a serious impact on the biomass efficiency cycle. Another aspect that requires serious consideration is the need for drying or other processing. A harvested wet willow crop will have a very low heat/calorific value per tonne and hence low economic value until its moisture content is significantly reduced. Kiln drying of biomass should be avoided and atmospheric drying should be used where possible.

In biomass power plants, wood waste or other waste is burned to produce steam that drives a steam turbine, which generates electricity. Biomass is often burned at combined heat and power plants. In recent times, small isolated islands and remote communities have been looking at developing small biomass plants to generate and supply electricity into a mini or micro grid, *i.e.* a localised electricity network augmented by wind generation and solar PV. Energy storage, employing either batteries or flywheels coupled with load management techniques, is used to provide system stability and security in these new self-sufficient mini/micro grids around the globe.

Biomass is not without its challenges, not least a long-term and consistent feedstock supply. For this reason, governments may introduce REFIT (renewable feed-in tariff) schemes to ensure that potential projects achieve sufficient financial support from investors. However, even with a REFIT scheme in place, biomass investors will be reluctant to come into the market if there is a concern regarding a stop–start fuel supply or if no long-term supply contract can be negotiated.

2.4.4 Waste Incineration. High-technology waste incineration or waste-to-energy (WtE) with sophisticated environmental controls has come to the fore in recent years, particularly for municipal solid waste, hazardous waste and medical waste. With increasing focus on the environment and on the planet's finite resources, waste reduction, recycling, reusing and composting, together with awareness of the circular economy where waste is a feedstock for other industries, are all contributing to a reduction in the use of both landfill and waste incineration. These are now seen as the final links in the waste chain, with waste incineration nevertheless playing a reduced but important part in waste management. All of this has resulted in a depleting feedstock, with some exceptions, and a low-level increase in waste incineration.

Fortunately, a significant portion of municipal waste – consisting of paper, cardboard, plastics, metals, glass, wood, rubber and textiles – can be recycled and reused, leaving a depleted feedstock for waste incineration. Similarly, increased emphasis on composting has led to more environmentally friendly uses of household and organic waste. For example, food waste from households and the food processing industries is being increasingly used as a feedstock for anaerobic digesters for the production of biogas. In many countries, national gas grids provide injection points where scrubbed biogas can be injected and sold to renewable generation and CHP plants through GPAs (gas purchase agreements, similar in structure to power purchase agreements).

Medical and biomedical waste requires special waste disposal practices owing to the risk of infections and other public health concerns. The increasing emphasis on healthcare, in both developed and developing countries, means that this waste stream is increasing. Until new methodologies are developed for recycling and reusing infectious waste, waste incineration will be the preferred method of disposal.

Hazardous waste is defined by environmental authorities in most countries as waste material that can be classified as potentially dangerous to human health or the environment on the basis of a range of characteristics, including its toxicity, flammability or corrosiveness. In the past, much of this hazardous waste was sent to landfill but, with concerns about contamination of air, surface water and groundwater from uncontrolled land-disposal sites, tougher regulations for this waste disposal have come into operation. As a result, advanced technologies have been employed for managing hazardous waste, including stringently controlled incineration.

As discussed earlier, waste incineration is frequently used in conjunction with waste heat recovery for district heating. This arrangement enables the WtE plant to operate as a high-efficiency CHP and hence deliver the greatest possible benefit to the local community and the environment.

3   Operation of Electricity Networks

In the early nineteenth century, Volta's battery and Faraday's electric motor, along with several other key inventions and theories, allowed the controlled conversion of different forms of energy to and from electrical energy.

Through the advancement of research and development in this area by Bell, Edison, Kelvin, Parsons, Tesla and many others, electrification of industry was made possible, resulting in the second industrial revolution. This electrical energy was and still is remarkable for its flexibility. It can be produced from a variety of energy sources, *e.g.* wind, coal and nuclear, and can be transported hundreds of kilometres to the point of consumption. Hence the generation of electricity can be placed near the source of the energy even if that is a significant distance from the ultimate user, *i.e.* it became no longer necessary to place large industries beside hydro dams or to transport coal inland to fuel power-intensive, mining *etc.*

By connecting multiple energy sources and loads to a simple grid, a reliable supply of energy is provided without ever, from the point of view of the consumer, having to be ordered, delivered or topped up. These large electricity grids can ride through the turning on and off of large appliances and can even maintain operation in the event of a significant fault, such as the unexpected disconnection of a major generator. In addition, extra-large grids, such as the synchronous grid of Continental Europe, also benefit from the spread of peak loads, *e.g.* since peak loads occur in Warsaw and Lisbon at different times, the overall fluctuations of power on the grid are reduced. In a similar way, although the average wind speed at a particular site varies very little from year to year, the instantaneous wind speed varies significantly; but, a distributed network of wind turbines and other intermittent energy sources across Europe would produce a very stable source of power.

Although the structure of the electricity supply industry (ESI), comprising large-scale generation plants, transmission and distribution networks and a supply function, has remained largely the same over many decades, developments in recent years have resulted in a more complex arrangement. The growth in intermittent generation (wind and solar), grid-connected storage, distributed generation and distribution-connected storage has added to more complex market arrangements in different countries. However, the supply of electricity is considered as primarily four distinctive sectors, as shown in Figure 3, with distributed generation, *etc.*, connected at the supply and distribution level. Bulk electricity is transported from power

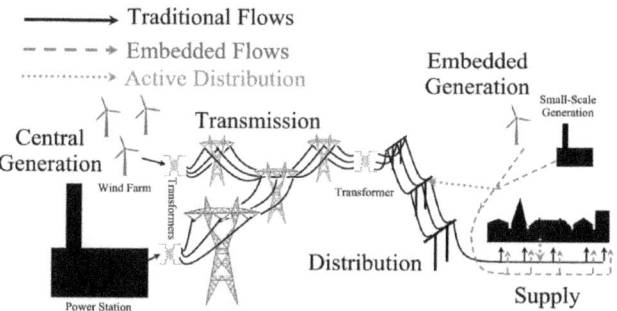

Figure 3 Schematic of distinctive sectors of supply of electricity.

stations (generation) through the high-voltage national grid (transmission) and through the low-voltage network (distribution) to end customers (supply). Coal-, gas- and oil-fired power plants usually compete for the electricity supply whereas more modern technologies such as wind and solar often have priority for energy supply. Transmission is the term used for the bulk movement of electricity, often over long distances, at high voltage from generating stations mainly to transmission substations where the voltage is stepped down to the distribution level through transformers. Transmission and distribution are in most instances monopoly sectors. Supply is normally a competitive sector that interfaces with consumers through a supply contract with metering and tariffs. Some of these tariffs, such as the wire charges and consumer levies, are normally regulated whereas the energy consumption tariff may be a competitive element that is contracted between the customer and supplier, usually for a specific period.

3.1   Transmission Network

A transmission system consists of a meshed network of high-voltage wires that transports bulk power from generation nodes to large demand centres (usually cities, towns and large consumers). The wires are normally arranged over a number of different voltages so that the grid may consist of three or four high-voltage (HV) voltages, *e.g.* 110, 220, 400 and maybe 700 kV; this meshed HV network is typically referred to as 'the grid'. Large power plants normally feed into the grid at high voltages consistent with their megawatt output; the higher the voltage for a particular output, the lower is the current and hence the lower are the losses and size of the conducting wires. For example, a large 400 MW plant would connect at 220 kV or above whereas a smaller plant could connect at a lower voltage.

The wires of each voltage network, *e.g.* 220 kV, are linked, coupling them to the next highest and next lowest voltage through what are referred to as tie-transformers, normally located in transmission substations. In the past, power flows were thought of as flowing from higher to lower voltages, passing through step-down transformers, and eventually stepped down further to what is commonly referred to as the distribution network. However, with the growth of embedded generation, where medium-sized generators such as wind farms, CHP plants and solar PV farms are connected at distribution voltages or at the low transmission voltages, power flows can frequently be reversed.

Each voltage network can be thought of as a particular floor in a large building: to get to the floor above or below, one must use stairs; in the case of the grid, electricity flowing from one voltage level to the next highest or lowest must flow through a transformer. Although these can carry power in both directions, power normally flows from higher to lower voltage. Network voltages below 100 kV are normally deemed to be distribution networks, although this can vary in different countries around the globe.

3.2　Distribution Network

A distribution system consists of a meshed network of low- and medium-voltage wires that transports power from large bulk nodes – referred to as substations – to individual customers. As with the transmission network, the wires are normally arranged over a number of different voltages so that the network may consist of three or four voltages, *e.g.* 38 kV, 20 kV and stepping down to 400 V three-phase supply to small business and 230 V single-phase supply to domestic customers. Again, in the past, power flows were thought of as flowing from higher to lower voltages, passing through step-down transformers and eventually connecting to the individual customer's meter point. In recent times, there have been several forms of micro generation from small, single-phase wind turbines, solar PV panels, micro CHPs or mini hydro schemes feeding into the distribution network at the lower distribution voltages, with medium-sized wind farms feeding in at the higher distribution voltages, resulting in reverse power flows.

Distribution networks encounter problems that are not normally seen in transmission networks. For example, because small generators do not have the same sophisticated protections as large conventional generators, there is an increased risk that they may feed into a 'dead' network during planned or forced outages of the network. This can prove to be a serious risk to maintenance crews working on the network, *e.g.* on 38 kV lines, as power fed in at low voltages, *e.g.* 230 V, can be stepped up to where repair crews are working. Distribution Codes normally require that all distribution-connected generators – large or small – are fitted with disconnection/trip mechanisms should the network voltage be lost. Another problem area is 'islanding' of distribution-connected generators, alone or while remaining connected to a small part of the network. For example, a circuit breaker connecting a part of the network where a small wind farm is located close to a factory could open/trip, leaving the wind farm as the sole feed to the factory. As these loads are unlikely to be matched exactly, the islanded frequency could increase or decrease dramatically, exposing equipment or personnel to risk. The need for reliable and sensitive detection and shutdown mechanisms is crucial to cope with such situations.

3.3　Distributed Generation

Currently, there is no consistent definition of distributed generation (DG) within the electricity sector literature. However, the Office of Gas and Electricity Markets (Ofgem) in the UK aptly describes it as 'embedded or dispersed generation, where the electricity generating plant is connected to a distribution network rather than the transmission network'. There are many types and sizes of DG, including CHP plants, wind farms, hydroelectric power, solar PV or one of the new hybrid generation technologies. Depending how distribution voltages are quantified, DG sizes may vary from fairly large wind farms down to small generators of less than 10 kW; all connected at different distribution voltages.

Over recent years, we have witnessed a dramatic growth in the number of distributed generators seeking to connect to the distribution network. Many DG developers are first-timers to the electricity market and frequently encounter significant difficulties in navigating their way through the planning, licensing and connection processes. The development of DG requires governments and regulators to ensure that planning, licensing and connection to the distribution network are such as to treat DG fairly. DG is expected to become more important in future generation systems as countries and citizens seek to contribute to combating climate change. From a national perspective, DG can greatly increase awareness of the environment in addition to supporting the meeting of national emissions targets. Furthermore, energy losses are considerably reduced as the electricity is consumed closer to the point of generation.

The development and increased availability of small-scale battery storage systems coupled with advances in wind turbine and solar PV technology are encouraging many householders, farmers and small businesses to have their own DG, while remaining connected to the distribution network. Many consumers are now able to supply their own energy and export excess energy to the grid, although the former may be the better of the two, depending on the regulatory framework that applies. For example, where the regulations allow net metering, there is a significant benefit to the customer/generator as any excess electricity exported onto the network during off-peak times can effectively be drawn down and used later at peak times. The benefit arises because electricity, which is normally expensive to store, has a different value at every point in time. Hence, if a customer/generator can export a surplus unit of electricity during off-peak times with a value of only 1 penny or 1 cent, and then draw it down and use it later in the day when valued at 5 pence or 5 cents, then he/she has gained considerably. However, this benefit has not been gained for free; rather, there is a cost that is spread across all electricity consumers so that in effect it is cross-subsidisation from the many to the few.

It is worth highlighting that customers with their own generation (including domestic generation with roof PV) purchase far fewer units of electricity and hence pay much less to the supplier and hence to the distribution and transmission operators. Since the fixed tariffs are usually not cost-reflective, with much of the fixed infrastructure costs and fixed billing and payments costs covered in the unit rates rather than in the standing charge or other fixed charges, when customers pay for fewer units (because they are generating their own energy) they are no longer paying their fair share of the fixed costs. Often the poorer customers who cannot afford PV are subsidising the wealthier who can afford it, raising an issue for the long-term sustainability of, in particular, the distribution operators under such regimes.[37]

3.4   Mini Grids

The World Bank states that 87.3% of the world's population in 2016 had access to electricity.[38] At a global population of 7.44 billion, this means that approximately 1 billion people do not have access to electricity. It was only in 2018 that the last remaining village not connected to the grid in India was connected. However, many people in villages in India still do not have access to electricity.[39]

Mini grids are normally associated with developing countries or small isolated communities. Such mini grids may be a long-term solution, as in the case of a remote island, or they may serve as an interim solution pending the arrival of the main grid at a future date. Decisions regarding their development are driven as much by social and political considerations as by economics.

A mini grid can be defined as a combination of generators, possibly coupled with an energy storage system and load management system, all connected to a distribution network to supply electricity to a local group of customers. The generation mix in such a mini grid may involve dispatchable generation such as diesel generator sets coupled with some non-dispatchable solar and wind, with a total capacity normally of between 50 kW and 10 MW. System stability is provided through the control system, employing energy storage such as batteries and/or flywheels and a load management system.

The control system ensures that frequency and voltage are maintained within specified limits by utilising the energy storage and, if these limits are reached, load control is employed. The mini grid normally serves a limited number of customers *via* a localised distribution network, which is operated in isolation from the main grid.

With the push towards the greater use of renewables and the higher availability of solar and wind energy on islands and in remote locations, there is a shift away from diesel engines. Many locations are seeking an all-renewable mini/micro grid with, for example, dispatchable biomass and non-dispatchable wind/solar supported by storage. This storage may well be hybrid storage, incorporating high-cycling lithium-ion batteries or flywheels and low-cycling lead–acid batteries. Issues to be addressed in these grids include load matching and grid stability from both voltage and frequency perspectives. The latter must be managed during both continuous frequency regulation and 'events' due to loss of generation or demand. Other areas that require attention are power quality, fault conditions and associated protection.

A *micro grid* should be seen as a small mini grid, where a similar definition applies, namely that it supplies a small group of electricity customers through an isolated network, employing all the elements of a mini grid, but with a total load between 5 and 50 kW. *Nano grids* are a step below this, with loads normally in the range 500 W–5 kW. *Pico grids* are the smallest level of all, where the load is below 500 W, and would normally be associated with a remote stand-alone dwelling.

There is a strong social drive behind these small grids as remote, isolated or developing communities no longer want to wait for several more years for electricity supply to reach them. Global access to the Internet, the availability of mobile and smart phones and a greater awareness of comparative deprivation are all driving the need for sustainable electricity systems among even the most remote of communities across developing countries. Where there are no plans to build a main grid and associated distribution network, then the push for mini grids increases. Some key considerations include the following:

- Are policymakers and regulators driving the construction and operation of mini grids within their country with the appropriate incentives for investors? These will only come into the market if the rewards are in line with the risks profile.
- Are major development entities such as the World Bank, Asian Development Bank, African Development Bank and European Union actively encouraging and promoting such projects across developing communities through the provision of adequate funding? An adaptive and replicative approach must be adopted so that projects can be replicated 100 times, having been adapted to incorporate the local conditions and sources of energy.
- Will regulators adopt a business-like approach to tariff setting by making provision for a subsidy, or through a regulated tariff, guaranteeing an acceptable risk, or will they opt for a tariff that is cost-reflective plus a margin?
- Will there be a sufficient uptake of the electricity supplies, particularly as the cost per kWh may be significantly higher than the cost to those who are grid connected? Also, remote and island communities tend to be poorer than consumers in cities, hence the possible need for subsidies and the need to contain costs through proven replicable projects.
- The envisaged time frame for access to the national grid, even though several years away, is a critical factor in the 'transfer' of the mini grid assets to the national grid/distribution network. If investors believe that there is a considerable risk of grid access coming earlier than expected or of having their assets grossly undervalued when subsumed into the national grid/distribution network, then they will be reluctant to invest.

In summary, national governments, policymakers and regulators and development/funding agencies all have a key role to play in bringing electricity to remote communities and thereby opening up a whole new world to them. The knock-on effects can be enormous: for example, facilitating further education, remote learning with foreign universities, developing programming and software skills and facilitating remote work. In terms of addressing the inequitable access of people to enjoying a better standard of life, access to a sustainable electricity supply can be viewed as a critical step. Furthermore, designing these grids to supply energy from renewable and ubiquitous sources can greatly reduce the running costs to isolated communities and curtail the increase in global GHG emissions through a reduction in the use of imported fossil fuels.

4   Stabilisation of the Electricity Grid

Electricity supply grids are in a transitional period, driven by national and international objectives with respect to renewable energy. The current state of evolution of a particular electricity grid is dependent on many factors, including its current generation portfolio, the availability of non-fossil fuel generation alternatives and geographical considerations. Such transitions have resulted and will further result in fundamental changes to the power system generation portfolio, leading to changes in the operational characteristics of each system under both steady-state and transient conditions and requiring significant transformation of the composition of system services required for the reliable operation of the grid.

4.1   System Support Services

System support services, also referred to as ancillary services, are services other than the supply of energy that are required for the secure and reliable operation of an electricity system. As the maximum supply of energy from wind increases on a grid, the need for these service and newly defined services, *i.e.* services that would have previously been supplied as an intrinsic part of conventional power generation, increases. Frequency and voltage control are the key criteria for the TSO in maintaining a secure and stable system. Frequency control is maintained through spinning reserve and standby reserve whereas voltage control is maintained mainly through automatic voltage control on generators that are running (see the section on synchronous machines in the chapter Electrical Storage), supported by tap changing on transformers.

Security of supply may be of particular concern to large multinational businesses, data centres and other industries located in flagship business parks. The advanced electronic equipment found in modern working environments often requires a high degree of reliability from the incoming electricity supply. For example, many sophisticated manufacturing processes rely heavily on microcomputers, variable-speed drives and robotic devices to achieve high levels of product throughput and product quality. Other areas of business, such as data centres and the bio-pharmaceutical and IT industries, will only invest where the energy and telecommunications infrastructure is strong and quality of supply is assured. Even the most robust grids have some outages, so large customers with critical processes must have uninterruptible power supplies (UPS) to allow them to continue operation during short outage periods. Additional system services will be required to facilitate a secure electricity supply with large-scale wind and solar penetration.

4.1.1   Spinning Reserve.   Spinning reserve is the increase in generating capacity that is available from generators that are running at times when the system frequency is below the normal operating level. This additional generation capacity is provided by dispatching a number of running plants below their full output so that in the event of the loss of a large generator and a consequential decrease in frequency, these plants can rapidly increase output to restore the frequency to normal.

The general rule that TSOs use for determining the level of spinning reserve required is that it should be equal to the largest infeed, *i.e.* generator or interconnector. Hence if the largest infeed was 400 MW then the total spinning reserves required would equal 400 MW to cover the loss of this infeed. An additional rule for prudent operation of the grid is that the largest infeed should be limited to 10% of the total demand. There can be a significant cost in providing spinning reserve as plants with low positions on the merit order must be dispatched to provide spinning reserve, displacing plants higher in the merit order (that are unable to provide the same level or type of spinning reserve).

Following an event on the system that leads to a sudden decrease in the voltage and/or frequency, the spinning reserve will slow the rate of change of frequency (RoCoF). The initial support to the grid is provided by an instantaneous synchronous inertia response where the RoCoF is determined by a combination of factors, including the magnitude of the supply/demand imbalance and the type of generators remaining on the grid after the event.[40–42] For such an event, the instantaneous response is augmented by the spinning reserve. Initially, the rotational kinetic energy (*i.e.* the 'inertia') of the synchronous generators coupled to the grid works to reduce the RoCoF. The next stage of the recovery comes from the primary reserve. This is provided by the ability of currently operating power plants to increase their output temporarily within a number of seconds, but this increased output is available for only a short period, of the order of minutes. Stage three in the recovery comes from the secondary reserve and is done by activating storage from hydro power plants or dispatching thermal plants from part-load to full-load output. This is done within minutes of the event and must remain in operation until the final stage of recovery can be activated, namely utilising the long-term reserve (*i.e.* at least one power plant that is shut down is brought on load), fully replacing the lost power plant with another one.

Spinning reserve rules would normally be more stringent for an isolated system to ensure that system security is not compromised. In large systems with ac interconnectors across several countries, such as in Europe, each country will provide spinning reserve at a level agreed between TSOs. However, this may be influenced as much by economic reasons as by system security; interconnection will prevent system collapse in any country across the system, but there are often penalty charges for excessive imports across interconnectors from neighbouring national grids.

The definitions of different reserves vary slightly between countries. The following definitions apply in Ireland, but are typical:

- *Primary Operating Reserve* (POR) is the additional power output (and/or reduction in demand) delivered compared with the pre-incident output (or demand), which is fully available and sustainable over the period from 5 to 15 s following an event. Note: the timing of the 'event' is defined in terms of the frequency nadir.

- *Secondary Operating Reserve* (SOR) is defined in a similar way, but it must be fully available and sustainable over the period from 15 to 90 s following an event.
- *Tertiary Operating Reserve* (TOR) is sometimes divided into two time frames, namely TOR1 and TOR2: TOR1 is from 90 s to 5 min and TOR2 is from 5 to 20 min following an event.
- *Replacement Reserve* is over the period from 20 min to 1 h following an event and it can be divided into synchronous reserve from plant that is running [Replacement Reserve Synchronised (RRS)] and desynchronised reserve from plant that is shut down [Replacement Reserve De-Synchronised (RRDS)].

4.2 Impact of Renewables on Operation of Electricity Grid

High penetration of renewable generation poses challenges for the TSO with regard to grid stability. The solution to managing these challenges will be met through the utilisation of traditional responses and reserves and a mixture of new system services, including backup from batteries and management of demand-side loads. As the maximum percentage of wind energy on the grid increases, the stability of the electricity grid can decrease significantly. This is primarily not due to the unpredictability of wind but rather is due to the displacement of conventional synchronous generators. All synchronous machines (generators, motors and condensers) provide instantaneous response to imbalances between electricity generation and load by using the rotational kinetic energy stored in the synchronous machines as a buffer. For example, if too little energy is generated, the rotational kinetic energy of the machines is converted to electricity, slowing their rotation and the frequency of the grid; if too much energy is generated, the reverse occurs. The instantaneous nature of this synchronous response cannot be replicated by batteries or other non-synchronous energy storage devices. However, the augmentation of synchronous response with conventional generators or synchronous condensers (*e.g.* generators spinning at synchronous speed with zero output but providing voltage support; see the chapter Electrical Storage for further details) increases the production of carbon dioxide and/ or reduces the efficiency of the electricity grid.[41,42]

Up to penetrations of about 40% renewables, grid stability is provided by conventional plant with heavy steam turbines and/or generators, which are synchronised to the grid, rotating at synchronous frequency (3000 rpm, *i.e.* 50 Hz, across Europe) or a factor of that frequency. If a generator fails, causing the supply of energy to the grid to decrease rapidly, then the system frequency will fall as the synchronous machines slow and automatically convert their kinetic energy into electrical energy (see the section on synchronous machines in the chapter Electrical Storage). This stored energy is injected into the grid, reducing the rate at which the frequency of the grid falls. After a short duration, this recovery is supported by other running generators that increase output arising from governor response to the decrease in frequency.

As more and more renewable generators are connected to the grid, there is less and less room for heavy conventional synchronised generators to provide this stability. With the reduction in the number of large synchronous generators running, the system has considerably less 'inertia' and hence the trip of a large generator can cause a rapid decrease in frequency. For example, electricity systems often operate up to a RoCoF threshold of $0.5\ Hz\,s^{-1}$ measured over 500 ms, requiring all generators on the system to ride through any RoCoF of up to this level. However, if the instantaneous system non-synchronous penetration (SNSP) is high, *e.g.* >40% of energy originating from wind and solar, then there is an increased risk that the RoCoF threshold will be breached and some generators may fail to ride through the fault and trip out, making a bad situation worse. For this reason, operators that are increasing their instantaneous SNSP to >40% for the whole system are seeking to increase the RoCoF or fault ride-through capability of their generating plant to $1\ Hz\,s^{-1}$ measured over 500 ms.

RoCoF has been identified as an issue only in recent years, but it will become a bigger issue for system operators in the future as penetration levels of non-synchronous intermittent generation increase towards and above 40%. It is worth pointing out that this non-synchronous penetration is for a whole system rather than one national grid within the wider system.

From a grid stability perspective, synchronous inertia is currently the most proven method of mitigating high RoCoF events and will continue to be the most relied upon method as its reaction is instantaneous. Currently, up to 4% of fuel consumed by conventional plant connected to the electricity grid is dedicated to system service provision. As the penetration of renewable energy (non-synchronous generators) increases on the system, the requirement for additional services with faster response times, which can compensate to ensure system stability and reliability, will increase. The stability of the electricity grid is determined over several time scales, the most important of which is the first few seconds, since any decrease in the RoCoF increases the time that other systems have to react to imbalances between generation and consumption.[43–45] Currently, significant development work is being undertaken on non-synchronous technologies for the provision of synthetic inertia, particularly with batteries, flywheels and demand-side response. However, their capabilities and value as sources of synthetic inertia have yet to be demonstrated.

Although synthetic inertia has the potential to provide a power response to help prevent high RoCoF events, there are significant question marks regarding whether it can be delivered in the appropriate response time. First, the accuracy of detecting a RoCoF for the purpose of triggering the synthetic device and, second, the response and ramp rates of the specific non-synchronous device have a serious impact on the performance of these devices. If the synthetic inertia cannot be delivered in the appropriate time frame, it will not provide an alternative solution to maintaining the RoCoF standard at 0.5 $Hz\,s^{-1}$ measured over 500 ms.

A significant concern for RoCoF alternative solutions is the commercial aspects of such projects. The revenue streams may not be sufficient to support the development of these emerging technologies in the mainstream electricity market. Serious thought needs to be given to this matter by regulators and system operators, otherwise none of these alternative solutions will materialise. Also, because these emerging technologies do not fit comfortably with existing distribution and grid codes (catering for conventional, hydro and wind generation plant), there is a need to develop an appropriate grid code for energy storage.[46]

4.3   Corrective Measures for Mitigating RoCoF

Most conventional power plants continuously supply system inertia response that slows the RoCoF and they then provide increased output, referred to as primary operating reserve (POR), after 5 s following a system event; this would be regarded as a normal governor response to a decrease in frequency. However, a much more rapid response is needed to limit the RoCoF and assist in its restoration when there is a high penetration of nonsynchronous generation on the system. For this reason, the concept of fast frequency response (FFR) – also referred to as enhanced frequency response (EFR) – has been developed so that rapid responses in 500 ms and less can be provided as a system service to the system operator to support the frequency. This rapid service can be provided by both synchronous and nonsynchronous generators by responding rapidly to changes in frequency and supplement any intrinsic inertial response.

In particular, FFR or EFR provides a power response faster than the existing POR times and may, in the event of a sudden power imbalance, increase the time to reach the frequency nadir and mitigate the RoCoF in the same period, thus lessening the extent of the frequency excursion. Very fast-acting energy storage can provide such grid stability. This application requires high power over short durations but with a very high number of charging/discharging cycles, and can be provided by supercapacitors, batteries, flywheels or other energy storage technologies. Some types of storage are economic for long-duration storage whereas others are more suitable for high-power but low-energy storage capabilities. Some have a limited number of charge/discharge cycles whereas other technologies have almost unlimited cycling but higher capital costs. Although 2 s may appear to be a fast response, considerably faster response times, *i.e.* less than 0.5 s, will be required from these systems. The fast reaction speeds of flywheels and electrochemical systems and the long operational life spans of many of these technologies make them ideal candidates for grid stabilisation and load levelling.[4] These systems with the appropriate frequency detection and control systems can respond within times as short as 200 ms, but such technologies cannot provide the stability currently provided by synchronous 'inertia', *i.e.* electromagnetic coupling of rotational kinetic energy to the electricity grid through the use synchronous generators/motors. If sufficient system inertia is not present, *i.e.* the kinetic energy stored in the synchronous machines that are coupled to the grid is below a minimum level, the RoCoF may be rapid during this initial duration and the pseudo-synchronous responses not fast enough to protect the grid, thereby compromising its stability.

However, the wind generators on the grid have similar characteristics to conventional synchronous loads and generators in that they possess considerable rotational inertia. At present, the rotational inertia cannot be used to stabilise the grid since the generators are not synchronously coupled to the grid. Similarly, transport networks, pumps and other systems that have stored kinetic energy when in operation are usually run on non-synchronous motors or are not connected to the grid. Thus, if practical means were developed to run a fraction of these generators and motors synchronously with the grid, the stability of the grid would be improved. Currently, wind generators aim to maximise their energy efficiency, since they are paid per unit of energy produced, but in the future if they are also paid for the supply of auxiliary services, such as synchronous response, they may operate synchronously with the grid, increasing grid stability (but reducing their energy efficiency). Overall, the energy efficiency of the supply of electricity may be improved by such actions since currently this synchronous inertia response must be provided by conventional generators, some operating inefficiently and significantly below their rated output power.

4.4   Demand-side Solutions and Smart Grids

A stable and efficient electricity grid requires precise matching of power supply and demand. This is typically achieved with the supply planned to respond to changing demand, where demand is modelled based on times when equivalent parameters were observed and the output of power plants is scheduled accordingly.

Traditionally, demand-side solutions have been employed to shift demand away from peak usage times in order to avoid the dispatch of expensive peaking plant and/or to relieve pressure on other grid infrastructure. These programs are all characterised by the use of incentives to promote favourable demand-side behaviour to support system operations. One such scheme involves contracts with large industrial users to reduce consumption at peak times during a particular season in return for preferable rates. From the user's perspective, this may be achieved by either reducing the load or turning on a local generator. For the same reasons, householders can have 'night rate' meters to incentivise them to shift their usage to times of reduced demand on the grid. This would typically be available to dwellings that use electricity for space and water heating.

Contracts are also available for companies that can reduce their demand at short notice in the event that, for example, unforeseen demand is present on the grid, a scheduled power plant will not be available or if wind energy forecasts have been inaccurate. More recently, networks have developed schemes that enable medium- to large-size electricity users to participate in demand management. Such a demand-side management unit consists of one or more demand sites that can reduce their demand when instructed by the TSO. For example, these units could have 1 h to reduce their demand and then be capable of maintaining the demand reduction for 2 h. Such units are usually operated by a third-party company that may contract with a number of demand sites and aggregate them together to operate as a single business unit, ensuring a high degree of implementation whenever the TSO issues instructions to reduce demand. Demand sites typically use on-site stand-by generation, plant shutdown or storage technology to deliver the demand reduction, and are required to be available 24 h per day, year-round. The ambition for demand-side management is that, in addition to it being a tool to promote behaviour conducive to the routine operation of the grid, it can evolve to the point that it can be used to form part of the response to frequency events on the grid.

A frequency event is typically caused by the unplanned loss of a power plant or part of the transmission or distribution network that causes an imbalance in the available supply and demand of power on the grid. The job for the TSO, as explained earlier, is to restore the grid frequency as quickly as possible.[47] However, as wind and solar electricity are increasing their penetration on the grid, this increase in asynchronous generation leads to less availability of all reserve types. This is a significant worry for the TSO in its role of maintaining a stable and efficient electric grid. Therefore, the role of demand response in emulating the current response to a frequency event is of great interest and significance in enabling the electric grid to allow greater penetration of low-carbon renewable energy.

The extent to which demand management can play a role in this is determined by how fast it can be deployed, in what magnitude and for what duration. To this end, different types of demand response should be mapped to their equivalent stage in the recovery mechanism from a frequency event based on the parameters with definitions similar to fast frequency response, primary operating reserve, *etc.* While a user's on-site, backup generation, for example at a data centre, will have an inherent delay in starting up and would likely map to primary or secondary reserve, the load reduction of a coordinated shutdown of large refrigeration warehouses should be deployable in a manner comparable to fast frequency response.

This will require the development of suitable technology, protocols and market mechanisms to do so, but none of this appears to be particularly challenging if demand management is pursued to achieve its greatest potential. Ultimately, it is possible to envisage a smart grid scenario where all appliances are cognisant of the operating conditions of the grid and can make choices to behave in a manner that supports a clean, stable and efficient electricity system (see the chapter Smart Energy Systems for further details).

4.5   Need for Energy Storage

As mentioned before, a stable and efficient electricity grid requires precise matching of power supply and demand. As the level of renewable energy on the grid increases, its non-synchronous and intermittent nature will result in instability in the supply of energy by the grid and mismatch between supply and demand. Many of the technologies, system services, grid codes and smart operation of the grid can improve stability and reduce mismatch. For example, load levelling, demand-side management, use of untapped reserves and provision of auxiliary system services can support a future grid that encompasses less synchronously coupled energy (*i.e.* inertia). However, there will always be occasions when further reserves are required or traditional plant will have to be turned back on. Therefore, a key to stable operation of the future electricity grid depends on the connection of substantial extra reserves to, at a minimum, allow enough time for traditional plant to be turned on and possibly to allow the grid to be operated from energy reserves for longer durations.

The energy storage methods required to provide these reserves may include any technology that can transform electricity to a form of energy that can then be retrieved and transformed back into electricity at some later time. It follows that flywheels, batteries and capacitors can be used for the provision of such services, but also pumped hydro, hydrolysis of water and forestry can be part of the set of solutions. However, all of these systems have parameters that determine their suitability for a particular support service, including response times, life span, cyclability and inefficiencies associated with the building, holding and retrieving of energy.

Flywheels, turbines and other rotating masses coupled to synchronous machines store rotational kinetic energy that is automatically and instantaneously fed back to the electricity grid following a fault or frequency excursion, providing synchronous inertial response. However, such systems consume energy during their operation to overcome losses to friction and drag and they only provide response as a function of RoCoF (see the chapter Electrical Storage for more details). Capacitors, supercapacitors, asynchronous mechanical batteries (*e.g.* flywheels) and superconducting magnetic energy storage (see the chapter Electrical Storage for further details) can provide fast frequency response and operating reserves for several minutes but cannot respond in tens of milliseconds and are not designed to provide cost-efficient reserves for sustained durations. Batteries can also deliver fast frequency response and, in addition, are designed to deliver reserves for up to several hours, depending on the battery chemistry. However, as explained in the chapter Electrochemical Energy Storage, the cyclability, required capability and usage, life span, life cycle and other factors must be considered when deciding which battery chemistry and design to use. In addition, there is a range of other slower reacting technologies such as pumped hydro and compressed-air storage (see the chapter Mechanical Systems for Energy Storage for further details) that have very high cyclability and long life spans and can deliver operating reserves cost-effectively for long durations. Therefore, although storage is the key to reliable future grids, not all energy storage technologies are equal and not all storage applications will be best served by the same technology. Rather, a mix of technologies is required where the response times, capacities, CO_2 production, efficiencies, and financial and environmental costs can be balanced to deliver an efficient and reliable electricity grid for the supply of energy for a low-carbon economy.

5 Electric Vehicles and the Electricity Grids

Transportation accounts for 14% of current anthropogenic GHG emissions,[1] as shown in Figure 1. Transport is an increasingly large user and driver of electrical energy storage technologies. Conventional petrol- and diesel-powered engines already make use of energy storage in a limited way. For example, they incorporate flywheels for smooth engine operation (see the chapter Electrical Storage) and use electrochemical energy stored in their (usually lead–acid) batteries for engine starting (see the chapter Electrochemical Energy Storage). Significant progress has occurred over the last 10 years in the development of batteries for stop–start technology,[48] which allows engines to be turned off when the vehicle is stopped, and for traction technology where the energy of a (usually lithium-ion) battery is used to drive an electric motor.[49] The term electric vehicle (EV) covers a range of different technologies, ranging from hybrid to fully electric vehicles. The energy storage in these vehicles uses flywheels, capacitors, fuel cells and batteries (including lead–acid, nickel–metal hydride and lithium-ion batteries and others).

The electrification of transport provides both opportunities and challenges for the electrical grid.[50] The deployment of EVs is growing in many countries, driven by climate change targets and air quality concerns in major urban areas. For example, cities such as London and Paris are already signalling plans to ban combustion vehicles and China is already the leading adopter of EVs in terms of absolute numbers deployed.

The most desirable operating mode for battery EVs is for overnight charging at a moderate rate, typically C/4 to C/8 (*i.e.* 4 to 8 h of charging time), and a daily driving distance that is well within the vehicle range. Longer commutes may require workplace or destination charging. However, short trips represent the greatest proportion of vehicle usage, with a large-scale survey[51] in the USA showing an average journey length of 8.9 miles (14.2 km) per trip, 95% of trips less than 30 miles (48.2 km) and less than 1% of trips over 100 miles (161 km).

Long distance travel and slow recharge times have always been, and still are, the key limitations of fully electric vehicles. Much research and development effort has been expended over a long period of time to address these issues. Tremendous progress has been made, but inherent fundamental limitations make it a difficult problem.

Public perception and expectations have generated a mind-set whereby long-range battery EVs are considered a necessity for the adoption of EVs on a large scale. This has driven the desire for increasingly high energy densities in terms of both weight (Wh kg^{-1}) and volume (Wh dm^{-3}). However, such high densities bring both safety and environmental concerns. The highest energy density batteries to date involve lithium-ion chemistries with electrode materials containing cobalt compounds;[52] however, the price and environmental issues around cobalt mining have raised some concerns (see the chapter Electrochemical Energy Storage for greater detail on lithium-ion batteries).

Plug-in hybrids provide an interim solution, but most manufacturers of mainstream battery EVs have adopted fast-charging strategies to address the long-distance travel limitation. Other strategies such as battery swap and mechanical recharging with metal–air batteries have been proposed and tested, but so far have not been adopted on a larger scale, although they may be in the future. The deployment of public charging points with fast chargers is being undertaken in many countries and, together with a strong ICT support infrastructure, can support EV long-distance travel.[53] It may be the case that safer and more environmentally benign battery chemistries with lower energy densities, together with public fast charging or battery swap facilities, may be a solution for the mass adoption of EVs.

Alternatively, plug-in hybrids may become a permanent solution. This could possibly occur if electrification of motorways *via* overhead cables and pantographs or *via* electrified tracks facilitated an extended driving range while requiring only sufficient energy for the journey to and from the motorway to be stored in the vehicle.[54] This would have significant advantages since each EV would need only a small energy capacity in its battery, reducing supply and demand issues and, most importantly, reducing security issues that occur due to war over limited resources, *e.g.* for cobalt in the Congo.[55] Furthermore, smaller energy capacity requirements would mean that many battery chemistries would be capable of providing the necessary capacity, reducing the cost per kWh of storage and facilitating a change in focus in research and development from maximisation of energy density to other considerations, such as lifetime, recyclability, efficiency and cost. Furthermore, even if such electrification were applied only to trucks, it would greatly reduce the GHG emissions from transport since, although trucks account for only 9% of the total vehicle population, they account for 17% of vehicle distance travelled and 39% of life-cycle road transport GHG emisions.[54]

Battery storage in EVs, currently dominated by lithium-ion technology, ranges from a few kWh for plug-in hybrids to tens of kWh for mainstream battery EVs.[56] High-end vehicles and future models can have over 100 kWh of storage. The peak power available from such batteries is typically at the

100 kW level (3 to 5C) for mainstream battery EVs, with several hundred kilowatts for high-end models.

The interaction of such vehicles with the electricity grid occurs when the vehicle is connected to the grid for charging and two distinct modes of charging can be identified: slow (or time-insensitive) overnight or workplace charging and fast (or time-sensitive) charging where the user is waiting for charge completion.

5.1 Slow Charging

Overnight home charging and destination charging are normally achieved with an on-board charger (OBC), built into the vehicle, and typically range up to 7.2 kW, *i.e.* 240 V single phase at 32 A. Such chargers are normally uni-directional and connect to the residential grid connection, typically through dedicated electric vehicle supply equipment (EVSE).

Bidirectional on-board charging equipment is technically readily achievable, with some additional complexity and cost in the associated power electronics. Often termed vehicle-to-grid (V2G), such equipment could provide a similar power level back to the electricity grid for load balancing or emergency stabilisation of the grid. Currently most EV manufacturers do not provide this as an option as there has been little demand from customers and an economic model to justify the additional cost and inconvenience to the vehicle owner has not yet been made.

Electricity utilities have identified potential capacity issues on legacy low-voltage distribution networks with future mass adoption of EVs along with mass adoption of electric heat systems, such as heat pumps. However, installation of smart grid support in domestic EVSE units could allow for smart charging algorithms to achieve maximum utilisation of local distribution networks without overload.[57] EV storage batteries and on-board chargers can readily support such variable charge rates and timing, but state-of-charge determination for the batteries may need to be improved to optimise charging rates.

With peak demand for electricity being normally in the daytime and early evening, the use of overnight charging is seen as a desirable feature, with many electric utilities already providing reduced tariffs for night-time electricity usage. Overnight or off-peak charging can directly contribute to a better load balance for electric utilities and effectively increase the utilisation of existing grid and distribution infrastructure.

The economics of EVs with overnight home charging are favourable, with capital and running costs for EVs in this mode being attractive, particularly in countries with high petroleum prices, *e.g.* in the European Union.

5.2 Fast Charging

The fast charging options available come in a number of standards and are listed under a range of terms such as rapid, quick or fast charge, but are probably best described as dc charging (although there is also a fast ac charge option from at least one manufacturer). Such dc charging options, in effect, present a port on the vehicle with direct connection to the terminals of the main traction battery, and also a signalling channel through which information such as battery voltage and current ranges and status can be exchanged.

The power electronics for conversion from the ac electricity grid to the dc source required for charging is then situated outside the vehicle and can be sized to a high power level. Power levels for mainstream battery EVs are typically up to 50 kW with proprietary systems available in excess of 100 kW. Increasing power levels up to and above 100 kW are being supported by existing standards and power levels of 350 kW are available as technical demonstrators.

As a direct connection to the main traction battery terminals is provided, it is possible to draw energy from the battery. With the power electronics conversion being done outside the vehicle, the additional cost and complexity of bidirectional power circuits are less of an issue and, for example, might be borne by the electric utility rather than the vehicle owner. In this manner, there is potential for the use of EVs as a distributed storage resource, subject to the development of a viable economic model. Indeed, if such chargers were used for slow charging whenever an EV was parked, the batteries would have longer life spans (compared with batteries that are always fast charged) and the batteries would be available for FFR and POR support of the electricity grid.

The fast charging capability of EVs was primarily intended as a method of mitigating their range limitation. To this end, the typical manufacturer claim is a 30 min period to fast charge the battery to an 80% state of charge from empty. For mainstream battery EVs, this requires a source and charger power of typically 50 kW with levels up to 100 kW expected. The expectation is that with increasing EV adoption, banks of such chargers would be located regularly along main transport arteries, typically every 50 to 100 km. Figure 4 shows an example of an eight-bay EV charge site located in Birdhill, Ireland. Each set of two bays has a power delivery capacity of 120 kW.

Figure 4   Eight-bay Tesla super charge site in Birdhill, Ireland, requiring 4×135 kVA grid connections, *i.e.* in excess of 0.5 MVA of grid connection. Photographer: Thomas Conway.

There is increasing research into higher recharge speeds, with claims such as 5 min recharge times. To charge a 50 kWh vehicle to 80% state of charge from empty in 5 min would require 500 kW per individual vehicle.

In the context of fast charging, the effect on the grid is potentially severe, as such loads represent a high load level with highly variable usage patterns and poor load factors. The demand for high power availability along major transport routes, where grid infrastructure may not have been provisioned, may also present a challenge for current electricity grids.

However, the demand for fast charging could be viewed as a high-value product with EV users prepared to pay a premium for energy availability at a high power level at a suitable public location. Such observation has led some to propose that such locations could be an ideal location for dual-purpose, grid-level, energy storage facilities.

The premium available for supply of energy to EV users may improve the economic viability of such deployments and the distributed nature of such locations may provide a synergy with the distributed nature of many renewable energy sources. The demand for high-power sources can then be met from local storage, reducing the need for upgrading of the existing electricity grid. Some fast charger facilities already incorporate local battery storage and local renewable energy sources such as solar panels.

5.3   End-of-life Usage

A further way in which EVs may impact the grid and electric energy storage is in the area of second use for EV battery packs. As the battery packs age through use in EVs, their energy storage capacity degrades and at some point may become too low for the users' range demands, or the vehicle may reach its end of life. Although battery material recycling may become realisable for EV battery packs, there are currently several factors – including a range of chemistries, structure of the electrodes and location of the active materials within other supporting materials – that contribute to making lithium-ion battery recycling more complicated than lead–acid or nickel–metal hydride battery recycling.[49,58,59] Furthermore, collection rates are currently very low, with only 5% of lithium-ion batteries being collected in 2010 in the European Union.[60] However, the cells of used EV battery packs still have a useful energy storage capacity and can be utilised for stationary storage.[61] It is too early at present to assess the impact of these effects and issues, but over the next several years a more significant number of vehicles are expected to reach their end of useful life and a battery pack reuse industry is expected to emerge.

5.4   Implications of Connecting Electric Vehicles to the Electricity Grid

The mass embracing of EVs will be a transformative change for the electricity grid. Their adoption, if not carefully managed, will require significant reinforcement of the grid and lead to further peaks in energy demand. These demands on the grid will require greater power generation and may lead to greater grid instability. However, the energy storage technology in EVs can also be viewed as a distributed network of energy storage devices that can be used for grid stabilisation, load levelling and smart operation of the grid. Furthermore, if this technology can be operated smartly, *e.g.* if EVs are charged slowly during the night, the variation in electricity supply during day-time peaks and night-time troughs will be reduced, increasing the use of wind energy whenever it is available and reducing the need for the electricity grid to invest in and own energy storage. However, since these EVs will probably be operating during day-time hours, their energy storage capacity will not be connected during peak operation times of the electricity grid.

If EVs are directly connected to the grid during operation *via* overhead cables or in-road tracks, they will require only small batteries and for significant durations, particularly in the case of freight transport, the motors in these vehicles could run directly from the grid. Although such systems will require significant investment,[62] this investment may reduce the need to reinforce existing grids to accommodate energy harvesting from renewables, charging of EVs and smart grid operation. Such a transport fleet would add stability to electricity supply at all times of the day since the batteries could be used for shedding of loads and backup of the grid, *i.e.* making them an integral and flexible part of a future smart grid. Therefore, the average power and number of generators connected to the grid would increase, increasing the reliability and stability of the grid. Furthermore, if the motors were synchronously coupled to the grid, the increased level of stored kinetic energy, and therefore the increased synchronous inertial response, would reduce the RoCoF and thereby increase the reliability of the overall grid. The increase in synchronous inertial response that could be achieved by such an approach would be small. However, as with synchronous inertial response delivered from any demand-side synchronous machine, where the energy losses to friction and drag occur regardless of whether or not the machine's operation is synchronised to the grid, support to the grid does not cost any additional energy.

The increasing adoption of EVs worldwide inherently provides a large electricity storage capability, and helps to load-level the grid by providing an off-peak demand for electric power. With more than 2 million fully electric vehicles on the roads already,[63] representing in the order of 40 GWh of storage, the widespread adoption of electricity storage is already under way.

6   Conclusion

The stored energy in coal, wood and hydrocarbon oils and gases originates from nuclear reactions in the Sun. This reaches the Earth as solar energy and has been stored in the form of chemical energy by living flora and fauna. Indeed, the energy in the wind originates from the Sun and, therefore, with the notable exceptions of geothermal, tidal and nuclear energy, most of the energy we utilise originates from solar energy.

The drive to harness and control energy is an inherent characteristic of the human race. From our ancient use of tools, fire and animals to today, every new technological advance has involved some new way of controlling energy. Each advance by itself may not have been ground-breaking at the time, but our history seems to be divided up into periods dominated by a new form of control of energy. Today it would seem that we are at the beginning of a new era where energy is no longer harnessed primarily from the burning of fossil fuels but from renewable sources of energy.

Stored forms of solar energy are being used up at a rapid rate and many of the most prosperous nations in the world now rely on significant imports of fossil fuels to ensure a continuous supply of energy so as to support their economies and maintain their ways of life. It follows that for many countries the burning of fossil fuels can be seen not just as depleting finite natural resources and producing significant anthropogenic GHGs and pollution but also as an extremely significant security-of-supply and political concern. For instance, 54% of the European Union's energy originated from fossil-fuel imports from non-EU countries in 2015, and the main supplier of oil, gas and solid fuels to Germany and Poland was Russia, on which these countries are therefore highly dependent for heat, transport and electricity.[64] It follows that the use of solar, wind and other ubiquitous energy sources to offset the use of fossil fuel imports is not only advisable for the care of our planet but also for increasing security of supply and possibly reducing international tensions.

However, the technologies that will support a future based on energy from such sources are not obvious and the transition from our current regime – supported by combustion engines and conventional electricity generators – is fraught with problems and technological obstacles. Nevertheless, these challenges present opportunities for individuals, companies and nations that can successfully transition to smart and reliable electricity grids powered by renewables. In such an ever-changing world, it is important for nations to be nimble and to peruse multiple new ideas. This is particularly true in the area of energy where renewable energy, energy storage and smart grids may greatly increase both our energy and political stability.

Reliable and stable electricity supply at a time of increased renewable energy penetration of the electrical grid has been highlighted as one of the key challenges in the drive for future prosperity and development. International agreements on reduction of carbon emissions and increase in the use of renewable energy sources mean that we will require technologies to mitigate the difficulties related to the supply of energy in this changing environment.

As smart grids are developed, large fluctuations between day-time and night-time electricity load can be greatly reduced, for example by charging EVs and performing other non-urgent tasks at night. Overall, although the fluctuations in consumption of electricity may be reduced, the total electricity consumed seems destined to increase, placing significant strain on existing electricity grid infrastructure. The expansion of the grid due to both distribution of generation, particularly to geographically isolated locations, and increased flows of electricity will require significant reinforcement of the existing grid infrastructure and may lead to the return of local grids. Such a transition will be helped by EVs, *etc.*, which, in addition to increasing average load, may lead to increased stability and significant changes in the operation of electricity grids.

References

1. Intergovernmental Panel on Climate Change in *Climate Change 2014: Mitigation of Climate Change: Working Group III Contribution to the Fifth Assessment Report of the Intergovernmental Panel on Climate Change*, ed. O. Edenhofer, Cambridge University Press, New York, NY, 2014.
2. C. Le Quéré, R. J. Andres, T. Boden, T. Conway, R. A. Houghton, J. I. House, G. Marland, G. P. Peters, G. van der Werf, A. Ahlström, R. M. Andrew, L. Bopp, J. G. Canadell, P. Ciais, S. C. Doney, C. Enright, P. Friedlingstein, C. Huntingford, A. K. Jain, C. Jourdain, E. Kato, R. F. Keeling, K. Klein Goldewijk, P. Levis, P. Levy, M. Lomas, B. Poulter, M. R. Raupach, J. Schwinger, S. Sitch, B. D. Stocker, N. Viovy, S. Zaehle and N. Zeng, *Earth Syst. Sci. Data Discuss.*, 2012, 5, 1107.
3. Intergovernmental Panel on Climate Change in *Climate Change 2007: Mitigation of Climate Change: Contribution of Working Group III to the Fourth Assessment Report of the Intergovernmental Panel on Climate Change*, ed. B. Metz, Cambridge University Press, Cambridge, New York, NY, 2007.
4. L. D. Mears, H. L. Gotschall, T. Key, H. Kamath in *EPRI-DOE Handbook of Energy Storage for Transmission & Distribution Applications*, ed. I. P. Gyuk and S. Eckroad, EPRI, Palo Alto, *CA* and the U.S. Department of Energy, Washington, DC, 2003.
5. *World Energy Resouces 2016*, World Energy Council, London, UK, 2016.
6. *BP Statistical Review of World Energy*, BP plc, London, UK, 2017.
7. *World Energy Outlook 2017*, International Energy Agency, Paris, France, 2017.
8. World Coal Association: Coal & Electricity, https://www.worldcoal.org/coal/uses-coal/coal-electricity (accessed 15 May 2018).
9. BP Global: Energy Economics, https://www.bp.com/en/global/corporate/energy-economics/statistical-review-of-world-energy/coal.html, (accessed 12 June 2018).
10. *DOE Fundamentals Handbook: Nuclear Physics and Reactor Theory: Reactor Theory (Nuclear Parameters)*, Department of Energy, Washington, DC, 1993, vol. 2.
11. C. Allardice and E. R. Trapnell, *The First Reactor*, US Department of Energy, Washington, DC, 1982.

12. *DOE Fundamentals Handbook, Nuclear Physics and Reactor Theory: Atomic and Nuclear Physics*, Department of Energy, Washington, DC, 1993, vol. 1.
13. IAEA Power Reactor Information System, https://www.iaea.org, (accessed 14 June 2018).
14. Reuters, https://www.reuters.com/article/europe-power-supply/germany-powers-france-in-cold-despite-nuclear-u-turn-idUSL5E8DD87020120214, (accessed 15 May 2018).
15. *Key World Energy Statistics*, International Energy Agency, Paris, France, 2017.
16. *The Nuclear Fuel Report*, World Nuclear Association, London, UK, 2017.
17. G. R. Keepin, *Physics of Nuclear Kinetics*, Addison-Wesley Pub. Co, Reading, MA, 1965.
18. United Nations and Scientific Committee on the Effects of Atomic Radiation in *Sources, Effects and Risks of Ionizing Radiation: United Nations Scientific Committee on the Effects of Atomic Radiation: UNSCEAR 2013 Report to the General Assembly with Scientific Annexes*, United Nations, NY, 2013.
19. L. Mathieu, D. Heuer, R. Brissot, C. Garzenne, C. Le Brun, D. Lecarpentier, E. Liatard, J.-M. Loiseaux, O. Méplan, E. Merle-Lucotte, A. Nuttin, E. Walle and J. Wilson, *Prog. Nucl. Energy*, 2006, **48**, 664.
20. *Thorium Fuel Cycle: Potential Benefits and Challenges*, International Atomic Energy Agency, Vienna, Austria, 2005.
21. *Technology Roadmap Solar Photovoltaic Energy: 2014 Edition*, International Energy Agency, Paris, France, 2014.
22. L. L. Ruttledge and D. Flynn in *IEEE Transactions on Power Systems*, IEEE, 2011, vol. 31, p. 1.
23. A. Ulbig, T. S. Borsche and G. Andersson, *IFAC-Pap.*, 2015, **48**, 541.
24. L. Ran, J. R. Bumby and P. J. Tavner, in *11th International Conference on Harmonics and Quality of Power*, IEEE, 2004, p. 106.
25. E. Muljadi, V. Gevorgian, M. Singh and S. Santoso in *IEEE Power Electronics and Machines in Wind Applications*, IEEE, 2012, pp. 1.
26. A. D. Hansen, M. Altin, I. D. Margaris, F. Iov and G. C. Tarnowski, *Renewable Energy*, 2014, **68**, 326.
27. O. Langhamer, K. Haikonen and J. Sundberg, *Renewable Sustainable Energy Rev.*, 2010, **14**, 1329.
28. E. Rusu and F. Onea, *Clean Energy*, 2018, zky003.
29. S. P. Neill, M. R. Hashemi and M. J. Lewis, *Energy*, 2014, **73**, 997.
30. S. P. Neill, M. R. Hashemi and M. J. Lewis, *Renewable Energy*, 2016, **85**, 580.
31. A. Westwood, *Refocus*, 2004, **5**, 50.
32. W. Borgquist, E. Meyer-Peter, T. Norberg Schulz and A. E. Rohn in *The Electrification of the Irish Free State: The Shannon Scheme*, The Stationary Office, Dublin, 1924.
33. J. W. Lund, *Geo-Heat Cent. Q. Bull.*, 2007, **28**, 1.
34. J. W. Lund and T. Boyd, *Geo-Heat Cent. Q. Bull.*, 2010, **29**, 12.

35. Y. T. Shah, in *Thermal Energy: Sources, Recovery, and Applications*, Chapman and Hall/CRC, 2018.
36. ECR Technologies, Inc., *EarthLinked® Ground-Source Heat Pump Water Heating System*, US EPA and Greenhouse Gas Research Center, Southern Research Institute, Frederick, MD, 2006.
37. *UK Electricity Networks*, Parliamentary Office of Science and Technology, London, UK, 2001.
38. The World Bank, Access to Electricity, https://data.worldbank.org/indicator/EG.ELC.ACCS.ZS?view=chart, (accessed 19 June 2018).
39. M. Safi in *The Guardian*, https://www.theguardian.com/world/2018/apr/30/india-fully-electric-after-last-village-connected-claims-government, (accessed 14 April 2018).
40. P. Daly, D. Flynn and N. Cunniffe, in *2015 IEEE Eindhoven PowerTech*, IEEE, Eindhoven, 2015, p. 1.
41. V. Knap, S. K. Chaudhary, D.-I. Stroe, M. Swierczynski, B.-I. Craciun and R. Teodorescu, *IEEE Trans. Power Syst.*, 2016, **31**, 3447.
42. *Electric Power Systems*, ed. B. M. Weedy, John Wiley & Sons, Ltd, Chichester, West Sussex, UK, 5th edn, 2012.
43. L. Bohan, *Advanced RoCoF Protection of Distribution Systems*, University of Nottingham, UK, 2012.
44. A. Dyśko, D. Tzelepis and C. Booth, *Assessment of Risks Resulting from the Adjustment of RoCoF Based Loss of Mains Protection Settings*, University of Strathclyde, UK, 2015.
45. T. Lobos and J. Rezmer, *IEEE Trans. Instrum. Meas.*, 1997, **46**, 877.
46. F. Díaz-González, M. Hau, A. Sumper and O. Gomis-Bellmunt, *Renewable Sustainable Energy Rev.*, 2014, **34**, 551.
47. *Electrochemical Energy Storage for Renewable Sources and Grid Balancing*, ed. P. T. Moseley and J. Garche, Elsevier, Amsterdam, Netherlands, 2014.
48. *Lead-acid Batteries for Future Automobiles*, ed. J. Garche, E. Karden, P. T. Moseley and D. A. J. Rand, Elsevier, Amsterdam, Netherlands, 2017.
49. L. Gaines, *Sustainable Mater. Technol.*, 2014, **1–2**, 2.
50. B. Bilgin, P. Magne, P. Malysz, Y. Yang, V. Pantelic, M. Preindl, A. Korobkine, W. Jiang, M. Lawford and A. Emadi, *IEEE Trans. Transp. Electrification*, 2015, **1**, 4.
51. U.S. Department of Transportation, Federal Highway Administration, 2017 National Household Travel Survey, https://nhts.ornl.gov/vehicle-trips, (accessed 15 May 2018).
52. N. Nitta, F. Wu, J. T. Lee and G. Yushin, *Mater. Today*, 2015, **18**, 252.
53. T. Conway, *IEEE Trans. Intell. Transp. Syst.*, 2017, **18**, 2311.
54. M. Moultak, N. Lutsey and D. Hall, *Transition to Zero-Emission Heavy-Duty Freight Vehicles*, International Council on Clean Transportation, Washington, DC, 2017.
55. A. King, in *Irish Times: Life Science*, Dublin, Ireland, edn. 10 May, 2018, p. 10.
56. E. A. Grunditz and T. Thiringer, *IEEE Trans. Transp. Electrification*, 2016, **2**, 270.

57. A. Dubey and S. Santoso, *IEEE Access*, 2015, **3**, 1871.
58. N. Natkunarajah, M. Scharf and P. Scharf, *Procedia CIRP*, 2015, **29**, 740.
59. X. Zeng, J. Li and N. Singh, *Crit. Rev. Environ. Sci. Technol.*, 2014, **44**, 1129.
60. The Guardian, https://www.theguardian.com/sustainable-business/2017/aug/10/electric-cars-big-battery-waste-problem-lithium-recycling (accessed 15 May 2018).
61. J. S. Neubauer, E. Wood and A. Pesaran, *SAE Int. J. Mater. Manuf.*, 2015, **8**, 544.
62. Siemens USA, http://siemensusa.synapticdigital.com/Featured-Multimedia-Stories/siemens-ehighway-roll-out/s/e9ff6e36-9c89-43b3-98e4-3a773420e98f, (accessed 17 June 2018).
63. *Global EV Outlook 2017: Two Million and Counting*, International Energy Agency, Paris, France, 2017.
64. Eurostat, http://ec.europa.eu/eurostat/statistics-explained/index.php/Energy_production_and_imports#More_than_half_of_EU-28_energy_needs_are_covered_by_imports, (accessed 19 June 2018).

Mechanical Systems for Energy Storage – Scale and Environmental Issues. Pumped Hydroelectric and Compressed Air Energy Storage

DAVID J. EVANS,* GIDEON CARPENTER AND GARETH FARR

ABSTRACT

This chapter introduces large-scale utility (bulk) energy storage in the form of pumped hydroelectric and compressed air energy storage. Both are mechanical energy storage technologies, converting electrical energy into potential energy, and both fall into the category of grid-scale energy management. Brief reviews and discussions relating to the general operational aspects and the legislative and environmental aspects of the two storage types are provided in the context of UK development. Both storage technologies offer the potential for better integration and penetration of renewable electricity sources and the reduction of greenhouse gas emissions.

1 Introduction

International, national and regional energy policies aim to promote the use of renewable-based electricity (RES-E) to reduce carbon emissions and secure local power supplies. The European Union has set ambitious greenhouse gas (GHG) emission reduction targets of 20% by 2020 and 80% by 2050.[1] In addition, the United Nations set a target of doubling the

*Corresponding author.

Issues in Environmental Science and Technology No. 46
Energy Storage Options and Their Environmental Impact
Edited by R.E. Hester and R.M. Harrison
© The Royal Society of Chemistry 2019
Published by the Royal Society of Chemistry, www.rsc.org

renewables share in the global energy mix by 2030, from the 18% seen in 2010 to 36% in 2030.[2] The transition to and increasing the use of renewable energies such as wind and solar, which are by nature intermittent, will introduce increased natural variability into electricity generation and capacity. To meet patterns of demand that do not follow such variations in generation, there will be a need for fast-ramping, backup generation, supported by reliable forecasting and, importantly, increased bulk, grid-scale storage capacity for electricity generated from renewables.

The UK has made considerable progress in decarbonising electricity generation as it moves towards a low-carbon energy system to meet climate change targets. This process has involved increased deployment of renewable energy technologies and changes in electricity consumption, alongside decommissioning of ageing coal-, nuclear- and gas-fired power stations.[3-5]

These rapid changes are presenting challenges for the future of the UK's electricity system. The increase in renewables is leading to an increasingly decentralised generating system and higher amounts of generation intermittency from natural energy sources. The increase in intermittency requires flexible generation to balance the system. In the past, fast-response thermal generation has largely met this need. The UK National Infrastructure Commission has set out priorities for a future energy system identifying the role of interconnectors, energy storage and flexible demand as being key innovations that will enable a smart power system to secure supply and meet carbon targets.[6] Importantly, it has been calculated that these innovations would save consumers up to £8 billion in system operating costs that would otherwise have been required for conventional investment in the energy system.

Although interconnectors will play an important role, increased energy storage will not only balance existing flexible generation, but will also enable greater amounts of renewables to be integrated into the system in the future. Technology developments and falling costs mean that battery storage is becoming increasingly common at a range of scales, although collectively their energy storage capacity remains small.

Energy storage involves converting energy from forms that are difficult to store to more convenient or economically viable storage forms. The energy can be stored in various ways, including chemical, electrochemical, thermal and mechanical storage. Some technologies provide short-term, small-scale energy storage options, whereas others represent grid-support, load-levelling and longer-term mechanical bulk (utility scale) energy storage options (see Figure 1). The two largest sources of mechanical energy storage are *pumped hydroelectric storage* (PHS) and *compressed air energy storage* (CAES):[7]

1. *PHS* – This is a type of hydroelectric energy storage used by electric power systems for load balancing. Energy is stored in the form of gravitational potential energy of water, pumped from a lower to a higher elevation reservoir. Typically, pumps are run by low-cost surplus off-peak electric power. During periods of high or peak electrical demand, release of the stored water through turbines generates electric

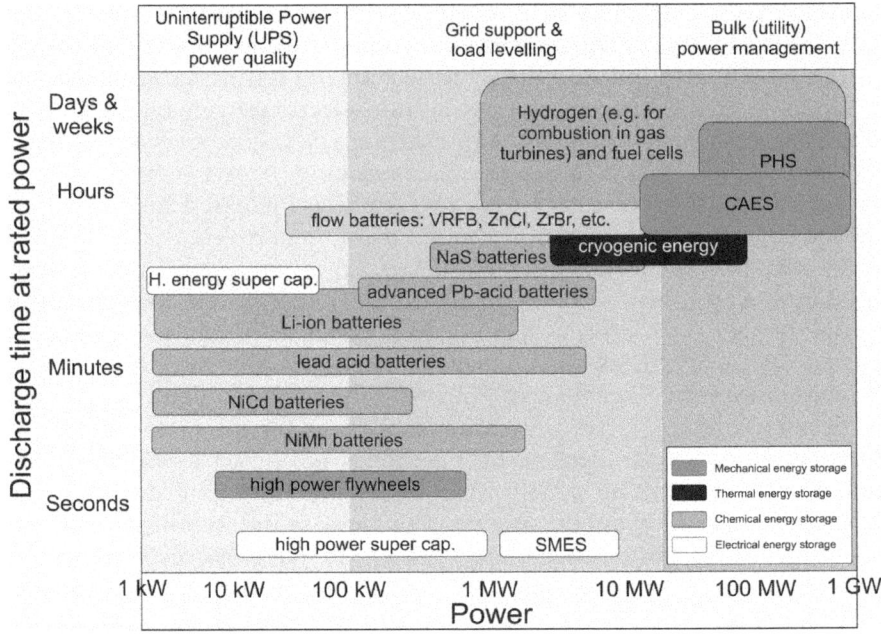

Figure 1   Schematic of main energy storage systems by duration and power. Compiled from ref. 15, 16 and 157.

power. Pumped hydroelectric storage is, however, a proven technology that can reliably deliver a large-scale and fast-responding storage capacity.[8] It is the most common form of energy storage, with a capacity of 188 GW, representing 96% of the total global energy storage capacity.[7] Geography, however, constrains the ultimate capacity for PHS and, as such, in most developed countries only limited prospects exist for further development, including in the UK,[9,10] although significant potential for hydro-storage capacity may still exist in many other areas around the world.[11]

2. *Conventional (diabatic) CAES, or adiabatic compressed air energy storage (ACAES)* – This is essentially equivalent to PHS in terms of applications, output and storage capacity. CAES is the second commercially proven, large-scale electrical energy storage (EES) after PHS, but instead of pumping water from a lower to a higher reservoir during periods (generally off-peak) of excess power or energy produced by base gas-, oil-, nuclear- or coal-fired units, air is compressed and stored in an air store 'reservoir' (currently salt caverns). As with pumped storage, release of power can be very quick and at peak times, as the air is withdrawn and used in the generation of electricity. CAES is viewed increasingly as offering bulk storage potential and a solution to levelling intermittent renewables generation [wind-power and solar photovoltaic (PV) plants] and maintaining a system balance.[12,13] CAES technology is based on the principle of traditional gas-turbine plants

and has some advantages over pumped storage, including the fact that its visible impact on the landscape is lower and there is more scope for building CAES facilities nearer the centres of wind-power production, especially in Germany and areas of the USA. Worldwide CAES capacity is 431 MW.[8]

CAES is not, however, an independent system, as it requires gas for heating of air and has to be associated with a gas-turbine plant during electricity generation – it cannot be used in other types of power plant during generation.[14] More importantly, the requirement for combusting fossil fuels and the associated emissions make CAES less attractive compared with other EES technologies.[15–17] However, advances in CAES technologies and linking to renewables generation (including offshore wind) offer the prospect of improved cycle efficiencies, with the reduction and possibly elimination of emissions.

The smaller contribution of grid-scale energy storage from CAES, with only two commercial facilities constructed and currently operational using solution-mined salt caverns, does not reflect physical or environmental limitations, more the lack of policy and economic drivers for energy storage in general and CAES in particular.[18]

This chapter considers the two main types of bulk, grid-scale, mechanical energy storage in PHS and underground (geological storage) CAES. Underground geological storage offers potentially very large volumes and is of five main types: salt caverns (solution mined or conventional mining), porous rock (aquifer or depleted hydrocarbon field), hard rock mined void, abandoned non-salt mine and lined-rock cavern.

In the UK, four PHS sites are operational and a further six planned, whereas only one CAES plant is planned in Northern Ireland, although others are being considered in Cheshire, all using salt-cavern storage (see Figure 2).

Non-geological CAES options not considered here include above-ground or shallowly buried containers or pipes.[19] Although such systems are attractive in that they permit the siting of CAES plants almost anywhere as no underground geological storage formation is required, they are estimated to be about five times more expensive than salt-based air storage caverns and porous aquifer-based air storage systems.[20] Other novel systems include energy bags secured to the sea bed,[21] schemes linking wind turbines with energy storage in supporting legs,[22–24] or where power is converted directly from the rotor by means of gas/air compression within the rotor blades.[25]

2 Pumped Hydroelectric Storage – Introduction to the Technology, Geology and Environmental Aspects

PHS is a well-established form of storage used by electric power companies in many countries for load-balancing, utility-scale electricity storage. Plants

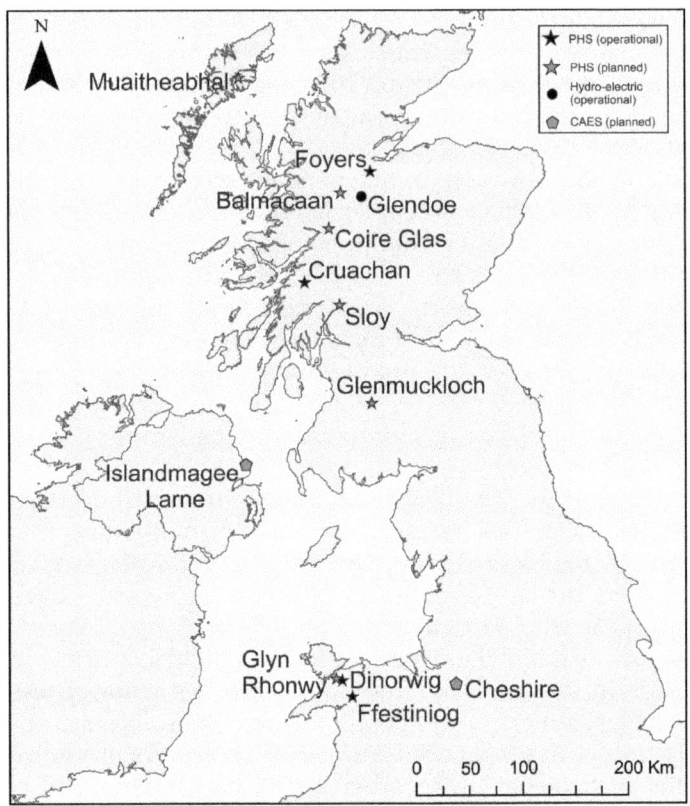

Figure 2　Location of operational and proposed PHS, hydroelectric and CAES facilities in the UK and referred to in the text.

that do not use pumped storage are classified as conventional hydroelectric plants. Essentially, in PHS water pumped from a lower to a higher elevation reservoir during off-peak periods stores energy in the form of the gravitational potential energy of water. In comparison with conventional hydroelectric dams and power generation, the reservoirs used with PHS are smaller and generating periods are much shorter, being measured in hours to parts of days. Pumped storage is the largest-capacity form of grid energy storage available, with more than 300 facilities worldwide.[26] However, the relatively low energy density of PHS/PHES systems requires either a very large body of water or a large variation in height between the reservoir and turbines.[27]

PHS was first deployed at sites in the Swiss and Italian mountains in the 1890s[28] and the first sites in Germany date back to the 1920s.[29] The first use of PHS in the USA was by the Connecticut Electric and Power Company in 1930, using a large reservoir located near New Milford, Connecticut. The water was pumped to the storage reservoir from the Housatonic River about 70 m below.[30] In the UK, the first PHS facility commissioned was at

Ffestiniog in North Wales during the 1960s. Dinorwig was the first hydro-power development built, using an old slate quarry in Elidir Mountain in the 1970s, with the modern Dinorwig plant having opened in 1984 and which has provided over 30 years of storage and supply.[31]

A number of PHS system configurations are possible, including open- and closed-loop, underground and pump-back systems (see Figure 3). Open-loop systems generally comprise an upper reservoir with no real natural inflow (inflowing river), which delivers water through tunnelling to the turbines underground and discharges to a river or lower reservoir. In some cases, the lower reservoir is the ocean, with the system using seawater. The upper reservoir may be an open natural or engineered body of water anywhere at the surface, provided that there is sufficient topography to provide the 'head' of water required for operation. This could be in mountain sites or along coasts with sufficiently high cliffs. The reservoir could be artificial or make use of an existing depression or mined void such as a former quarry, as at operational Dinorwig and the proposed Glyn Rhonwy plant in North Wales.[31,32] Such sites tend to be located in areas of massive, homogeneous, hard rock types including igneous and crystalline metamorphic rocks such as granites and gneisses, or some hard sedimentary rock types like dolomites. These rock types are more likely to contain fewer fractures and be impermeable, preventing leakage. However, grouting and lining of the storage and any dam structure may be required. Hence sites are likely to be in mountainous areas and potentially in areas of outstanding natural beauty, introducing social and ecological issues for proposed developments. There are examples where delay or abandonment of pumped hydro construction has occurred because of environmental concerns.[33]

Underground PHS operates in a similar fashion to conventional PHS (see Figure 3), with the exception that, although involving a surface reservoir and an underground powerhouse, the lower reservoir is underground and mined from hard rock.[34] It converts the energy of a water column flowing from a surface reservoir to the underground reservoir and then pumps the same water mass back to the surface. The system eliminates dependence upon natural topography and permits design for higher head operation. In this sense, it carries a lesser environmental impact in comparison with conventional PHS.

Closed-loop systems not involving connection to surface waters involve both reservoirs being excavated underground, with a number of projects considered in old mines, including a 1500 MW project at Norton, Ohio, and a 1300 MW project in California.[35,36] Pump-back plants operate as a hybrid combination of pumped storage and conventional hydroelectric plants, with an upper reservoir that is replenished in part by natural inflows from a stream or river. There has been modification of some standard hydroelectric power plants to operate with a pump-back system, such as the Grand Coulee Dam in the USA. This was expanded in 1973 with water pumped back up from below the dam,[37] and similar conversions of other such plants are under consideration in other countries.

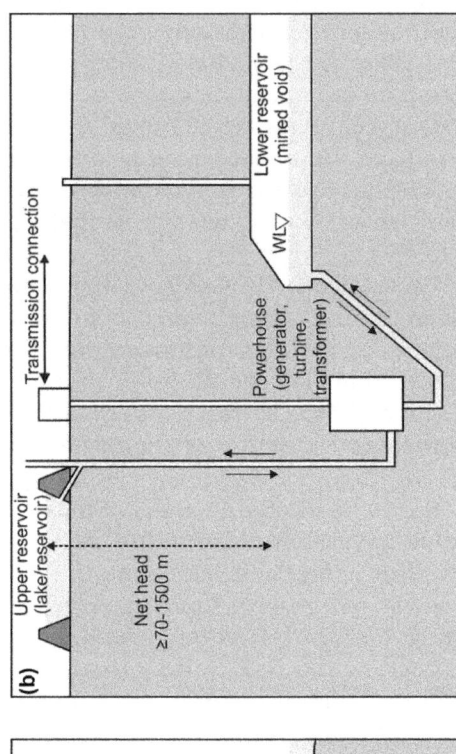

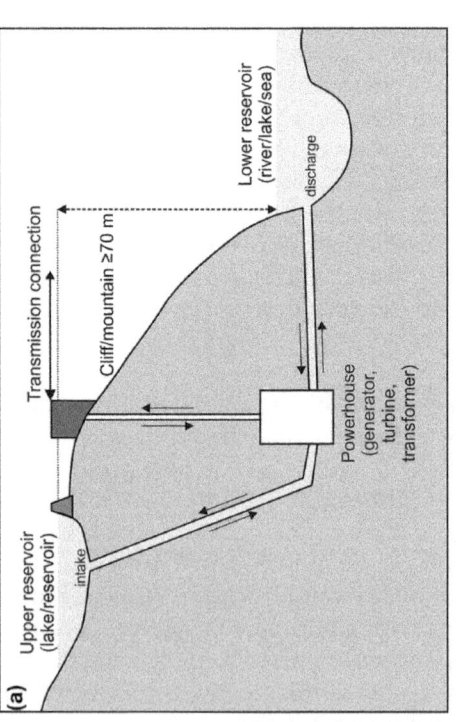

Figure 3 Schematics of PHS schemes: (a) conventional PHS with surface reservoir and discharge to lower water body, either river, lake or sea; (b) underground PHS with lower reservoir created by mined void below ground.

The design of almost every PHS/PHES power plant is highly dependent on the site characteristics: there needs to be sufficient water available and both the topography and geology of the area must be favourable. Suitable sites are, therefore, likely to be in hilly or mountainous regions, which are potentially in areas of natural beauty and which would pose additional social and ecological issues. All schemes will require very site-specific investigations to determine rock mass qualities and whether major permeable zones, such as faults or fractures in the rock, exist. Most reservoirs will require some sort of engineered lining (often clay) to reduce leakage rates and risk. Evaporation may be a problem in some areas.

In the USA, a growing number of proposed projects are avoiding highly sensitive or scenic areas, with some proposing to take advantage of 'brownfield' industrial sites such as old mines, using different mined levels as the reservoirs. One such 1300 MW scheme is being considered by Eagle Crest Energy in California, which would reduce environmental impacts to a minimum.[36] Although additional challenges exist compared with using fresh water, PHS plants can operate with seawater, with the world's first demonstration of seawater PHS at the 30 MW Yanbaru project in Okinawa, Japan, commissioned in 1999.[38] Built on top of a cliff, it used seawater; however, owing to a lack of anticipated local electricity demand, the plant was dismantled in 2016.[39]

2.1 Efficiencies and Economics

PHS presents a number of benefits:[7]

- It is a mature technology where installations have a long design life, enabling it to play an important role in the long-term planning and management of energy supply systems.
- Its large scale of storage allows it to deliver for long discharge periods, contributing to security of supply.
- Provision of frequency response and short-term operating reserve services.
- Energy arbitrage or the time shifting of electrical energy by consuming the surplus from renewables at times of high availability or use of lower cost, baseload electricity power to pump into storage and delivering fast response discharge at higher-value periods of peak or system demand.
- The high-flexibility response of absorbing and generating electricity smoothes the demand profile between highly variable renewable electricity source (VRES) generation and the response by conventional thermal plants, enabling them to operate with improved efficiency and at lower cost.

Although PHS stations are net consumers of energy, they remain an important component of an integrated, renewable energy system. They have system efficiencies of 70–85%, which compares favourably with other mechanical energy storage systems such as compressed air energy storage (CAES), which has efficiencies in a similar range. In contrast, hydrogen fuel cells have an efficiency of just 20–50% and lithium-ion batteries an efficiency of nearly 100%.[14,40]

The construction and installation costs of PHS are estimated to be twice those of conventional hydropower plants with similar capacity while operating costs are almost equal.[41] With capital costs estimated at between £500 and £2000 per kilowatt of installed capacity,[7,42] it has been shown to be the most cost-efficient storage technology, with a levelised cost of delivered electricity of £105 per megawatt hour (MWh).[42]

Despite the storage benefits that PHS schemes can deliver for energy systems, they currently face a number of economic challenges, where the greatest risk is uncertainty in long-term revenue. PHS schemes are complex infrastructure projects requiring extended periods for development and construction, have high capital expenditures and have a long design life of 40–80 years.[43] This type of investment is, therefore, not ideal for cost recovery in market-driven energy systems.

Most historical PHS has developed where electricity markets have a degree of public ownership. Analysis has shown that a reduction in the global development of PHS coincided with large-scale deregulation of electricity markets during the 1990s.[14] The technology requires long-term investment and some certainty on return, where a stable regulatory and financial environment will, therefore, be a precondition to attract private funding. Modern liberalised energy markets tend not to provide this certainty where they are difficult to forecast, contracts are short term and future policy decisions can be unpredictable, leading to increased risks associated with developing this type of hydropower scheme.[44]

Revenue streams for PHS will, therefore, need to come from the range of services that the technology can deliver, although each of these streams is subject to considerable uncertainty. Although the technology also delivers a number of external benefits, such as reduced carbon emissions, lower systems costs and energy security, there is currently no mechanism to internalise their value.

The UK National Infrastructure Commission concludes that current regulatory conditions unintentionally disadvantage storage providers, preventing them from participating across the various electricity markets, but that much greater use could be made of the technology where better regulation creates a level business environment between generation and storage.[6]

The UK government has developed interventions in the energy market to provide greater certainty for investors in some sectors, although these are either not currently viewed to favour PHS or actively favour other technologies. The Capacity Mechanism was introduced to provide support for backup generators to help balance the energy system, but the time scale on which it operates is considered too short to attract investment in PHS.[45,46]

Development of the Cap and Floor mechanism was specifically to support investment in interconnectors to the UK.[47]

Although a number of PHS developments are currently at advanced stages of planning, there remains some uncertainty as to whether they will progress to construction unless appropriate measures are put in place to provide favourable investment conditions for private finance.[7]

2.2 UK Deployment of PHS

2.2.1 Existing Facilities. There are currently four PHS schemes operating in the UK (see Figure 2) and, with a capacity of 2828 MW, they represent 1.5% of global capacity. These are Dinorwig[48] (Gwynedd, North Wales) at 1728 MW, Ffestiniog[49] (Gwynedd, North Wales) at 360 MW, Cruachan[50] (Dalmally, Scotland) at 440 MW and Foyers[51] (Highlands, Scotland) with a capacity of 300 MW. The Ffestiniog, Cruachan and Foyers schemes were commissioned in 1963, 1965 and 1975, respectively. The Dinorwig power station, commissioned in 1984 and having taken 10 years to complete, is one of the largest PHS schemes in the world, where a flow of 60 m^3 s^{-1} is released from the upper Marchlyn Mawr reservoir, through turbines set in the former Dinorwig quarry and discharging into the lake Llyn Peris.

2.2.2 Potential for Future PHS. The increasingly important role of storage in our energy systems has led to renewed interest in new PHS schemes. Although topography, water resource availability and environmental and landscape protection are principle constraints to the development of new schemes, a Geographic Information System (GIS)-based study concluded that there is still considerable potential for the development of more pumped-storage sites in the UK.[52] A 2010 study for the Scottish Government identified there being a role for 7000 MW of new pumped-storage hydropower to meet Scotland's energy system storage needs by 2030.[53] Based on a typical scheme capacity of 400 MW, this means a requirement for 17 new schemes, and the study concluded that it is unlikely that this number of sites is available. In reality, the number of potential sites for the construction of large, new PHS schemes is likely to be limited and more comprehensive, site-level studies will be required to increase confidence in estimates of future UK capacity.

There is, however, interest in developing a small number of new sites in the UK and upgrades to existing sites to increase capacity. Collectively, they could more than double the UK's current PHS capacity. A summary of these is as follows:

- Coire Glas (Scotland) – consents secured for a new 600 MW scheme, although studies are being carried out to increase this to 1500 MW.[54]
- Balmacaan (Scotland) – proposals to develop a new site for a 600 MW scheme.

- Glenmuckloch (Scotland) – proposals to develop a new 400 MW scheme.[55]
- Glyn Rhonwy (Wales) – consents secured for a 50 MW scheme, although applications are being made to increase this to 100 MW.[32]
- Cruachan (Scotland) – proposals to upgrade the current scheme to add up to 600 MW of additional capacity.[56]
- Sloy (Scotland) – Conversion of existing 152.5 MW hydropower scheme through addition of 60 MW pumping capacity.[57]
- Muaitheabhal (Isle of Lewis, Scotland) – proposals for a new 300 MW scheme.

2.2.2.1   Traditional Sites and Potential Problems

Historically, PHS schemes were constructed in mountainous or upland areas utilising existing but modified natural lakes for upper and lower reservoirs. This type of scheme, described as an open-loop system, has the reservoirs in hydrological continuity with existing water catchments. Enlargement of an existing natural lake generally creates the reservoirs through construction of an impoundment to provide the increased storage volume required for power generation. Hydraulic controls are established on the lower lake (reservoir) to regulate outflows and retain necessary storage volumes for pumped-hydropower operation. The Dinorwig scheme is typical of this arrangement. The connectivity of open-loop schemes with natural hydrological systems means that their operation is likely to have impacts on the wider water environment. Conversely, a closed-loop PHS scheme is one in which there is limited hydrological connectivity beyond the boundary of the engineered scheme and operates, from a water resources perspective, largely independently of its hydrological setting. The proposed 100 MW Glyn Rhonwy scheme, near Llanberis in North Wales, is a good example of a closed-loop scheme that would utilise two redundant slate quarries to create the upper and lower reservoirs, each with a storage volume of 1.1×10^6 m^3. The mountainous topography is necessary to provide not only the landform suitable for dam construction and reservoir storage, but also the hydraulic head necessary for power generation. To illustrate, the Cruachan scheme has a reservoir storage volume of 10×10^6 m^3 whereas the hydraulic head for the four UK pumped-storage schemes ranges between 300 and 500 m.

Given the scale of conventional PHS schemes in upland areas, they are typically constructed using tunnels and chambers to convey flow, house turbines and provide access and service routes instead of using buried or overground infrastructure. There are risks associated with extensive use of tunnels in such schemes and tunnel collapses have occurred in many developments, both in the UK and elsewhere. Although a conventional hydropower development, the tunnel collapse on the newly constructed Glendoe scheme in Scotland put the scheme out of action for 3 years at an overall cost of £100 million (see Figure 2).[58] Numerous other collapses have been recorded in countries with major hydropower development programmes, such as in Ethiopia,[59] Vietnam[60] and China.[61] Causes have been attributed to poor construction practices, poor or inadequate geological assessment and unfavourable geological conditions.

2.2.2.2 Cliff-top Sites

Recognising the many constraints of typical inland, mountain settings for PHS, there is increasing interest in the use of seawater and cliff-top sites. These types of scheme would operate in the same way as traditional PHS schemes but would require the construction of only the upper reservoir on land at elevation and the use of seawater for power generation, discharging into the sea, which acts as the lower reservoir. As already described, the first experimental seawater scheme was on Okinawa Island, Japan, in 1999.[38,39,62] There were a number of challenges in using seawater for the scheme. To prevent leakage of salt water into the terrestrial environment, the upper reservoir was lined, and the quality of the water created difficulties in operation and maintenance owing to its corrosive nature. To reduce the impact of the scheme on the marine environment, environmental mitigation measures were installed.[63] Increasingly, cliff-top sites are seen as potential alternatives in places with limited scope for conventional upland schemes and with less impact on lake and river ecosystems. Investigations into the feasibility of new cliff-top schemes are ongoing in a number of countries, including Australia[64] and Hawaii, USA.[65] The use of seawater is also under consideration in a low-head context, with the conclusion that it is a technically and economically feasible option.[66]

2.3 Environmental and Regulatory Factors in PHS

2.3.1 Overview of Environmental Regulation for PHS. PHS schemes are large, complex civil engineering projects that are likely to have significant impacts on their local environment during the construction, operation and decommissioning stages of their design life. Although the general design principles of PHS schemes are similar, their environmental impacts will vary according to where they are located and how they are arranged and are operated.

Extensive legislation is in place in the UK to protect the natural environment. This ranges from targeted conservation of high biodiversity value habitats and rare species of flora and fauna to protecting the quality of our wider environment, including soils, air, water and landscapes. In some cases, such as water, a key purpose of the law is for the proper and equitable use of a common natural resource.

Regulation of activities affecting the environment is achieved through a system of permits, licences and consents issued and enforced by range of public sector agencies and authorities across the UK. The principal environmental regulators are Natural England (NE) and the Environment Agency (EA) in England, Natural Resources Wales (NRW), Scottish Environment

Protection Agency (SEPA), Scottish Natural Heritage (SNH) and Department of Agriculture, Environment and Rural Affairs (DAERA) in Northern Ireland. Local Planning Authorities also have a statutory duty to consider the environmental impacts of development in decision making for Town and Country Planning.

The underlying approach for environmental regulation requires that development activities are assessed for their impacts, leading to the issue of a consent or assumed permission, authorising an activity where the impact of that activity on the environment can be controlled to meet standards of protection set out in legislation. The consent will include specific conditions on how that activity must be carried out, and where failure to comply with these conditions is an offence and can lead to enforcement action.

The level of regulation of controlled activities generally varies according to environmental risk. Some activities meet low-risk criteria, which, although controlled in law, may be undertaken without the need for a specific consent if, when undertaking that activity, a set of generic rules are followed. Conversely, bespoke licences may be required for high-risk activities where extensive analysis of potential environmental impacts must be submitted to the regulator to inform a licensing decision.

It should be noted that differences exist in environmental legislation between the devolved administrations in the UK. Although the overall objectives of national legislation remain similar, the processes involved in obtaining authorisations and conditions associated with them may vary, although much of our domestic environmental legislation has developed in response to European Commission directives with duties transposed into UK law.

Environmental regulation covers an extensive range of complex subject areas, from impacts on internationally rare bryophytes or invertebrates to effects on noise, landscape and management of wastewater. The aim of the following sections is not to provide complete and exhaustive detail on all aspects of environmental regulation but to provide a brief introduction and overview of some key areas relevant to the development of PHS, with a focus on ecological impacts. Although there are engineering and operational differences between PHS, storage reservoir and run-of-river hydropower schemes, they pose common environmental risks where the development of any large hydropower scheme should follow a standard process of environmental assessment and will need to meet similar regulatory requirements. Any assessment is required to consider a range of regulations and legislation, as illustrated in Figure 4.

These sections are presented with the caveat that if readers wish to further their understanding of any subject areas described in this chapter, then they should seek detailed advice and guidance from the relevant environmental regulator. Finally, this text was written prior to the UK having left the European Union and some uncertainty remains over the long-term impacts of Brexit on environmental regulation.

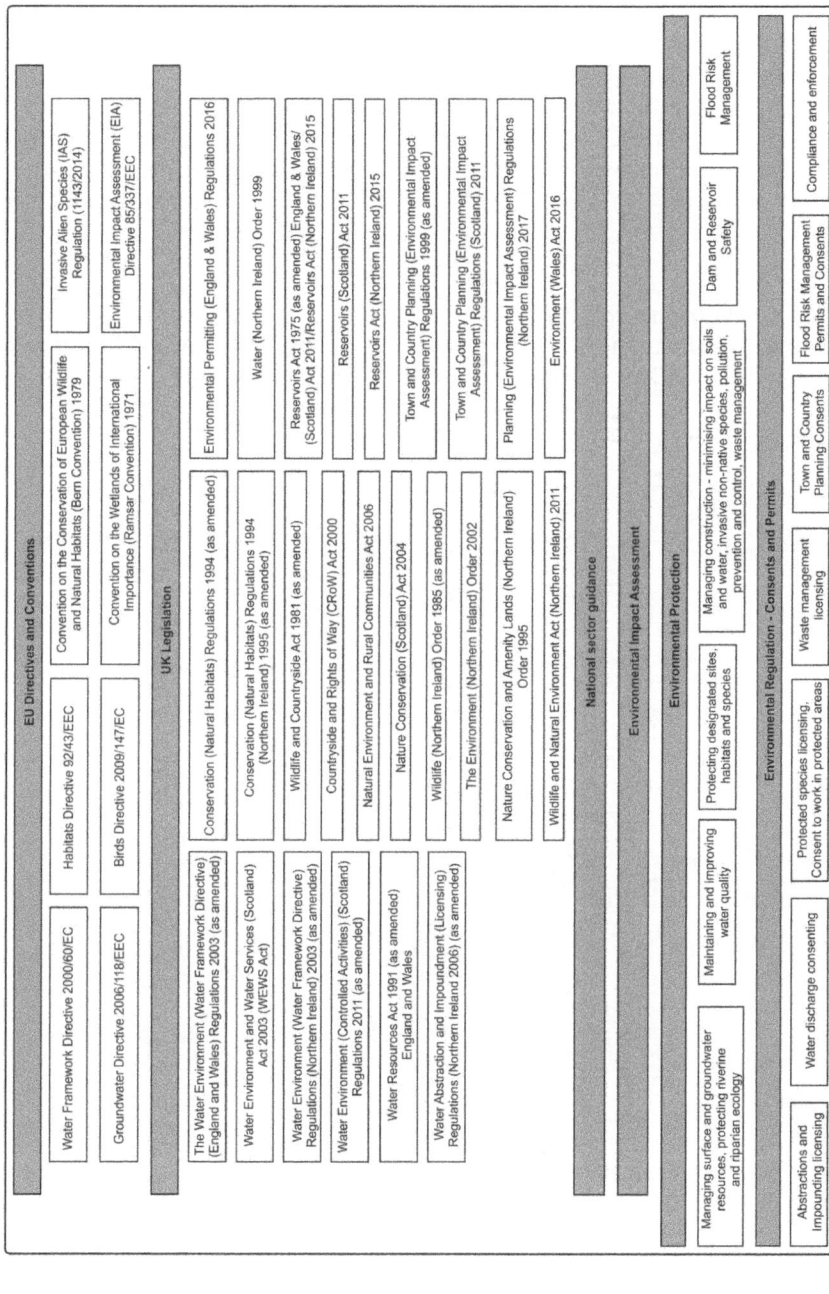

Figure 4 Schematic to illustrate the principal EU and UK regulatory and legislative frameworks against which a PHS scheme would be assessed in the UK.

2.3.2   Environmental Assessment Process.   An Environmental Impact Assessment (EIA) is the core tool that enables a consistent and structured approach to be taken when assessing how a development will affect the environment, the significance of potential impacts and the options available for mitigating those impacts. It is likely that a statutory EIA will be required for all PHS schemes where the generation capacity is greater than 0.5 MW and/or the development is located fully, or partly, in an environmentally sensitive area.[67–74] The process can lead to improved environmental design and provides decision makers with the information they need for Town and Country Planning and environmental regulation (see Figure 5).

2.3.3   Abstraction and Impoundment Licensing.   Water resources are managed across the UK using a regulatory system of abstraction and impoundment licensing, which is central to the development and operation of

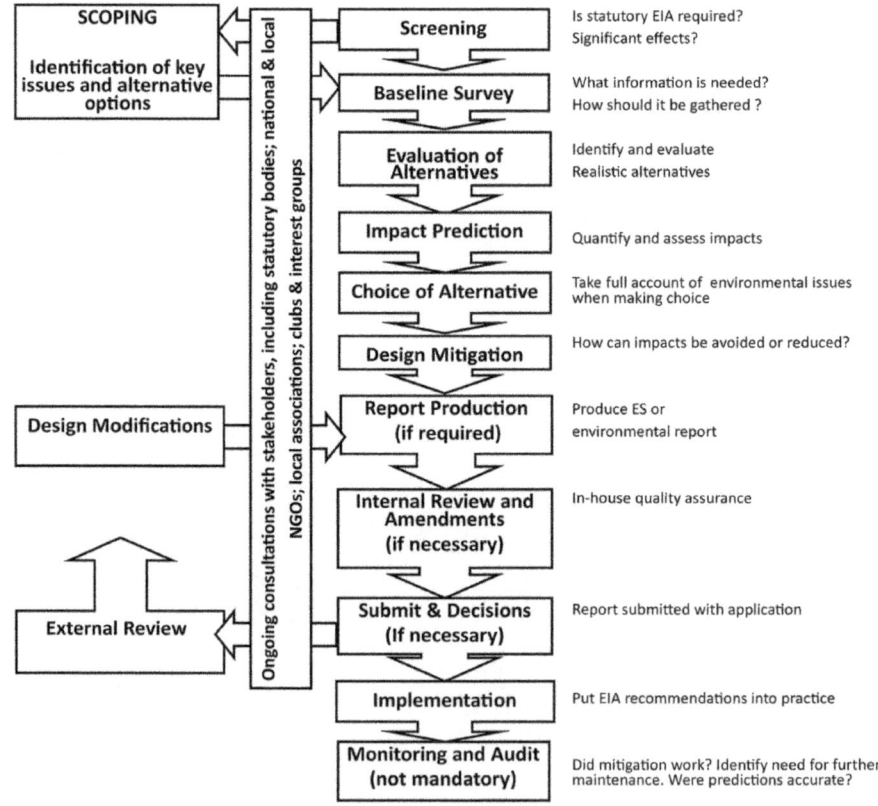

Figure 5   Key steps in the environmental impacts assessment process.
Source: *Environmental Impact Assessment (EIA): A Handbook for Scoping Projects*, Environment Agency, 2002 (contains public sector information licensed under the Open Government Licence v3.0).

pumped-storage hydropower schemes.[75–78] It is highly likely that these licences will be required for future PHS, although the actual requirements will be dependent on the specific arrangements of individual schemes. The licensing system is in place to ensure the sustainable use of water resources for society at a national scale where this includes, for example, public water supply, industrial, navigation and amenity purposes. The licensing system also protects the ecology of our lakes and rivers by controlling the amount of water available for abstraction and governs the construction of impoundments in watercourses where there is an obstruction, impedance or diversion of flow.

Although Water Resources legislation or Controlled Activities Regulations are the principal legal tools used to implement abstraction and impoundment licensing, a raft of wider environmental law exists that is specific to the protection of certain sites, habitats and species. These must be considered within the licensing process and be fully addressed before a licence can be issued.[79–82] One of the most important of these is the Water Framework Directive (WFD).

The WFD targets the protection of inland surface waters (rivers and lakes), transitional waters (estuaries), coastal waters and groundwater. It is based on a spatial structure of River Basin Districts, each of which is subdivided into water bodies representing the water environment at catchment and subcatchment levels. The WFD has introduced a cyclical process of river basin planning, involving the characterisation and assessment of impacts on River Basin Districts, environmental monitoring, setting environmental objectives and the design and implementation of a programme of measures needed to achieve them.

Environmental regulators must take the WFD into consideration when determining abstraction and impoundment licences, ensuring that authorisations meet environmental standards and do not compromise environmental objectives.

Article 4.7 of the WFD does, however, include provision for derogation of its objectives in certain circumstances. This recognises that some impact on the water environment may be necessary for society, but only where a development meets the criteria of providing benefit to human health, human safety, for sustainable development or other overriding public interest. In such cases, it must be demonstrated that all steps will have been taken to mitigate the adverse impacts of the activity on the status of the water body and that it cannot be achieved by a significantly better environmental option for reasons of technical infeasibility or disproportionate cost. Compliance with objectives and standards of other European Community legislation such as the Habitats Directive will still be required.[83]

Environmental legislation is more extensive now than in the 1960s and 1970s when the Dinorwig and Cruachan schemes were constructed and, as described earlier, our water environment is now expected to meet the environmental standards of the WFD in addition to there being more protection for designated sites, habitats and species. The implications of this are that open-loop schemes that have hydrological connectivity beyond the development boundaries will present a risk of wider ecological impact and face greater challenges in meeting the regulatory requirements for abstraction and impoundment licensing than a closed-loop scheme that has a higher level of hydrological independence from its surrounding drainage network.

Any raised dam structure or engineering works proposed to enhance the storage capacity of a natural lake will require an impoundment licence. Where a natural lake is utilised for a lower reservoir there may be less need to increase the storage volume, although it is likely it would be necessary to install some form of hydraulic control on its outflow to manage levels for hydropower scheme operation. Like the upper reservoir in an open-loop system, a lower reservoir utilising an existing lake would have its own natural upstream catchment and pattern of inflows, with subsequent outflows reflecting the storage and attenuation characteristics of the lake. Impoundment licensing would be required to permit the construction of outflow control structures from the lower reservoir, as their operation would impede natural outflow and alter the hydrological regime in the downstream watercourse.

Legislation recognises these storage reservoirs as inland waters or sources of supply, where any water taken from them would require authorisation *via* an abstraction licence. It is likely, therefore, that for any new scheme a licence would be required for the release of water from the upper reservoir for power generation and conversely to allow water to be pumped from the lower reservoir back into the upper storage. Abstraction from one reservoir to another within a closed-loop system such as the quarry-based Glyn Rhonwy scheme, however, would probably be exempt from licensing, with the exception of the need to abstract from a source of supply outside the system to import 'infill' and 'top-up water'. It is also likely that closed-loop, quarry-based schemes using existing voids would be exempt from the need for impoundment licensing if their engineering works were not creating a raised structure that obstructs or impedes surface water flows within an open hydrological system.

Where storage reservoir catchments are small and have limited inflows, it might be necessary to augment the water resource in the pumped storage system with a transfer from an adjacent catchment, which in turn would require a transfer licence. Like an abstraction licence, a transfer licence authorises abstraction of flow from one source of supply with subsequent discharge to another source of supply without intervening use.

Inter-catchment transfer of water for operational purposes, or to increase water resource availability for PHS, presents several environmental risks that require careful assessment. Of primary concern is the risk of spreading invasive, non-native species.[84] There is also a risk of distributing existing native species of fish, macrophytes and invertebrates between catchments *via* mass water movement, with the potentially undesirable consequence of introducing them into an ecosystem in which they may currently be absent. Biosecurity Risk Assessments are, therefore, essential components of the

EIA. Inter-catchment transfer should also be assessed in the context of water quality in order to avoid the introduction of flow into a riverine or lake ecosystem that disrupts the natural nutrient, chemical and quality balance in which that catchment's aquatic organisms have evolved. Fish screening will be required on abstraction intakes where fish populations are present in the open-system reservoirs but are less likely to be needed in closed systems if fish are absent, preventing their ingress to the penstock and turbines.

Natural lakes used as reservoirs and actively managed for power generation will have a significant impact on both lake and stream hydrology and where potential changes will require analysis to inform the licensing process. Changes to natural outflows might include an overall reduction in flow volumes within the downstream channel and loss of flow variability, particularly in the low- and medium-flow ranges. This will affect the timing, frequency, magnitude and duration of components of the natural flow regime. Alterations to these ecological flow components change the hydrological, hydraulic and geomorphological parameters of the river and riparian environment, modifying the condition of the habitat upon which river biota are dependent to progress through their life cycles.[85]

Fluctuations in lake water levels (where the lake is an actively managed reservoir as part of a hydropower scheme) have also been shown to have adverse effects on littoral invertebrate and macrophyte communities.[86–88] Although the flora and fauna of the littoral zone of lakes are dependent on a range of physical characteristics of the shoreline, including slope, substrate and exposure to wave action, studies have shown that the degree of control and consequent duration, frequency and extent of water-level changes are of fundamental importance to littoral communities.[89,90] In upper reservoirs of pumped-storage schemes, large fluctuations in water levels occur on a frequent basis and have a significant impact on littoral flora and fauna,[88] whereas lower reservoirs experience fluctuations within naturally occurring ranges that are less likely to affect lake ecology adversely.[87]

It is likely, however, that to meet current regulatory requirements, particularly the WFD, future open-system schemes would need to mitigate their impacts in the form of more dynamic operating regimes and greater complexity in compensation-flow releases from storage reservoirs than in the large hydropower schemes of the past.[85] The detail of mitigation measures would be dependent on the ecological and geomorphological sensitivity of the downstream watercourse to changes in flows and of the effects of fluctuations in water levels on lake flora and fauna.[89,90] They could, for example, take the form of seasonally varying compensation flows, actively managed, periodic reservoir releases and seasonal restrictions on timing, volume and frequency of abstractions to retain lake water levels within natural ranges. Such operational controls would be incorporated into abstraction and impoundment licence conditions, or set out within a formal operating agreement, against which an assessment of regulatory compliance can be made.

Potential changes to surface water flows and groundwater levels resulting from scheme development should be analysed for their connection with other existing water features such as springs, flushes and mechanisms of water supply to wetland areas. Dewatering is likely to be a key component of construction works. An abstraction licence may now be required for this activity in England and Wales. There are, however, exemptions for temporary, small-scale, low-risk dewatering activities from both surface and groundwaters, subject to their meeting restrictions based on duration, site sensitivity and volumes. An application for an abstraction licence in this situation will require support from a Hydrogeological Impact Appraisal.[92] It may be necessary to apply for a Consent to Investigate a Groundwater Source, which authorises test pumping and permits monitoring of the effect of any drawdown on local surface water features, the results of which should inform the Hydrogeological Impact Appraisal. In Scotland, abstraction of water from quarries or dewatering is a controlled activity covered by General Binding Rules or licensing, depending on the nature of the operation.[91]

2.3.4 Regulation of Dam and Reservoir Safety. In the UK, reservoirs created by structures designed to hold water above natural ground level are controlled by legislation. The roots of the legislation date to early twentieth century dam failures and the loss of life caused by dam failure, and the subsequent catastrophic flooding that can occur. The laws vary between England, Northern Ireland, Scotland and Wales,[93–96] but revolve around the common principle that the design, construction, alteration, maintenance and eventual removal of raised reservoirs is done under the scrutiny of qualified civil engineers. The point at which regulation applies is identified by the capacity of water retained above natural ground level, which is either $10\,000$ m^3 or another figure provided by specific regulations. The required civil engineers are specialists within the field of reservoir construction and may act as reservoir panel engineers only if appointed by government on recommendation by the Institution of Civil Engineers. Before construction of any reservoir, advice should be sought from the relevant regulator regarding specific laws applicable and the steps to be taken to comply with them.

2.3.5 Water Quality. Open-loop system pumped-storage hydropower schemes have the potential to affect water quality through the principal mechanisms of

- increasing reservoir storage capacity through enhancement of natural lakes by dam construction or increased hydraulic control of outflows;
- introduction of regulated/compensation flows into watercourses downstream of impoundments;
- mixing of reservoir water during pumped storage operation.

Changes to water quality can adversely affect the conditions for aquatic flora and fauna in reservoirs and downstream rivers and streams. Lake water bodies naturally become thermally stratified where solar heating warms the surface water layer, known as the epilimnion, whereas water at the bottom becomes cooler and denser, known as the hypolimnion. Surface water will heat up and cool down more rapidly and over a greater temperature range than deeper water. In natural lakes, the extent to which the water layers become mixed is caused principally by disturbance of the surface by wind and other local factors such as tributary inflows.

When the surface water cools to below the temperature of the water below it, it becomes denser and sinks, causing the water column to mix towards an isothermal state in which the water temperature is the same from surface to bed. During winter months, therefore, lakes tend to be isothermal, whereas during summer months they tend to be stably stratified, with warm water at the surface and cooler water beneath.[97]

Enhancements to natural lakes to increase storage capacity will alter the physical characteristics of the water body and potentially lead to changes in thermal stratification, particularly where there is a low surface area-to-volume ratio, rendering it less responsive to fluctuations in solar radiation.[98]

Changes in water temperature are significant to aquatic ecology, not only through the direct effect on biological processes and metabolic rates of aquatic organisms, but also indirectly, where it affects other parameters of water quality and chemistry such as dissolved oxygen, conductivity, pH and density. Ecological impacts may occur where the quality of water is substantially altered from that occurring naturally. Whereas outflows from natural lakes occur from the lake surface, discharges from impoundments are frequently released from depth, leading to cold-water compensation flows or reservoir releases that affect the thermal and chemical characteristics of the downstream watercourse.[98,99] In summer months, where thermal stratification has occurred in upper reservoirs, degasification of cold water can occur as flow passes through the turbine, leading to low dissolved oxygen levels in discharge water.[100,101] Regular operation of a PHS scheme can, however, disrupt natural patterns of circulation within both upper and lower reservoirs, leading to increased vertical mixing and disturbance to natural thermoclines and processes of stratification, with potential consequences for lake ecology.

From a regulatory perspective, a full assessment of how any new PHS scheme would affect water quality in connected water bodies would need to be undertaken to meet the requirements of the WFD.

2.3.6 Designated Sites, Protected Species and Habitats. The Environmental Impact Assessment for any proposed PHS development should identify the risk and significance of any likely impact on protected sites, habitats and species. Several European- and UK-level designations provide statutory protection for sites of high conservation importance across the UK. The principal designations are Special Areas of Conservation (SACs), Special Protection Areas (SPAs), Ramsar wetland sites and Sites of Special Scientific Interest (SSSIs). The Joint Nature Conservation Committee website provides comprehensive information about these designations.[102]

SACs are designated under the commonly termed EC Habitats Directive.[103,104] The Directive lists habitats and plant and animal species of international importance within its Annexes. It is the duty of competent authorities (CAs) across the UK, including but not limited to environmental regulators and Local Authorities, to maintain these features at, or restore them to, favourable condition status.

The Birds Directive[105] sits parallel to the Habitats Directive, with the objective of specifically protecting wild birds and associated habitat. Similarly to the Habitats Directive, there is a requirement to designate and maintain sites of importance for the conservation of the rare and vulnerable species within SPAs. SACs, together with SPAs, form the Natura 2000 network of areas protected for nature conservation across member states of the European Union.

Ramsar sites are specific areas of marsh, fen, peatland or water habitats of high conservation value designated under the Convention on Wetlands of International Importance.[106] They include onshore, transitional water and coastal sites and, with relevance to hydropower developments, many upland lochs/lakes and wetlands.

SSSIs (ASSIs in Northern Ireland) are protected by law to prevent operations and activities from being carried out that might damage or destroy the scientific interest of part or all of the site.[107–113] Consent must be sought from the relevant environmental regulator to carry out any controlled activity from which there may be risk of damage to SSSI features. Undertaking controlled activities without consent is an offence, which could lead to an unlimited fine and payment of costs necessary to repair any damage to the site.

The Habitats Directive and 1981 Wildlife and Countryside Act largely protect individual species. The Act lists protected plants and animals under various schedules to which varying degrees of protection apply. The local regulating authority can grant licences under the Act to allow certain activities to be carried out that would otherwise be offences under the legislation, provided that the activities being undertaken are for non-developmental purposes such as scientific, research, education and conservation.

Lists of priority species and habitats identified as being threatened and requiring action for their conservation have also been drawn up under the UK Biodiversity Action Plan (UK BAP).[114] UK BAP priority species and habitats are incorporated into statutory lists for domestic or UK legislation.[110,115–118] Following devolution of many environmental duties, individual UK nations have developed their own biodiversity strategies. These reflect national priorities, requirements and responsibilities, although the principal underlying objective remains to halt the loss of biodiversity and continue to reverse previous losses through targeted measures to conserve species and habitats.[119,120]

The network of internationally important SACs was chosen to represent best the features requiring protection under the Habitats Directive where many of those features are found in riverine, lake/loch and upland environments. Given that sites suitable for future PHS developments are likely to be in similar remote upland areas, there is a high likelihood that the construction or operation of any new PHS scheme would affect a Natura 2000 site and would need to address the requirements of the Habitats Directive as a central part of the development process.

The Habitats Regulations require that CAs, which in the case of hydropower developments will principally be the Local Planning Authorities and environmental regulators, must carry out a formal assessment of the implications of any proposals. This will be individually or in combination with other plans and projects that may affect the designated interest features of European sites before deciding whether to permit or authorise any such proposals. These plans and projects in the case of PHS might refer to Town and Country Planning Consents, abstraction and impoundment licences or other environmental permits associated with specific construction activities. The assessment process is usually termed the Habitats Regulations Assessment (HRA), which requires a precautionary approach to be taken in any assessment as to whether a proposal is likely to have an adverse effect on the integrity of a site and thereby compromise the ability of the site to meet its conservation objectives.[121] If it cannot be concluded beyond reasonable scientific doubt that a plan or project will not result in an adverse effect on the integrity of a European site, then the permission or consent being sought should be refused by the CA.

There is, however, provision within the Habitats Directive for a derogation that under certain circumstances would allow a plan or project to be approved, even though the HRA has shown that there is potential for it to have an adverse impact on site integrity.[122]

Significant long-term impacts on designated sites or protected species are more likely to be associated with open loop type PHS schemes than closed ones, where both their construction and operation risks have more geographically extensive impacts due to their high connectivity with adjacent terrestrial, wetland and riverine ecosystems.

A key component of the overall EIA process is the Ecological Impact Assessment (EcIA), which provides a framework through which designated sites and protected species that may be affected by a proposal can be identified, and assessment of development impacts completed. The Chartered Institute of Ecology and Environmental Management (CIEEM) has produced comprehensive guidance on this.[123]

Protected habitats may also form a component of protected landscapes, including those in National Parks, Areas of Outstanding Natural Beauty (AONBs), Special Landscape Areas (SLAs), National Scenic Areas (NSAs) and Areas of Great Landscape Value (AGLVs) in Scotland. An assessment of projects, such as large pumped hydropower on landscapes, will form a key part of any Town and Country planning application and may be included in the EIA.

2.3.7 *Managing Construction to Protect the Environment.* Construction of a PHS scheme will involve key activities including the construction of temporary access tracks, working and lay-down areas, construction of permanent access roads, tunnelling, trenching, site dewatering and land drainage and use of construction plant. Translocation of species and habitats is often considered as an option to mitigate impacts on ecology, although *in situ* conservation is preferred by environmental regulators.[124,125] Reduced impact on the environment can be achieved through comprehensive planning and sensitively managed site operations, which can substantially reduce the impact of construction activities on the environment. This can be achieved through the development of, and adherence to, site-scale Environmental Management Plans (EMPs) and activity-specific method statements.[126] The underlying principles within an EMP involve understanding the risks from construction activities, identifying the environmental receptors and the potential pathways in which they are connected. The EMP and method statements must have regard for industry best practice and any regulatory requirements that apply to a given construction activity.[127,128]

These documents can have a high level of importance where their development, and the operational procedures set out within them, can be secured as conditions in licences or consents by a Regulating Authority and used to demonstrate compliance with environmental legislation. Failure to comply with them may lead to a breach of consent and an environmental offence with potential for enforcement action. The EMP should include a programme for environmental monitoring pre-, during and post-construction. Associated with this should be an ongoing audit during construction activities in which checks are made to ensure that environmental mitigation measures and sensitive work practices are being implemented. It is usual, and often a condition of licence on large civil engineering projects, that an Ecological Clerk of Works (ECoW) is employed with specific responsibility for site environmental management.

Preventing pollution of the water environment is central to good site environmental management, where the water environment includes both surface and groundwater. Potential causes of pollution from construction sites include leakage/spillage of fuels, oils and fluids associated with machinery, use of construction materials, *e.g.,* concrete, and uncontrolled discharge of site run-off that is high in sediment. A range of rules, consents and permits apply to activities in which there is a risk of pollution and a detailed assessment of consenting requirements should be included within the EMP.[87,129–134]

Regulations covering waste management are complex and are best addressed through the development and application of a Waste Management Plan. The plan should detail who is responsible for managing waste on-site, what types of waste will be generated, how it will be stored and handled, the options for recycling and disposal, which contractors will be used to recycle or dispose of the waste legally and how waste generated will be measured. Advice should be sought from national regulators in developing the plan and securing waste management consents or permits.[135–137]

The construction of access roads and trenches for pipes and cabling needs careful planning to minimise the risk to site hydrology. Access roads should be sited to avoid sensitive habitats such as peatland or wetland areas and maintain natural pathways for surface water drainage.[138] Construction of trenches risks introducing preferential flow pathways that intercept or disrupt natural subsurface drainage patterns and risks either reducing the availability of flow for existing wetland features or increasing discharge into an alternative site. Risks and measures to mitigate the impact of construction activities on site hydrology should be included in the site drainage plan described earlier.

The importance of a Biosecurity Risk Assessment has already been described in the context of PHS design and operation. The Risk Assessment should extend to the construction phase to establish construction site biosecurity control measures, preventing the spread of invasive non-native species and diseases from site operations.[139,140]

Although the principal aim should be to avoid impact on habitats during the construction of infrastructure, it is likely that some loss of habitat will occur in large projects. This loss might be short term owing to the need, for example, for construction of temporary access tracks, lay-down and working areas, or permanent for roads or where increases in reservoir or lake capacity inundates terrestrial habitat. A Habitat Management and Restoration Plan allows for sensitive habitats to be identified and enables careful planning and design of infrastructure, using techniques such as micro siting to minimise impact on them. Environmentally considerate construction practices and restoration techniques can be set out in method statements to support the plan. Post-construction reinstatement aims to reverse the impacts of temporary disturbance and recover the original condition and functioning of the habitat.[141,142]

3   Compressed Air Energy Storage – Introduction to the Technologies, Geology and Environmental Aspects

Compressed air energy systems have been in wide use since the late nineteenth century, with tanks often used for storage. Various cities, including Paris, Birmingham, Dresden and Buenos Aires, had systems where the delivered power was used in homes, industry and transport (*e.g.* trams). The mining industry utilised rock caverns for small-scale compressed air storages, enabling them to power mine trains, pumps, pneumatic drills and hammers.[143]

The use of the underground to store fluids and gases under pressure is also not new, with patents issued dating back to the turn of the twentieth century. One of the first applications of the use of compressed air was for mining operations, as illustrated by the Ragged Chute Compressed Air Plant in Northern Ontario, Canada. Built in 1910, at the peak of the Cobalt silver boom, it was the world's only water-powered compressed air plant.[144,145] With only two brief maintenance interludes, the plant ran continuously for just over 70 years until destroyed by fire in the 1980s, after which it was converted to a 7 MW hydroelectric power station in 1991.[145] Compressed air storage was used at Striberg Mine in Sweden from 1910. First references to electrical energy storage by means of compressed air storage for utility application were included in patents issued in the USA,[146] to the turbine manufacturer Stal Laval in Sweden and Great Britain (1952) and to Djordjevic in Yugoslavia (1950).[143] The first patent applications, reviews and assessments of large-scale CAES potential appeared from the 1960s, with small air-storage chambers also developed alongside power stations in Scandinavia (see Figures 6 and 7). The technologies behind current and future electrical energy storages based on compressed air, hydrogen or synthetic methane (SNG) storages are derived from proven storage technologies developed for underground natural gas storage, dating back as far as 1915 (see, *e.g.*, ref. 147).

The value of CAES was recognised in the early 1980s and the US Department of Energy initiated a broad CAES research programme with efforts directed at evaluating the potential for grid-scale CAES in many configurations, including caverns in bedded and domal salt, excavated openings in hard rock environments (using water-column pressure compensation or not) and aquifers.[148] Expectations then, and at times since, were high that CAES would gain utility acceptance,[149,150] but there has not been the take-up expected, with only two commercial projects operating, both hosted in salt caverns. Commissioned in 1978, Huntorf in Germany was the world's first commercial CAES plant, driving a generator to contribute 290 MW for 3 h every day to meet local peak electricity demand.[151,152] It remains operational almost 40 years later. The 110 MW plant at McIntosh in Alabama, USA, was the second commercial CAES facility, which also is still operational today.[153] Since then, other tests and demonstration plants have appeared briefly, but nothing has progressed to commercial operation. These have included aquifer storages (Sesta, Italy, and Pittsfield, USA, in the 1980s and Iowa in 2010), a salt-cavern storage (Gaines, Texas, USA, in late 2012) and the world's first pilot 0.5 MW test plant with thermal storage in 2016, using the abandoned unlined Pollegio–Loderio tunnel, north of Biasca in the Swiss Alps.[148,154,155]

In the late 1990s, concerns grew surrounding GHG emissions from fossil fuel use and interest turned to the large-scale deployment of renewable energy sources such as wind and solar. The intermittency of these energy sources and other factors, including the high price of natural gas fuel for peaking plants, and increased concerns about grid reliability, have resulted in renewed interest in CAES technology.[17,156] From the early 2000s, a number of active proposals for CAES development in the USA were under consideration, including a 260 MW aquifer storage facility in Iowa and a 2700 MW CAES plant in an abandoned limestone mine at Norton in Ohio (see Figures 6 and 7).[157,158] Comprehensive reviews of early and subsequent CAES R&D efforts can be found in the literature.[143,148,156,159,160] Interest in CAES has continued in the USA, with a number of projects supported by the Electric Power Research Institute (EPRI) and funding from the American

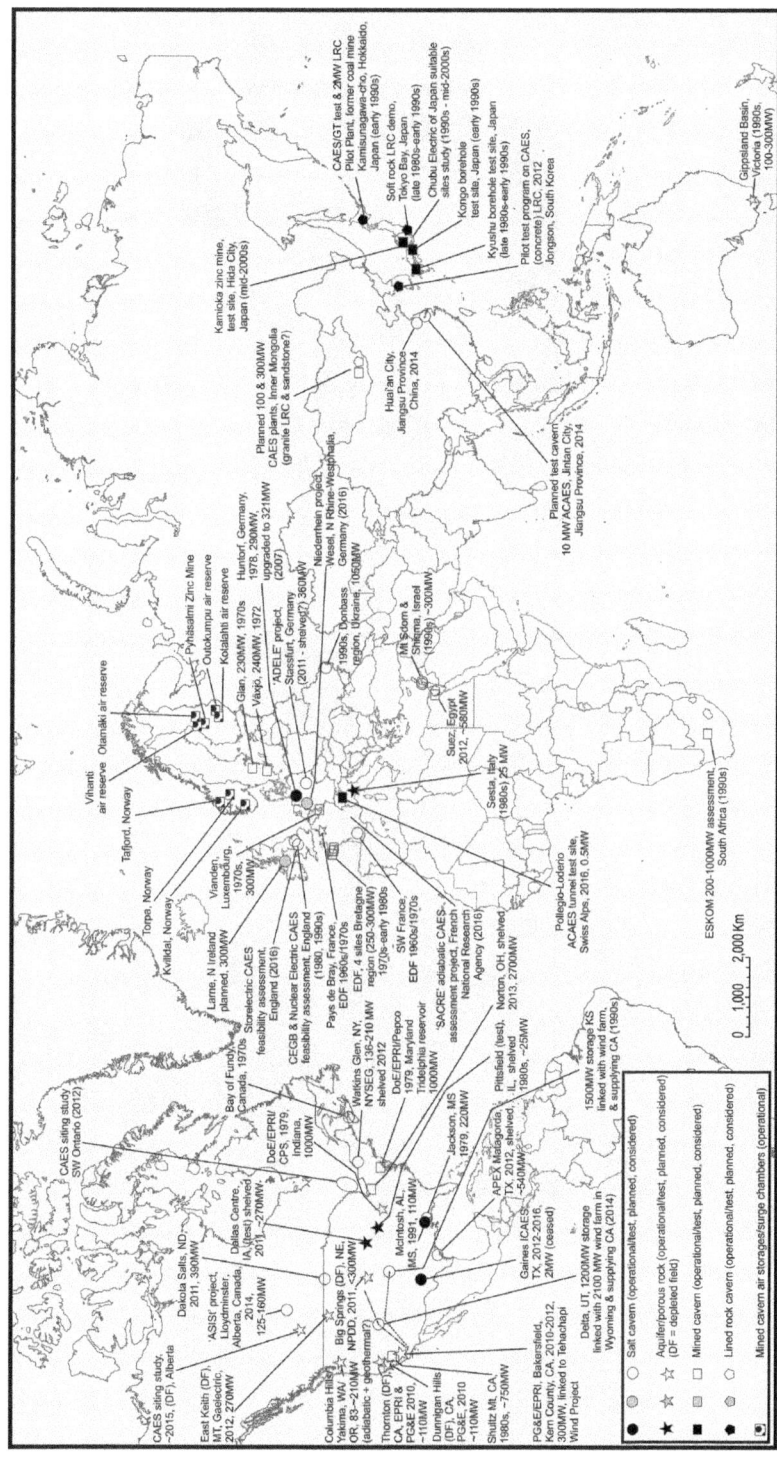

Figure 6 Map of compressed air facilities, CAES assessments, projects and operational facilities around the world. Sample only shown for the USA, but including many of the more recent assessments jointly undertaken with EPRI and ARRA funding.

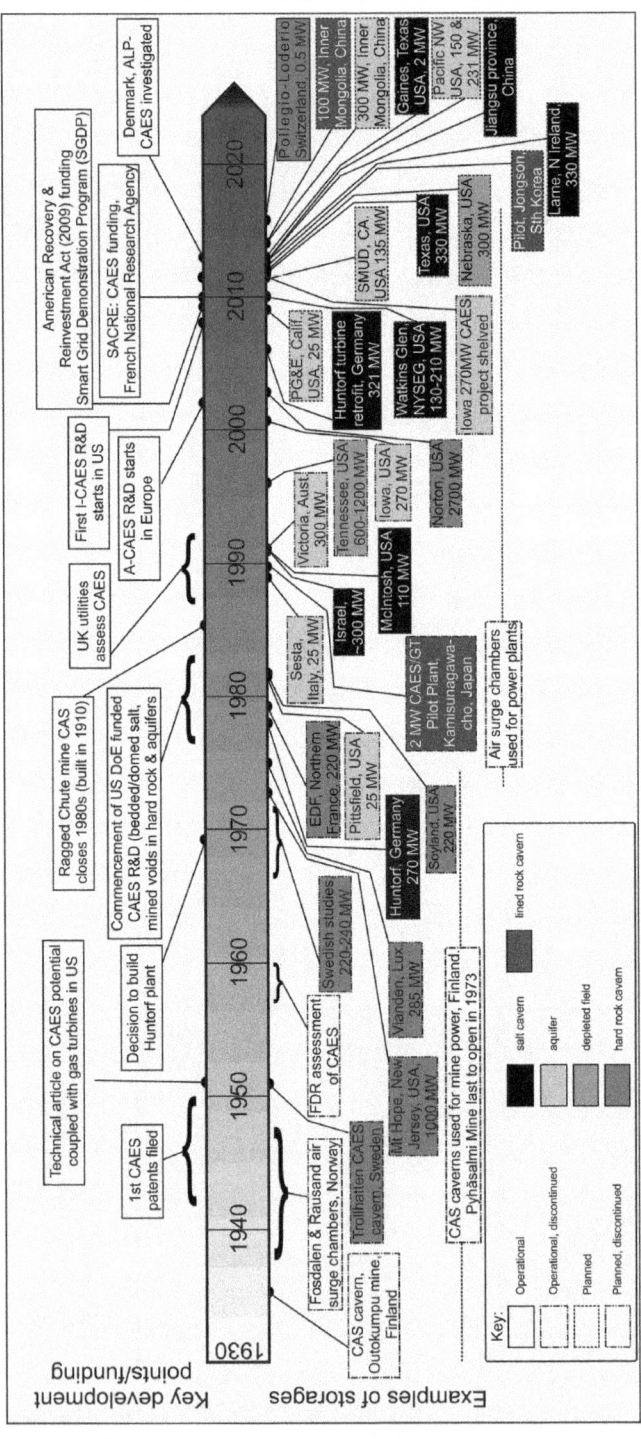

Figure 7 Timeline of compressed air usage, CAES R&D and principal industrial efforts. Based in part upon ref. 156.

Recovery and Reinvestment Act of 2009 (ARRA). Many US states have, over the years, looked into CAES potential, *e.g.* the PG&E/EPRI 300 MW CAES depleted field assessment in southern California, linking storage to the large Tehachapi Wind Project[161] (see Figure 6). The timeline summarises compressed air usage and CAES development to the present day.

Current operational CAES facilities at Huntorf and McIntosh prove CAES to be a commercially viable and mature, cost-competitive technology to meet grid-scale storage requirements (see Table 1). However, there are geological and geographical limitations on the distribution of these rocks and the ultimate volumes available to support salt-cavern CAES. It is widely accepted that the potential exists to advance CAES technology beyond caverns to use pore space in aquifers and depleted hydrocarbon reservoirs as the storage volume. This section outlines the basics of CAES in terms of the applications, geological storage options, potential regulatory requirements and the environmental impacts and considerations when considering CAES technology.

3.1 Applications of CAES

Increasingly, the applications and potential of CAES are recognised in terms of supporting power system operation and CAES will play an ever-growing and important role in energy management between generation and consumption. Potential applications and benefits of CAES include the following:[164]

- Price arbitrage.
- Peak shaving and demand-side management – using CAES to manage the demand-side energy by storing at the lower price and releasing the stored electrical energy during peak-demand times.
- Integration of more renewable power generation plants – enabling the seamless integration of renewable power generation plants into the existing power network to provide stable power grids while solving the problems of the inherent intermittence and instability of renewable power generation.
- Applications to smart grids and wind energy networks – playing a role in both supply and consumption.
- Applications to compressed air engines – compressed air energy being converted into other forms of mechanical energy through compressed air engines.
- Applications in other fields – including in the event of power supply failure, CAES acting in black-start capacity, rapidly providing power to important users.

3.2 CAES Configurations – DCAES, ACAES/AACAES, ICAES

CAES is a form of mechanical storage, the basic concept of which is simple: electrically driven compressors convert electric energy into potential energy

Table 1 Summary of technical and economic characteristics and maturity level of electrical energy storage technologies. Based on ref. 14, 40, 42 and 157, and others where noted (BES = Battery Energy Storage, VRFB = Vanadium Redox Flow Battery, PSB = Polysulfide Bromine flow battery).

Technology (source)	Power rating (MW) / Cycle efficiency (%)	General storage duration	Lifetime (years) / Cycling times (cycles)	Discharge time	Power capital cost ($ kW^{-1})	Energy capital cost ($ kWh^{-1})	Operation ($ kW^{-1}) / Maintenance ($ kW year^{-1})	Technology maturity level
PHS	100–5000 / 70–85	Long term, hours–months	40–60 / 10 000–30 000	1–24 h+	600–4300	5–100	0.004 / ~3	Commercialised, mature
Large-scale below-ground CAES	5–300 / 42–54	Long term, hours–months	20–40 / 8000–12 000	1–24 h+	400–1000	2–<50 (?up to 120)	0.003 / 19–25	Commercialised, mature
Above-ground small-scale CAES	0.003–3 / —	Long term, hours–months	~30 000+ /	30 s–3 h	517–550	200–250	Very low	Immature, test/development stage
ACAES with thermal salt TES[162]	128 / 60–70	Long term, hours–months	— /	1–10 h	1720	—	—	Immature, test/development stage
GE ACAES[163]	100 / 66–70	Long term, hours	— /	1–4 h	1070–1200	—	—	Immature, concept
GE ACAES[163]	80 / 70–72	Long term, hours	— /	1–4 h	1150–2000	—	—	Immature, concept
Flywheel	0–0.25 / 90–95	Short term, <1 h	~15–20 / 20 000–21 000	Milliseconds–15 min	250–350	1000–5000	0.004 / ~20	Early commercial
BES (VRFB, ZnBr, PSB)	30 kW–15 MW / 60–85	Long term, hours–months	5–20 / 2000+ – 12 000+	Seconds–10 h	600–2500	150–1000	~20 / ~80	Demo and commercialised

(more precisely, exergy) of pressurised air, which charges a storage that can be of many potential types. During compression from atmospheric pressure to a storage pressure of ∼70–80 bar (1015–1160 psi), the air temperature rises to ∼600 °C (the heat of compression). Processing and geological storage of compressed air at such high temperatures are not possible and the heat must be reduced to around 110–120 °F (43–49 °C) prior to storage, which is achieved by a series of intercoolers. The handling of the heat during compression and prior to expansion provides the main CAES configurations (see Figure 8), which can be in the form of conventional, diabatic CAES (DCAES), (advanced) adiabatic (ACAES/AA-CAES), or near-isothermal CAES (ICAES).

When electricity generation is required (*e.g.* when power generation cannot meet load demand), the stored pressurised air is withdrawn (and expanded) in the turbine/generator unit. However, during expansion, the air cools and prior to entering high-pressure turbines must be heated, either by the combustion of fossil fuels or the heat recovered from the compression process.

In the current operational DCAES (hereafter CAES) plants, the heat of compression is lost to the atmosphere as waste (reducing system efficiencies), and there is a need for additional fuel in the expansion process (see Figure 8a). The air then passes through a series of high- and low-pressure turbines, where it is mixed with fuel and combusted, with the exhaust expanded through low-pressure turbines. The high-pressure turbines capture some of the energy in the compressed air (see Figure 8a and b) (see, *e.g.*, ref. 14). Both the high- and low-pressure turbines are connected to a generator to produce electricity (see Figure 8). The exhaust is hot and, prior to release, waste heat may be captured *via* a recuperator and used during the air withdrawal phase, as at the McIntosh plant (see Figure 8b). At the Huntorf facility, reheating occurs in combustors using natural gas fuel. The overall efficiency of CAES is approximately 42%, as for the Huntorf power plant,[151] which is low in comparison with other EES technologies (see Table 1). Following 28 years of operation, the retrofitting of the whole expansion train at Huntorf provided an increase in the output power from 290 to 321 MW.[152] However, the exergy of the exhaust is still not utilised. The addition of a recuperator to recover the waste heat from the gas-fired expansion process provides improved efficiencies (∼54%) at the McIntosh CAES plant.[157]

The burning of fossil fuel produces CO_2 emissions, reducing the efficiency of the storage-recovery cycle and adding to the cost of the recovered electrical energy. This compromises the ecological benefits that most would associate with renewable energy sources and technologies. The current focus of research is mainly on the so-called 'second-generation' ACAES/AACAES, the design of which is to deliver higher efficiencies *via* a zero-carbon process that eliminates the supplemental gas-firing process. In an adiabatic (or advanced adiabatic, ACAES/AA-CAES) process, the heat of compression is captured and stored in a thermal energy storage (TES) plant. It is later used to heat the air during withdrawal and expansion, increasing round-trip efficiencies.

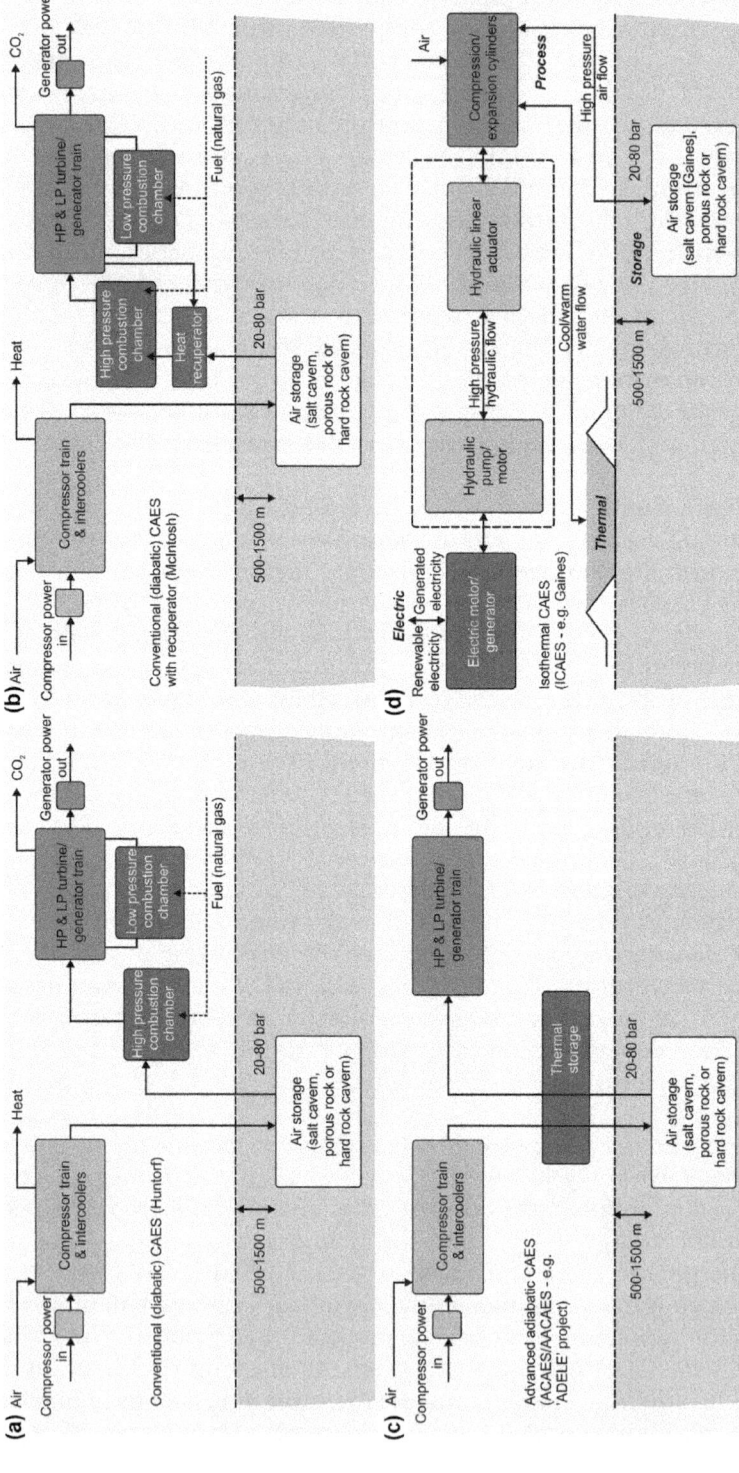

Figure 8 Basic CAES configurations: (a) conventional diabatic CAES (*e.g.* Huntorf), (b) conventional CAES with recuperator (*e.g.* McIntosh), (c) adiabatic CAES and (d) near-isothermal CAES (see ref. 166 and 167).

During withdrawal operations, the air leaves the storage cavern and passes through the TES before being applied to an expansion turbine coupled to a generator, without the need for any co-firing of fuel. Improved round-trip efficiencies of >70% are expected,[40] with 70% reported in RWE's 'ADELE' AACAES test project.[165] Predicted system efficiencies at the ALACAES AA-CAES tunnel test facility, commissioned in 2016, are expected to approach 90%.[154] Emerging from RWE's work, General Electric (GE), together with EPRI, looked at two designs, for building and possible performance testing of a large-scale ACAES plant.[162] One design was to develop an ACAES plant using existing commercial GE equipment (not requiring the development of a high-temperature compressor system) or, alternatively, to design, develop and manufacture an entirely new high-temperature compressor for application to ACAES.

Other emerging technologies include isothermal or near-isothermal CAES (ICAES), which also offer the potential for increased efficiencies over CAES and ACAES/AACAES (see Figure 8d). The process requires the continual removal of heat from the air during the compression cycle (interstage cooling) and its continuous addition during expansion to maintain an isothermal process. To date, technology testing has generally involved above-ground processes, including storage.[156] However, General Compression (GC) completed the design and installation of a General Compression Advanced Energy Storage (GCCAES™) plant, integrated with a 2 MW Gamesa G97 wind turbine, a pre-existing former fuel storage salt cavern, a surface brine pond (for thermal storage) and connection to the Lea County Electric Cooperative electricity grid (see Figure 8d).[166,167] Such configurations do not require gas line connections or produce air pollutants, including coal ash, CO_2 emissions or other risks, and would be ideally suited to areas remote from conventional power plants and associated transmission lines, or where conventional power projects would not gain approval. Commissioned in late 2012, the plant operated until it ceased operation in 2016,[168] but the results will prove invaluable in moving the technology forward.

Further efficiencies would be possible if, instead of using energy derived from fossil fuels during compression, the compression is achieved using renewable energy sources (excess wind or PV; see Section 3.7). Other studies aimed at increasing efficiencies and lowering costs include high-temperature ACAES, where prior to compression the air is preheated, which could be linked to PV, or low- to medium-grade waste heat.[169] To date, however, CAES remains the only commercially implemented system.

3.3 Geological Storage Options

The following sections review briefly the main geological storage options, together with indications of costs and possible issues during storage. Underground gas storages operate by compressing gas during injection and which, on withdrawal, decompresses. The gas storages operate between maximum and minimum pressures and set pressure rate changes. This is in order to maintain the geomechanical stability of the reservoir to avoid overpressuring and fracturing, or underpressuring and collapse. Rock strength and storage depths (equating to lithostatic pressure – the pressure or stress imposed on a layer of rock by the weight of overlying rock) determine the operational pressure ranges. The gas between the minimum and maximum pressures is the working gas volume, and the gas that must remain in storage to maintain minimum pressure is known as cushion gas.

3.3.1 Solution-mined Salt Caverns. Thick, deeply buried (>250 m) bedded or domal halite/rocksalt ('salt', NaCl) deposits exist in many countries[170] and provide the host rock to the two commercial CAES facilities. The favourable physical properties of halite means that the construction and operation of very large unlined caverns are possible, which offer highly flexible, inexpensive storage in comparison with other geological storage options (see Table 2). They are not only stable for very long periods of time, but also halite is gas tight with no cavern pressure losses over often extended periods of storage (months), and is also inert with respect to liquid and gaseous hydrocarbons and to oxygen and hydrogen.[171] Salt caverns are constructed by a process known as solution mining: water is pumped *via* cased boreholes into salt deposits deep underground, dissolving the salt to create a cavity and brine as a by-product. The brine is extracted, with the process continuing until the cavern is the required size and shape. The brine produced may be used for industrial applications or is disposed of at sea or by injection into saline aquifers. The latter is often the case in many US gas and hydrocarbon storage operations. Globally, salt caverns rank third in terms of natural gas storages, accounting for 6% of facilities as at 2014,[171] and represent the cheapest main storage option (see Table 2).

To date, both of the two operational commercial CAES systems use solution-mined salt caverns. Huntorf operates with two air caverns (140 000 and 170 000 m^3) and one gas storage cavern,[151,173] whereas McIntosh has one cavern (566 000 m^3).[174] At Gaines in Texas, Texas Dispatchable Wind LLC (a subsidiary of General Compression) commissioned and briefly operated a small-scale near-isothermal 2 MW plant using a reconditioned

Table 2 CAES capital costs of storage. Compiled from ref. 42, 157 and 172.

Geology	Storage/reserve capacity cost ($ kWh^{-1})	Cost of storing air adjusted figures (€ kWh^{-1})
Salt cavern – solution mining	1–5	1.01
Salt 'cavern' – dry mining	10	9.71
Porous rock – aquifer, depleted field	0.1	0.1
Hard rock – existing mine	10	
Hard rock – excavated cavern	30	29.55
Abandoned mine – limestone, coalmine, *etc.*	10	9.71

former liquid hydrocarbon storage cavern ($\sim$31 100 m^3 or 265 000 barrels) between late 2012 and 2016 (see Figure 6).[166,167] The proposed ADELE adiabatic CAES demonstration plant in Germany was scheduled for completion in 2016 but has not progressed further.[165] Other salt-cavern CAES sites assessed around the world include ones in Israel, Canada and the UK (see Figure 6), with Gaelectric's current application to construct a 200–300 MW plant near Larne in Northern Ireland.[175,176]

3.3.2 Porous Rock – Saline Aquifers and Depleted Fields.

3.3.2 Porous Rock – Saline Aquifers and Depleted Fields. Dependent upon the existence of a suitable caprock and trapping structure, porous rock formations with sufficient porosity and permeability offer potential for storage, with many gas storage facilities having been constructed using both depleted hydrocarbon reservoirs and aquifers (water-filled reservoirs). In the porous-medium storage concept, injection of air through a borehole displaces the water, creating an air 'bubble' within the pore space in the near-well region, creating a 'gas cap'.

Globally, with proportions of 81 and 13%, respectively, for stored natural gas, depleted field and aquifer structures represent the first and second most important storage types (see, *e.g.*, ref. 147 and 171). EPRI suggest that aquifer storage is the least expensive geological storage option (see Table 2). However, a number of things must be taken into account when considering CAES aquifer storage that add to the costs of developing an aquifer storage. First, the geological and physical characteristics of aquifer formation, such as suitable trapping structure plus caprock and adequate porosity/permeability, are not known ahead of time and significant investment has to go into investigating these and evaluating the aquifer's suitability for gas storage. The same would be true for CAES, with air also representing a lower energy density and value compared with natural gas or hydrogen. Second, all of the associated infrastructure may have to be developed, including installation of wells, extraction equipment, pipelines, dehydration facilities and specialised equipment near the wellhead, and possibly compression equipment. This increases the development costs compared with depleted hydrocarbon reservoir sites. The aquifer initially contains water with no naturally occurring hydrocarbon gases (or air) in the formation and, of the gas injected, a significant proportion will be physically unrecoverable. Consequently, aquifer storage typically requires significantly more cushion gas than depleted reservoirs: up to 80% of the total gas volume. Therefore, developing an aquifer storage facility for gas storage is usually time consuming and expensive, making it the least desirable type of natural gas storage facility.

With depleted fields, proven porosity, permeability and trapping structure exist, but some potential technical challenges and problems arise when considering CAES. First, residual hydrocarbons remain in the reservoir after production ceases. The injection of air can lead to co-mingling with the gas and give rise to the formation of ignitable gas mixtures during the withdrawal phase. Formation damage may also have occurred during production due to the depressuring and collapse of the reservoir. In a similar fashion, the caprock may have suffered fracturing during pressure decrease and its integrity may need consideration. Further risks arise with potential reactions between oxygen and the mineral constituents of the reservoir rock, which can lead to oxygen depletion and the potential for bacterial/microorganism growth due to the introduction of warm air/oxygen into the reservoir.

CAES in depleted fields has been, and is currently being, assessed with a number of sites located in the USA (see Figures 6 and 7). As part of the 2009 ARRA Energy Smart Grid Demonstrations Program and storage demonstrations, PG&E together with EPRI have considered projects in California for an advanced 300 MW, 10 h storage CAES plant using a depleted gas reservoir.[177] Borehole tests at candidate sites have been undertaken but, to date, no test plants have been developed.

Potential benefits with aquifers may include enhanced deliverability rates if an active water drive exists (the natural, regional flow in an aquifer, or as water in the reservoir expands during hydrocarbon production), which supports the storage reservoir pressure through the injection and production cycles. However, the aquifers have not previously been pressurised to the levels likely to occur during storage and reached when conditioning the reservoir – the aquifer contains only water and to begin displacing the pore water and replacing it with natural gas or air, high injection pressures are required. Water coning could occur, where water infiltrates the near-wellbore region and is produced instead of hydrocarbons (or air). In addition, once injected, a certain amount of natural gas/air is physically unrecoverable, being trapped in the pore spaces. Since aquifers are naturally full of water (and not gas), cushion gas requirements may be as high as 80% of the total 'gas' volume. This results in less flexibility in injecting and withdrawing the stored gases and higher costs.

Testing of aquifer storage sites has taken place (see Figures 6 and 7). Much early aquifer assessment took place in Illinois, USA, with the Pittsfield structure tested (~200 m below ground level)[148,160] and which demonstrated the feasibility of daily cycling of compressed air.[148,160,161] However, reservoir pressures were lower than expected, with 'fingering' of air within the aquifer found (the unequal movement of air in the porous rock, producing uneven or 'fingered' profile), both of which affect the growth of the air bubble. Also, over longer storage periods, oxygen depletion occurred, related to the oxidation of minerals, particularly iron sulfides (pyrite).

Another early aquifer storage assessment was a 25 MW test facility at Sesta, Italy, which ran between 1987 and 1991.[178] Again, the project encountered less than ideal reservoir conditions and a 'geothermal anomaly', and was shut down. The most recent aquifer storage assessment was the innovative 270 MW Dallas Center CAES project, coupled with wind generation near Des Moines, Iowa, USA (~890 m below ground level).[179,180] The project ceased following 8 years of study and development owing to geological limitations, including reservoir heterogeneity (thickness, permeability and porosity) and high sulfate concentrations in water.

Hence, despite a large US government research programme over 30 years ago, more recent testing of an aquifer site in Iowa and modelling studies that suggest that aquifer CAES is very feasible, fundamental questions remain regarding the hydrological and energetic performance of CAES in porous media.[17,18]

3.3.3 Mined Voids – Abandoned Mines and Hard/Competent Rock Caverns (Unlined). This category includes two main types of unlined rock caverns (cavities) that are generally conventionally mined (excavated) in competent rock formations: abandoned mines and specifically mined caverns for storage, both of which have been used as gas storages. Some rock masses are gas tight, but generally the host rock, although stable, has fractures and joints and is not tight with respect to liquids and gases. Hydrostatic pressure of the water in the overlying rock mass provides containment: groundwater above the storage percolates through the cracks and into the storage, preventing the gas from leaking outwards and upwards. The water collects in sumps at the base of the cavern, to be pumped to the surface. Water curtains are sometimes constructed, comprising a system of holes drilled above the storage and into which water is pumped to increase the hydrostatic pressure.

Gas storage has been undertaken in abandoned coalmines at Anderlues and Peronnes in Belgium[181] and at Leyden in the USA.[182] However, owing to leakage, the storages closed after a short period of operation. Liquid hydrocarbons have been stored in specifically mined unlined rock caverns in the USA and Europe since the 1950s. One former salt mine located at Burggraf-Bernsdorf in Germany is currently operating as a natural gas storage.[171] The feasibility of storing wind energy as compressed air in the Lyons Salt Mine in Kansas, USA, was investigated in 2009[183] but never progressed. Investigations took place into constructing a 2700 MW CAES plant in a former limestone mine (10×10^6 m^3) at Norton, Ohio, USA.[158] Interest started in 2001, but the project stalled and appears to have been abandoned:[171] ACAES is being tested in a tunnel in the Swiss Alps.[154,155]

The construction of rock caverns is associated with a higher level of risk of accident compared with salt caverns and porous storages due to excavation of the caverns by mining techniques (drilling, blasting and clearing the fallen rock). Unlined rock caverns have the highest initial capital costs for geological storage options (Table 2) and these would be increased if the construction of water curtains were needed to provide containment. Although there have not yet been any commercial compressed air storages, small unlined caverns, operated at air pressures of around 0.8 MPa (116 psi/ 8 bar), were used from 1936 to support electricity supply in Finnish mine operations (including the Pyhäsalmi zinc mine[184]). Compressed air cushion surge chambers attached to hydropower plants use unlined rock caverns as part of power generation operations at Fosdalen and Rausand in Norway.[185] Elsewhere, in Japan, a series of borehole tests were conducted for CAES potential in unlined caverns in a range of rock types during the 1980s, although no sites were developed (see Figure 6).[186–188]

3.3.4 Lined Rock Caverns. Lined rock caverns (LRCs) are a variation on the mined void storage type, with rock caverns constructed and lined with an artificial, gas-tight barrier, generally comprising concrete and stainless-steel sheet. The technology was proved for gas storage following the construction and testing of a pilot plant between 1988 and 1993 at Grängesberg in south-central Sweden and the opening of a demonstration cavern (40 000 m^3) at Skallen near Halmstad in southwestern Sweden (see, *e.g.*, ref. 189). Construction of several small CAES test facilities using LRCs took place in former mine tunnels in Korea and Japan during the 1980s (see Figure 6).[190,191] Tests were successful, but in general short, and no facility was developed.

Investigations into the feasibility and stability of shallow CAES LRCs in cavities and tunnels are ongoing, with studies on the coupled thermo-dynamic, geomechanical behaviour and stability of lined caverns located at depths of 60–120 m.[189,192,193] Maximum air pressures of 20 MPa may be possible, depending on the geotechnical conditions.

Although of smaller volume and higher cost than other storage options, LRCs offer a number of distinct advantages over existing conventional underground storage types. They have greater geological flexibility, are a technology that provides adequate degrees of tightness and high storage pressures, even at relatively shallow depths, and have a capability for high-frequency cyclic operations. They can therefore create conditions for the storage of large volumes of 'gas' in regions where more favourable rock types do not exist, or those present are unsuitable for geological CAES storage. As such, this storage concept is particularly well suited where the challenge for energy systems will be the management of electricity, because the major sources of renewable energy such as wind or solar do not always occur in regions with favourable geology.

Costs are the highest of all geological storage operations due to mining operations and the construction of the cavern lining (Table 2). Capital expenditure (CAPEX) of the Skallen storage was €27 million, or €675 per m^3.[194]

3.4 Operational Modes of CAES 'Reservoirs'

The operational nature of cavern and porous media storage differs. For salt caverns and mined voids, two operational modes that are generally considered, with compressed air in both systems stored at up to 70–80 bar pressure, are the following (see Figures 8 and 9):

1. Fixed-volume, variable-pressure system (isochoric) – the most common mode, with a cavity or cavern of fixed volume operated over an appropriate pressure range, increasing as air is pumped in, and *vice versa*. Both the Huntorf and McIntosh facilities operate in this mode and require throttling to maintain constant turbine inlet pressures.

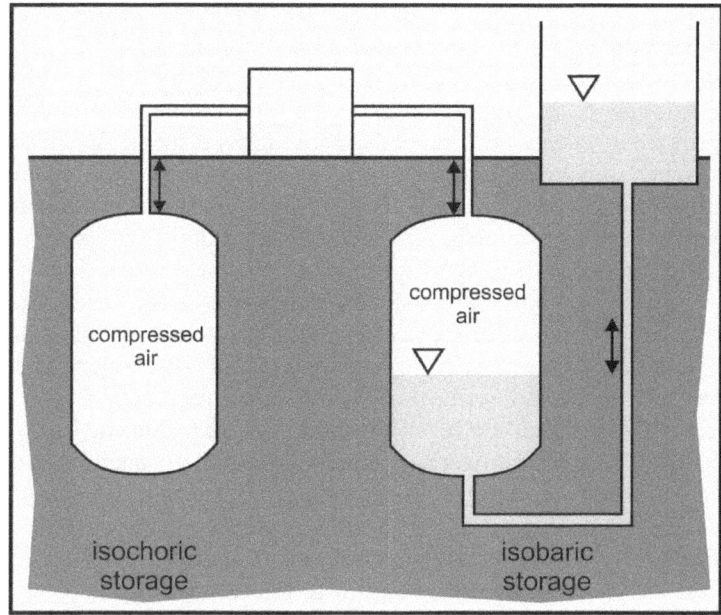

Figure 9 Schematic of operational isobaric (constant-pressure) and isochoric (constant-volume) pressure modes in underground storage.

2. Constant-pressure but variable-volume storage cavity (isobaric) – the underground storage cavity is linked to a water (or brine) reservoir at the surface; as compressed air is pumped in, water is displaced from the cavity (brine compensation mode). The pressure in the cavern remains almost constant, despite the increase in volume of the air, which represents a more efficient system than the constant-volume method.[195] Examples of liquid hydrocarbon (propane) salt-cavern storages operating by brine displacement are found on Teesside in the UK.[196]

The two modes present different environmental considerations (see below).

The literature (see, *e.g.*, ref. 197) suggests that the operational window for CAES salt caverns lies between 500 and 1300 m (see Figure 8), based upon operating pressures being directly dependent on depth and power-plant components. However, breakthroughs in compressor and turbine technology could allow CAES deployment to greater depths than previously possible, with 1500 m being considered at Larne in Northern Ireland.[176] To maintain the mechanical stability and integrity of salt caverns, they operate within a minimum and maximum pressure range and pressure change rate during injection or withdrawal. These limits are set by the mechanical properties of the salt, depth of storage, and borehole size, being 30–80% of lithostatic pressure. However, CAES caverns may occasionally require lowering to atmospheric pressure during maintenance periods, although the lower pressures should be for the minimum time possible.[151,171]

Porous media storage presents different conditions and problems from those of salt-cavern or mined voids, with significant differences existing between an open cavern and the distributed pore space in an aquifer (or depleted field), and how energy is stored (spatially). Pressure evolution in a cavern shows nearly uniform pressure throughout the cavern because in an open cavern there is very little resistance to air flow, permitting very rapid pressure propagation. In contrast, a porous rock provides a large resistance to airflow, which results in pervasive pressure gradients established by the injection of air.[18] In an aquifer, the reservoir pressure increases during the initial fill period and then oscillates during each production, recharge, shut-in and production period, with much of the variation occurring in and near to the well. This is due to the reservoir itself, which limits the flow of air and thus the extent to which injection-related pressure propagates over time.[17] Although aquifer storage may exhibit a water compensation mode over the long term, it acts as a constant-volume, variable-pressure system for short-term cycling,[160,198] with pressure gradients remaining throughout the injection period.[18] During shut-in, the gradient tends to diminish but does not disappear entirely in the reservoir, hence variable-pressure (pressure gradients) rather than single-pressure values dominate energy storage in an aquifer. Added to these physical constraints are the facts that system efficiencies and dynamics are affected by a large loss of heat to the formation found through the wellbore, and the loss of oxygen in the stored air, probably due to oxidation reactions with minerals in the formation.

Other considerations may include aquifer water drive arising from regional flow and unconfined or confined aquifers due to structural compartmentalisation. The latter, together with poor permeabilities, could lead to isolated pressure regimes, such that water drive does not exist, or is much reduced. These issues may lead to both overpressuring during charging or underpressuring during air withdrawal, as the influx of water to replace the air is not possible. Both situations may lead to loss of reservoir performance and permanent aquifer/reservoir damage. Gas storage experience in aquifers suggests that maximum charging pressures should not exceed ~80% of lithostatic pressure, and maximum reservoir storage pressure should not exceed 50% of lithostatic pressure.[160] Pore pressures exceeding lithostatic pressure could lead to fracturing or parting along bedding planes in the reservoir and overburden rocks, with the possibility of air loss, displacement of groundwater and damage to surface structures. Depleted reservoirs may act in a similar fashion, but experience from production will assist the understanding of reservoir, pressure, recharge and management.

3.5   UK Potential for Deployment of CAES

Shell and the Gas Board assessed the CAES potential in the UK the late 1980s and 1990s.[199,200] ICI (now Ineos) looked at CAES prospects for their salt field in Cheshire and, although the salt beds have been proved for gas storage caverns, no CAES project has been progressed. Interest has once again returned, with sites under consideration in Northern Ireland and in England, both part funded by EU programmes.

Following lengthy investigations, in 2015 Gaelectric submitted planning application documents for a deep (~1500 m) salt-cavern hosted CAES facility in the Larne area of Northern Ireland.[176] Storelectric was planning a 40 MW, 800 MWh pilot plant using a salt-cavern air store in northwest England,[162] although little information is available regarding progress.

3.6 Planning and Regulatory Environment for CAES

In the UK, halite beds occur only onshore in England and Northern Ireland (NI) and currently, salt-cavern storages seem the most likely for a CAES facility. To date, no planning applications for CAES have been submitted in England, with the only one in the UK being for the Gaelectric Project-CAES Larne, NI, submitted in 2015.[201] NI legislation differs in some details from that in England (The Planning Act 2008) and the separate planning legislation for Wales and Scotland. However, it provides evidence of the status of such a project in terms of national planning policies and applicable legislation. Thus, together with underground gas storage applications in England, in which similar storage technologies are employed, and existing oil and gas legislation, it is possible to highlight the main legislation and regulations that will likely be applicable to applications relating to large-scale energy infrastructure such as CAES in England (see Figure 10).

The following overview attempts to place a CAES application in context with what is a complex EU and UK legislative and regulatory environment likely to affect CAES development in the UK (see Figure 10). It is necessarily brief and cannot address all issues. A number of Competent Authorities (CAs) will be involved in the process, including the relevant environmental regulator, the Health and Safety Executive (HSE) and, for some operations in offshore areas, the Maritime Agency (MA) together with The Crown Estate (TCE). The differing storage types will likely come under different areas of legislation/regulation, particularly in the offshore areas.

For the Gaelectric Project-CAES Larne, NI, proposal, the Planning Act (Northern Ireland) 2011 is the primary planning legislation and is equivalent of the Planning Act 2008 (plus amendments) in England. This gives the NI government responsibility for regional planning policy, the determination of regionally significant and called-in applications, and planning legislation. The EU Trans-European Energy Infrastructure (TEN-E) Regulation (347/2013 EU) sets out rules for the timely development of energy infrastructure across Europe and streamlining the permitting processes for major energy infrastructure projects [designated projects of common interest (PCIs)] that contribute to European energy networks among member states. PCI status demonstrates that projects conform to a range of criteria, including having significant benefits to at least two member states, contributing to market

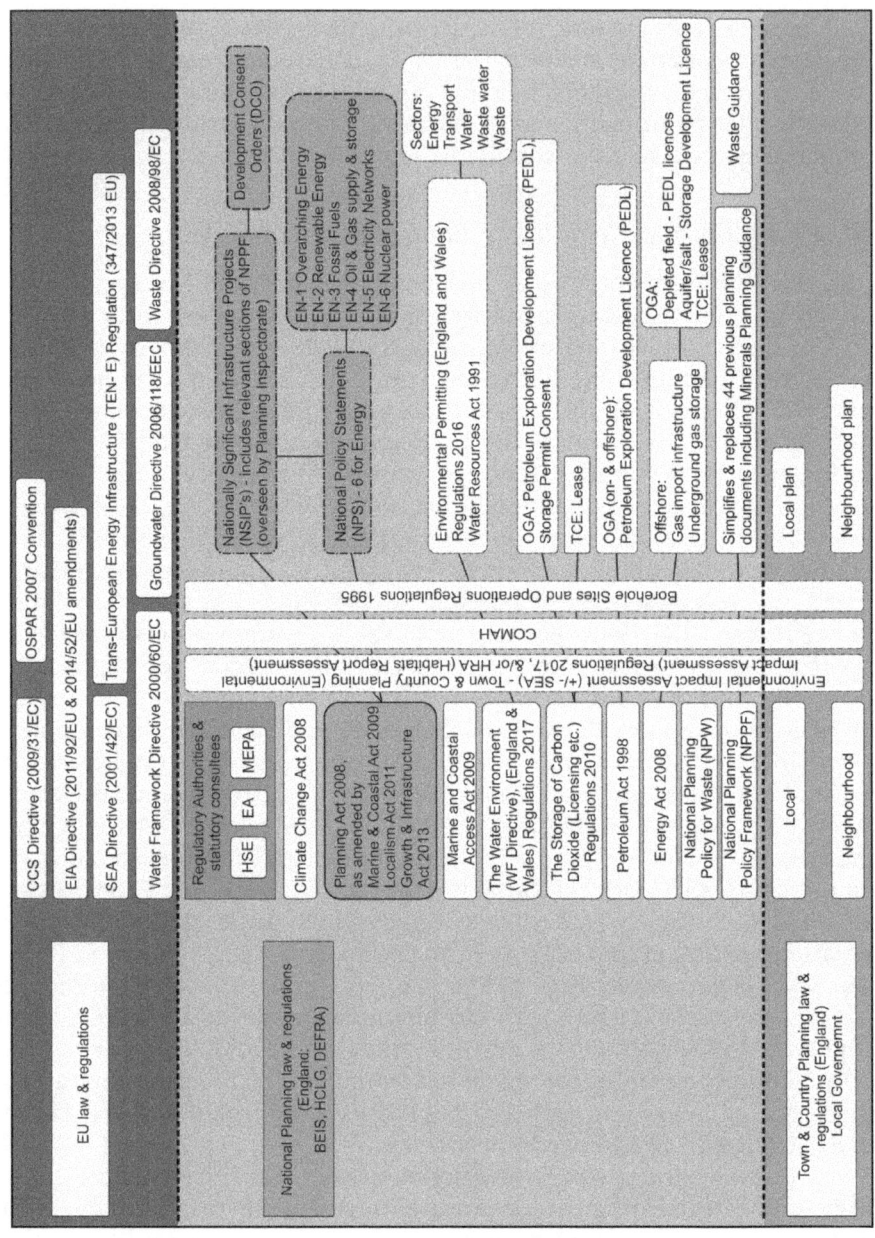

integration and further competition, enhancing security of supply and reducing CO_2 emissions. In October 2013, Project-CAES Larne, NI, was designated a PCI and in July 2015 the Strategic Planning Division of the Department of the Environment confirmed it as a 'Development of Regional Significance' under the new Planning (NI) 2011 Act and 2015 (Development Management) (DM) Regulations. The Department thus holds jurisdiction over the proposed development and the planning application.

This established the precedent in the remaining UK onshore areas for a CAES application to proceed through the main national Planning Act 2008 (as amended by the Marine and Coastal Access Act 2009, the Localism Act 2011 and the Growth and Infrastructure Act 2013). Offshore CAES will be subject to different legislation and probably will follow that around natural gas storage and carbon capture and storage (CCS).

The Planning Act 2008 and relevant National Policy Statements (NPSs), and also any other matters that are considered both important and relevant, established that large, nationally significant infrastructure projects (NSIPs) are now considered separately by the Planning Inspectorate [originally the Infrastructure Planning Commission (IPC)]. The Act created a new Development Consent regime [with the application for, and issue of, Development Consent Orders (DCOs)] for NSIPs in the sectors of energy, transport, water, wastewater and waste (see Figure 10). It includes associated developments, which in the case of underground gas storage facilities comprises both surface works such as pumping/compressor stations, boreholes and pipelines to storage facilities, and monitoring boreholes.

Eleven NPSs support this legislation, setting out government policy on different types of national infrastructure development. Six of the NPSs are in Energy (EN-1 to En-6), all of which could have relevance to CAES, depending on the source of electricity and later generation. As Figure 10 suggests, other legislative and policy areas lie behind and provide reference during the DCO application process. This includes the National Planning Policy Framework (NPPF), which establishes policy for local planning authorities in various areas, including mineral planning and the sustainable use of minerals, to the preparation of environmental impact reports, adherence to water and borehole regulations and HSE requirements. The EA and HSE are statutory consultees in the process. A number of underground gas storage projects have now successfully obtained a DCO, including the Preesall (2014) and

---

Figure 10   Illustration of the key EU and UK (England) legislation, regulations and regulatory bodies likely to apply to an application to develop a CAES plant. Main Planning Act (2008) and supporting policies are the likely primary legislation under which a DCO planning application would be submitted for a CAES facility onshore (highlighted). Also indicated are relevant Competent Authorities involved in a DCO application and assessment, with likely applicable legislation for storage in the offshore area (Energy Act, 2008), overseen by BEIS (OGA) and The Crown Estate (TCE).

Keuper Gas Storage (2016) applications, both constructing salt caverns. Preesall was subject to EA requirements and obtaining a consent to discharge trade effluent (brine), which at the time in England fell under the Water Resources Act 1991 (as amended in the Environment Act 1995[202]). Applications today would also need to consider the Water Environment Regulations (2017) and National Planning Policy for Waste, the latter including waste (brine or rock) generated during the construction of mined voids (caverns).

Additional legislation would likely apply for porous rock CAES storage where depleted hydrocarbon fields are considered. For onshore areas, alongside the Planning Act legislation, it is likely that certain other regulations applying to the water and oil and gas industries will apply to proposed developments. This is because any air injected to storage will likely co-mingle with residual gas and be produced together with the air during withdrawal cycles. The two important areas of legislation are the Petroleum Act 1998 (which vests all rights to the nation's petroleum resources in The Crown, including the offshore continental shelf area) and the Energy Act 2008. The Oil and Gas Authority (OGA), on behalf of the Department for Business, Energy and Industrial Strategy (BEIS) and The Crown, grants a Petroleum Exploration and Development Licence (PEDL) that confers exclusive rights to 'search and bore for and get' petroleum in a defined area. Northern Ireland issues its own licences to cover its onshore area independently of the OGA. The EA has issued guidance on the environmental permits required for onshore oil and gas operations in England.[203]

The use of aquifers for CAES, both on- and offshore, would be subject to legislation and a number of EA regulations relating to both surface and subsurface infrastructure or operations, many of which are applicable to oil and gas activities.[204] The primary aim of the EA is the prevention of pollution of groundwater and protection of it as a resource. But it is also to manage water supply, which might be required during the construction of salt caverns, by abstraction from either rivers, aquifers or the near-offshore area. Regulations would apply to the disposal of the brine in such operations. As part of aquifer protection, any boreholes drilled to develop and support the project would fall under the Borehole Sites and Operations Regulations (1995), which require boreholes to be properly completed to avoid communication and contamination between formations.

Offshore, CAES applications are likely to face slightly different legislation owing to rights relating to minerals, the seabed and subsurface space invested in The Crown. Again, there is no experience relating to CAES but that of the offshore gas storage and CCS industries provides guidance. The Energy Act 2008 covers a number of areas, with the following likely to be applicable to offshore CAES: offshore gas infrastructure, carbon dioxide storage, Renewables Obligation, decommissioning of energy installations (nuclear, offshore renewables and offshore oil and gas) and offshore transmission. In terms of natural gas storage, anyone who wants to explore for, drill for or use a site in the UK's offshore area for that purpose must hold a licence issued by the OGA under Section 4 of the Energy Act 2008. In addition, for storage activities in all offshore areas, including the territorial sea, a lease is required from The Crown Estate (TCE), as Section 1 of the Energy Act 2008 vests all rights to store gas in the offshore area in The Crown. Northern Ireland's offshore waters are subject to the same licensing system as the rest of the UK continental shelf area (UKCS). The lease and licence provide the framework for regulatory consent for the physical activities at the site, *e.g.* intrusive drilling and the subsequent submission of a Gas Storage Development Plan (GSDP) for OGA approval. If it is intended to store gas in a petroleum reservoir (whether fully or partially depleted), the law requires a petroleum Production Licence (PL) because native gas will inevitably be produced from the reservoir when gas is recovered for use.

Storing gas in a non-petroleum-related geological feature such as a salt bed or dome, requires no PL and only an application for a Gas Storage Development Licence is required. Aquifer storage would likely have to refer to legislation relevant to CO_2 storage, which requires safeguarding of the aquifer and preparation of a GSDP. As such, the use of the subsurface storage space for CAES would likely fall under this legislation.

Should gas pipelines be required at a CAES site, then offshore oil and gas pipeline consents are also the responsibility of the Secretary of State for BEIS and are issued in accordance with the Petroleum Act 1998. Any CAES assessment is also likely to reference relevant British and European Standards on the construction and operation of underground (gas) storage facilities (BS EN 1918-1 to 5:2016) and the HSE SPC/ENFORCEMENT/185 document.[205]

Projects may lie across on- and offshore areas and thus more than one consenting jurisdiction and specifically across two main consenting regimes, which have overlapping regulations on EIA and HRA (Habitats Report Assessment) that apply, respectively. Project-CAES Larne, NI, falls within this category.[201] In a similar manner, the Preesall Gas Storage development in northwest England involved on- and offshore components, the latter requiring marine consents issued under Part 4 of the Marine and Coastal Access Act 2009.[202]

3.7 Environmental Performance, Emissions, Sustainability and Economics of CAES Systems

A series of papers have reviewed the environmental performance, applications, outlook, potential and costs of various EES in power system operations.[16,19,40,42,206] Clearly, no single EES technology can meet the requirements for all power system applications. However, CAES is recognised as an important technology with which to compensate for the intermittency and fluctuations that occur with renewable sources and thereby increase the penetration of renewables (wind power and PV). CAES requires suitable geology and has site-specific technological requirements, which, when combined with a lack of grid-scale adoption, means a lack of data relating to the economic performance of sites. CAES, generally regarded as a storage technology to assist in reducing emissions, is also not a 'pure' storage system. It requires fuel for combustion in the turbine during the generation phase, which both increases operating costs and is associated with GHG emissions, although less so than normal gas-turbine generators.[16] These factors further hamper the assessment and comparison of CAES with other EES technologies for grid operators, power suppliers, energy system analysts and policy makers developing regulation strategies.[42]

Many past environmental assessments have looked at diabatic CAES. Increasingly, research on the performance of ACAES systems[207] and linking CAES with electricity production from wind power (which was discussed as early as 1976[208]) and using other fuels such as biogas as substitutes to reduce further the environmental impacts are under consideration.[209] There is now a growing literature on the economic and environmental performance and benefits of ESS technologies and co-locating renewable generation and CAES in addition to those relating to increased grid stability.[9,15,16,210–212]

3.7.1 Economics of CAES – Capital Costs. Costs of EES are discussed in the literature (see Table 1).[14,40,207,213] Capital cost is one of the most important factors for the industrial take-up of EES.[14] However, a comparison of the merits and costs of the various technologies is not entirely straightforward. Levels of uncertainty and complexity hamper estimations and comparisons of the cost of large-scale EES systems: apart from PHS, their deployment is scarce, they are in differing stages of development, storage sizes can be different for the same power rating technical diversity, differences in application exist and the economic performance of existing sites is not widely reported. CAES has only two long-term commercial plants and the need to cost in the value of gas with all its market and supply variations further complicates assessments. That said, in general, EES technologies include two main sections: a power conversion system (PCS), which may consist of two separated units for charging and discharging, with different characteristics, and an energy storage section (that containing the storage medium, *e.g.* the cavern in CAES[42]).

Various studies have shown CAES and PHS to have certain economic advantages in terms of capital costs and construction energy requirements over other EES technologies (see Tables 1–3). In terms of construction, salt solution mining for a CAES cavern is relatively low in energy intensity compared with building structures to house battery components and electrolytes. In addition, air and water CAES and PHS represent very cheap energy storage media compared with batteries, which require intensive mining, refining and manufacture.[16]

The power and energy densities of different EES technologies play a significant role in determining the applicability and cost-effectiveness of EES and supply. For a given amount of energy, the higher the power and energy densities are, the smaller is the volume of the required energy storage system.[40] PHS and CAES have low densities, requiring large reservoirs, and are thus most suitable for use mainly in stationary, bulk (utility)-scale facilities, with discharge times of 10 h or longer.[40]

Also important in determining the economic performance are cycle (round trip) efficiencies,[40] with CAES having a low cycle efficiency (<60%), which improved from the Huntorf (42%) to the McIntosh plant (54%). The next generation of ACAES plants promise greater efficiencies (~70%). Lifetime and cycling times affect the overall investment cost, with low lifetime and low cycling times increasing the cost of maintenance and replacement.

A growing literature exists on the levelised cost of electricity (LCOE) and levelised cost of storage (LCOS) of CAES and other EES technologies in three main applications: energy arbitrage, transmission and distribution (T&D) support and frequency regulation. Typically, two main approaches exist when assessing the cost of EES technologies:[42] calculating total capital cost (TCC; €/£/$), or life-cycle cost (LCC; cost per kW). The majority of studies report the costs of CAES as TCC.[42]

TCC evaluates all costs covering the purchase, installation and delivery of an EES unit, generally expressed in terms of the cost per kW (power), cost per kWh (energy) and cost per kWh per cycle. TCC comprises two main areas: first, above-ground equipment, and second, the cost of developing the below-ground energy storage (solution mining of a cavern, preparation and conditioning of a porous storage, or conventional mining of a hard-rock cavern). Above-ground costs include costs of the power conversion system [PCS, associated with the interface connecting the storage to the utility (typically cost per kW)] and balance of power (BOP) costs (including costs for project engineering, grid connection interface and integration facilities, *e.g.* transformers, and construction management that includes the cost of land and accessibility).

Inexpensive storage media (air and water) mean that CAES and PHS have low energy costs, which for CAES, where components have long lifetimes and do not require frequent replacement, reduces total storage costs.[214] Capital costs for EES technologies show that in terms of the bulk, utility-scale storage unit, underground CAES (all geological options) represents the lowest costs in comparison with all systems, other than metal–air batteries;[14,40,214] calculated power and energy capital costs for CAES range typically between $400 and $1000 per kW, or $2–50 per kWh, respectively (see Table 1). Calculated capital costs per cycle kWh of PHS and CAES show they are among the lowest of all storage technologies, at 0.1–1.4 and 2–4 US¢ per kWh per cycle, respectively.[14] Elsewhere, 18 storage technologies were reviewed,[42] with the calculated average TCC of CAES being the lowest at €893 per kW or €92 per kWh (see Table 3). However, the costs of CAES plants are themselves strongly influenced by the geology of the storage,[157,215] with energy costs ranging from $1–5 per kWh for solution-mined salt caverns to $30 per kWh for excavated hard-rock caverns (see Table 2). Owing to their low construction costs and widespread availability, many regard aquifers as suitable sites for compressed air storage.[148,160] However, plant siting is critical to the success of a CAES plant construction effort: the McIntosh plant

Table 3 Main capital cost items and levelised costs breakdown for grid-scale EES systems competing with CAES. Compiled from ref. 42.

Bulk EES type	Cost of PCS (average/ range) (€ kW⁻¹)	BOP (average/ range) (€ kW⁻¹)	Cost of storage section (average/ range) (€ kWh⁻¹)	Fixed O&M costs (average/ range) (€ kW year⁻¹)	Variable O&M costs (average/ range) (€ MWh⁻¹)	TCC, per unit of power rating (€ kW⁻¹)	TCC, per unit of storage capacity (€ kWh⁻¹)	Annualised LCC (€ kW year⁻¹)	LCOE (delivered electricity) (€ MWh⁻¹)	LCOS (1–8 h periods) (€ MWh⁻¹)
NiCd	239/206–329[a]		780/564–1120	11/4–24	N/A	1093	699	842	421	~500–350
Fe-Cr	362/326–523[a]		145/64–156	3.3/2.7–6.9	0.4/0.1–1.0	1132	569	417	209	~675–140
VRFB	490/472–527[a]		467/433–640	8.5/3.4–17.3	0.9/0.2–2.8	2512	307	706	353	~690–290
NaS	366/241–865[a]		298/180–563	3.6/2.0–17.3	1.8/0.3–5.6	2254	343	487	244	~410–190
Lead-acid	378/95–594	87/43–130	618/184–847	3.4/3.2–13.0	0.37/0.15–0.52	1923	437	646	323	~575–270
Flywheel	287/263–470[a]	N/A	2815/865–47764	5.2/4.3–6.0	2.0/0.2–3.8	867	4791	210	N/A	N/A
CAES (above ground)	846/804–887	N/A	109/86–131	2.2/2.2–3.7	2.2/1.9–3.0	1315	263	319	159	~355–100
CAES (under ground)	843/543–1014	N/A	40/4–64[b]	3.9/2.0–4.2[c]	3.1/2.2–2.5[d]	893	92	269	134	~315–71
PHS	513/373–941	15/3–28	68/8–126	4.6/2.0–9.2	0.22/0.19–0.84	1406	137	239	120	~250–54

[a]Including BOP costs.
[b]Mainly for storages with 8 h discharge time.
[c]Major fixed O&M expected every 5 years, totalling €67 per kW of installed capacity.
[d]Note: natural gas prices are not equal in the different studies, but on average are in the range €8–20 per MWh with emission costs €18–22 per ton CO_2.

spent less than $10 million (in 2010 dollars) to develop a 566 000 m³ salt cavern.[216] Interest rate (%), energy cost ($ per kWh) and power cost ($ per kW) are the top three factors affecting the total storage costs.[214]

3.7.2 Economics of CAES – Life-cycle Costs. LCC analyses, which may be annualised (ALCC; cost per kW year), consider investment and operational costs over the entire life cycle of electricity storage [fixed operation and maintenance (O&M), variable O&M, replacement, disposal and recycling, in addition to TCC]. It depends directly on the characteristics of the service (*e.g.* number of cycles per year), the power market (interest rates, price of power) and technological features. The method allows a broad overview of different technologies with different system designs and various operation modes, permitting evaluation and comparison of technologies with different cost structures. Commonly used parameters in economic evaluations of electricity storage, expressed as cost (€/£/$) per kWh of electricity output, are the LCOE and the levelised cost of electricity storage (LCOS or LCOES).

The following summarises briefly recent LCC analyses of different CAES configurations undertaken in various studies. LCOS values of 18 EES technologies were calculated by including all capital and operational expenses, discounted to today's values.[42] The main conclusions were that both PSH and CAES have the lowest LCOS of all technologies for energy arbitrage (up to 8 h) at €54 and €71 per MWh respectively (see Table 3). The data for bulk EES show the LCOE delivered by PHS as the most cost-efficient technology (€120 per MWh), with CAES second at €134 per MWh. In terms of the ALCC of EES systems applicable to bulk energy storage (energy arbitrage), PHS offers the minimum cost (~€240 per kW year), whereas for CAES, the cheapest technology in terms of capital costs, the costs of fuel and emissions decrease its profitability (~€270 per kW year).[42]

A US DoE survey calculated LCOS for PSH, CAES and a number of battery technologies for the US market with different plant designs and cycles.[217] The results were $150–220 per MWh for PHS, $120–210 per MWh for CAES and between $60 and 6000 per MWh for battery technologies.

The LCOS for PSH, CAES and five battery technologies in 11 operational modes were compared with those for selected fossil alternatives.[218] The results illustrated that CAES, with an LCOS of $192 per MWh, and PSH, with an LCOS of $188–274 per MWh, are near cost-competitive with gas turbines ($165–218 per MWh). All studies show that large increases in LCOS occur with decreasing amounts of yearly discharged energy (longer storage times), owing to the same CAPEX and operational expenditure (OPEX) costs relative to a smaller amount of discharged energy.[219]

CAES and PHS technologies use conventional machinery, with the lifetime of the plant and its operational economics largely determined by the mechanical components. The long lifetimes and numbers of cycles possible with both PHS and CAES plants, the latter of which can operate for 40 years or more with many thousands of cycles, mean that PHS and CAES represent the lowest costs per cycle kWh among all the EES technologies.[14] CAES, with low replacement needs, compares well in terms of the LCC of those EES systems subject to significant replacement costs during the lifetime, which are more sensitive to interest rates.[42,214]

Long-term storage costs of PHS and both conventional CAES and ACAES/AACAES systems have been assessed,[219] with the author concluding that for a storage charged and discharged once per year (one cycle per year), PHS is the most cost-efficient technology [between 93 and 185 €ct per kWh (€ct = euro cent)]. The LCOS of CAES at one cycle per year is significantly higher at above €2 per kWh. For short-term storage operated almost daily, with ~150 GWh of energy discharged per year, the PSH systems show the lowest cost at 5–9 €ct per kWh. The LCOS of ACAES and CAES systems are similar at 7–11 and 10–12 €ct per kWh, respectively. However, the differing cost structures of these two technologies suggests that the ACAES system has lower cost at a higher number of cycles, whereas the CAES system is more competitive in the range of less full-load hours. A large share of the LCOS of CAES arises from the need for fuel, with the cost of the technology depending heavily on the cost and stability of natural gas prices. Whereas CAES has a high gas cost, the ACAES system has a high insurance cost. The estimated costs of the heat storage for ACAES are in the region of 25% of the total CAPEX.[219]

LCOS for a number of systems, including both PSH and ACAES, have been calculated.[220] PSH was found to have the lowest LCOS of 2.5 €ct per kWh, excluding the cost of charged electricity, but ACAES can operate at 5.3 €ct per kWh. An LCOE assessment of various utility-scale EES technologies including PHS and CAES was undertaken, using a simple-cycle combustion turbine (SCCT) as the baseline technology for comparison and against which EES must compete.[221] SCCT LCOE values were $86–125 per MWh, which compares with the mean LCOE for PHS and CAES of $46 and $55 per MWh, respectively. However, LCOE values for storage systems are very sensitive to costs associated with reservoir types and siting needs. These are increased by costly 'scrubbing' equipment required to clean the air of gases picked up during storage in depleted reservoirs, including potentially harmful H_2S.

Regarding the electricity source, for the UK, the levelised cost of generating electricity from wind farms is higher than that of fossil fuel, which in part is linked to the costs of ensuring standby generation during periods of little wind or sun.[222] An assessment of the Danish electricity system suggests that although CAES is technically capable of increasing VRES penetration, even in energy systems with very high shares of wind power and combined heat and power (CHP) production, the investment in CAES systems cannot be justified and must be combined with other functions.[223] A profitability analysis for PSH and CAES for the UK market suggests that neither of them is profitable without subsidy.[222]

3.7.3 Environmental Performance and Safety of CAES. The application of EES promises to reduce the curtailment of renewable electricity sources and the need for fossil power, providing clear global benefits. However, depending on the storage type, negative net effects on the environment may arise during the life cycle of electricity storage. The use of energy storage with electricity generation increases the input energy required to produce electricity, and also the total GHGs.[16,207] The issues are numerous and this review can only provide an outline of the chief environmental issues when considering the development of the technology. Many of the environmental concerns are also engineering, operational and economic.[148] General considerations are discussed, followed by more specific issues associated with specific CAES storage types.

3.7.3.1 *General*

CAES, based on conventional gas-turbine technology, involves the combustion of fossil fuels, a consequence of which is that conventional CAES has considerably higher emissions during operation than the other storage-only technologies.[16] Total GHG emissions, excluding primary electricity generation, have been calculated at ~288 tonnes CO_2 equivalent (CO_2e) per GWh (see Table 4), most of the direct emissions arising during operation from the combustion and transportation of natural gas to the CAES site.

Using life-cycle assessment (LCA) methodology, often involving LCC analyses, a number of studies have assessed the environmental performance of electricity storage technologies and, in particular, grid applications.[16,211] The

Table 4 Life-cycle energy inputs and GHG emissions related to plant construction and operation. Compiled from ref. 16. [O&M = Operating & Maintenance, SCR = Selective catalytic reduction, EE_{op} = operational energy (excluding stored electricity) requirement per unit of energy delivered by storage system (GJ_t GWh^{-1}), Ef_{op} = operational emissions factor per unit of energy delivered by storage system (tonnes CO^2e GWh^{-1}), ER_{net} = net energy ratio (kWh_{in}/kWh_{out}), GJ_t = Gigajoule, thermal energy]

	Component	Life-cycle energy (GJ_t MWh^{-1} storage capacity)	GHG emissions (tonnes CO_2e MWh^{-1} storage capacity)
Fixed components – construction	Cavern development	16.2	1.2
	Site and buildings	36.7	3
	Plant electrical	65.9	4.7
	Subtotal – plant	*102.6*	*7.8*
	Electrical T&D	14.2	1
	Gas infrastructure	130.5	9.2
	Decommissioning	2.3	0.2
	Total	*265.7*	*19.4*

		GJ_t GWh^{-1}	Tonnes CO_2e GWh^{-1}
Variable components – operational	Fuel	4649	234
	Fuel delivery	518	51
	O&M&SCR	42	3
	Total variable – EE_{op}, EF_{op}	5210	288
	ER_{net}	0.735 times primary energy	0.735 times source emissions

'cradle-to-grave' approach takes into account all energy and emissions related to an energy plant, including the raw material extraction, construction, processing, manufacturing, distribution, use phase, repairs and maintenance, and decommissioning (disposal/end of life and/or recycling of materials). The life-cycle GHG emissions from electricity delivered by an energy storage facility originate from three major sources: generation of electricity to be stored, storage plant operations and construction of the energy storage facility.[16] The authors reported GHG emissions in terms of tonnes of CO_2e emissions and included non-CO_2 GHGs, such as methane, sulfur and nitrogen oxides.

However, emissions from primary electricity generation vary considerably, being dominated by the primary electricity generation emissions, particularly when the generation energy source is fossil. If net emissions of EES technologies are considered, then CAES can perform more favourably. Nuclear and most renewable technologies have low life-cycle CO_2 emissions (typically in the range of 10–100 tonnes CO_2e per GWh), whereas fossil generation produces between 475 and 1300 tonnes CO_2e per GWh[224] – considerably higher than the 288 tonnes CO_2e per GWh for CAES.

CAES, requiring natural gas combustion, exhibits higher GHG emissions than PHS or BES when coupled with low-GHG electricity generation. However, as emissions from primary electricity generation increase, CAES becomes more competitive; when coupled with fossil fuel generation (gas, oil or coal), CAES is associated with significantly lower life-cycle GHG emissions than PHS or BES. The effect of this 'fuel switching' in the CAES system therefore makes it the preferred technology to store electricity generated from coal.[16] The performance of the storage systems is reportedly tied to the electricity 'feedstocks' used, with the use of renewable energies decreasing the combustion of fossil fuels and hence reducing CAESs impact on the environment.[211]

Conventional CAES is now a mature technology and its limiting factor is its dependence on natural gas. Major improvements in either reduced energy input or the efficiency of these systems in the near future are unlikely. The use of renewable energy sources such as carbon-neutral biomass fuels during generation would lead to significant GHG emission reductions from the CAES systems.[209]

ACAES or AACAES, with predicted increased efficiencies of 70–75%, promises to help reduce direct emissions further, possibly to near zero. Considering both LCOES and life-cycle GHG emissions, AACAES has relatively low life-cycle GHG emissions and, together with PHS, shows the best performance for large systems.[207] However, its development has not occurred, although a small ACAES test facility is currently operational in Switzerland (see Section 3.3.3).

Development of isothermal CAES systems would lead to major reductions in direct emissions. The GCAES test facility at Gaines, Texas, linked to a wind turbine, operated for about 3 years, supplying electricity back to the electricity grid using heat stored from compression in a pond at the surface. There was no link to fossil fuels.[166,167]

Using LCA, the environmental impacts associated with both CAES and ACAES systems as a means of balancing the electricity output of an offshore wind farm have been addressed.[212] For the CAES system, wind power production and natural gas combustion are the main contributors to the assessed life-cycle environmental impacts. CAES increased acidification, particulate matter, photochemical oxidant formation and climate impacts significantly, but there were decreases in eutrophication, toxicity and mineral resource-depletion impacts. Significant contributors for the ACAES system were wind power production and thermal energy storage, with moderate increases in environmental impacts found across all assessed impact categories, which are sensitive to the required capacity of the thermal energy storage.

More general and indirect environmental benefits of EES and CAES include the following:

- Storage may lead to the integration of more renewable energy into the electricity grid.
- Reduced use of less efficient fossil-fuelled power stations with relatively high air emissions per kWh, reducing air emissions from central generation.
- Time shifting of electric energy from intermittent renewable generation makes renewables more viable alternatives to fossil-fuelled generation. Depending on how it is used and where it is located, electricity storage could reduce the need for additional generation, transmission or distribution facilities, and the land on which they are built. Furthermore, if storage is located at or near loads, then reductions in energy losses associated with T&D could be achieved.
- Reduced need for generation to provide spinning reserve and load following.
- Assist generation facilities to operate at optimal levels, reducing the use of less efficient generating units that would otherwise run only at peak time.
- Added capacity provided by electricity storage can delay or alleviate the need to build additional power plants or transmission lines.

3.7.3.2 Specific Issues Relating to Geological Storage

Early studies of CAES in the USA during the mid- to late 1970s and early 1980s identified the areas of main environmental concern, which remain and are perhaps more relevant today in terms of environmental and climate change issues. At the time, many areas were poorly understood, but have subsequently been considered and reported on during the development of the operational CAES plants and by the underground gas and fuel-storage industry. Investigations into CO_2 disposal in porous strata (saline aquifers) continue to inform some areas. A problem determining and quantifying problems is that they are also site specific.

This section can provide only a short review of very differing storage environments (see Section 3.3) and results of differing construction methods and generated waste products. The following areas are among the main factors when considering CAES deployment:

- air quality and climate – pollutant emissions and meteorological impacts;
- water quality – consumption, discharge and hydrological impacts;
- geological structure and seismicity;
- land requirements – both at site and for disposal of wastes;
- fuel requirements – including required pipelines for gas or water/brine;
- noise and safety;
- aesthetic – including the impact of multiple surface facilities and collection/supply lines.

Of interest in the Iowa investigations was the negative public perception of a CAES plant. As a result, the project name was changed from 'Iowa Stored Energy *Plant*' to 'Iowa Stored Energy *Park*', and the project website appearance and text were revised from an engineering, power-plant orientation to conform with the various important environmental reliability and energy-security messages identified in the research as important to community members.[179]

3.7.3.2.1 Salt Caverns. Environmental issues for the development of salt-cavern storages exist and are different from those posed by porous rock storages. During solution mining, significant volumes of low-salinity water are required, abstracted from local rivers or from the sea. Abstractions from a river can adversely impact the river system and the associated flora/fauna. Early considerations of CAES using salt caverns considered that the operational mode would be uncompensated (isochoric), as excavation costs are low and brine-compensated operation has significant design and environmental issues.[148]

During cavern creation, if no use exists for the brine produced (*e.g.* as a chemical feedstock in the petrochemicals industry), then it becomes a waste product, the disposal of which requires careful consideration.

Apart from the Portland gas storage project, where storage of some brine during storage operations was considered in a deep saline aquifer,[225] there have been no plans to inject brine from solution mining into porous rocks. During the construction of gas storage caverns at sites close to the shore, the brine has been disposed of offshore, as at Hornsea and Aldbrough in East Yorkshire,[226] as would be the case at Preesall and Portland. The EIA requires modelling of the brine plume and dispersion, plus the impacts on the local seabed and communities. At King Street in Cheshire, no local water supply or brine use was available and construction of a 61 km twin pipeline from the Cheshire Basin to an outflow point in the Mersey Estuary was proposed. This was both to bring in seawater and to take away brine from the solution mining of the caverns and required additional environmental analyses regarding the pipeline route.[227] Such operations raise questions about the sustainable use of a mineral resource. But in all cases, the examining authorities recognised that no markets existed in the UK or indeed internationally, effectively deeming the brine a waste product (see, *e.g.*, ref. 228).

Cavern storage operations using the brine-compensated (isobaric) mode would partially reduce the sustainability issue, as brine would be recycled to some extent. However, it may pose environmental risk as a large surface shuttle pond/reservoir is required for the brine, which must be secure from leakage and/or failure. The deep ($\sim$2.2 km) Portland gas storage project was to operate caverns in the brine-compensated mode, for which the surface storage of brine would not be practicable. Instead, brine produced during the gas injection mode was to be pumped *via* a pipeline and injected into a deep underground aquifer about 18 km to the north.[225] The environmental issues of both pipeline and aquifer storage had to be considered by the Council Planning Officers during the application.

3.7.3.2.2 Porous Rock Storage (Aquifers and Depleted Hydrocarbon Fields).
Porous rock storage presents a complex series of problems and environmental concerns. The first is to ensure that there is no contamination of an aquifer system. The injection of air above ambient temperature may affect *in situ* subsurface conditions, mechanically, geochemically and biologically. Local aquifer flows might be perturbed and, in the presence of fluids, geochemical changes could occur, leading to hydrolytic and oxidation reactions and mineral dissolution and transport, all of which could lead to permeability changes, weakening and increased corrosion. In addition, oxidation could lead to depletion of oxygen in withdrawn air, which could result in impaired combustion efficiency in the gas turbines.[229] Similarly, gypsum scale could be precipitated during oxidation, occluding porosity and impairing CAES performance.

The drilling of boreholes requires completions that prevent leakage of air or water between aquifers that may lead to pollution of water supplies. In addition, injection of air to form a bubble within the aquifer may affect local aquifer pressures, water production and adjacent groundwaters. Water levels in existing wells may fluctuate owing to long-term formation pressurisation that can occur several years after operations begin. This could lead to the need to limit the use of aquifers due to sensitivities of the operation to drawdown. Additionally, CAES systems themselves may be adversely affected by unscheduled or scheduled withdrawals of water from other wells. Storage pressures would have to be determined to prevent the increased pressures in the storage reservoir causing increased pore pressures in fault zones in contact with the reservoir that may cause re-activation and movement.

Injection into depleted hydrocarbon sites will carry health and environmental risks different from those with aquifer storages. Injection into depleted hydrocarbon sites will require a number of considerations. First, caprock integrity and possible damage will require assessment. During production, the reservoir pressure declines, which may lead to settling of the caprock and overburden, causing fracturing. Upon repressuring of the reservoir, the opening of fractures may result, permitting the escape of either native gas or air that might migrate to the top of the reservoir. This could lead to fugitive gases finding their way not only into overlying strata with potential aquifers, but also eventually to the surface. In both scenarios, potential future hazards exist.

Furthermore, oil and gas production does not lead to the removal of all the hydrocarbons and remnant oil and/or gas (methane, but possibly CO_2 and H_2S) is present in the reservoir, with which the injected air will co-mingle. Consequently, during withdrawal cycles, the air will contain some native gases, which will require separation and subsequent disposal. In the case of gas, there is a real danger of explosion, and H_2S poses major health risks.

Wellbore desaturation and reversible air–water interface movements provide the potential for geochemical reactions and the generation, transport and deposition of fine particulate matter. Particulates generated around the wellbore could be carried in the airflow to the CAES turbomachinery, where they might damage the turbine blades and other sensitive equipment. Airflow rates mean that the CAES air stream would be able to pick up particles of nearly any size generated within a few feet of the wellbore.[229,230]

Despite the potential problems associated with porous rock storage, reviews of the proposed Iowa Stored Energy Park project concluded that investigations into the aquifer storage site suggested that it represented a well-researched and planned 'environmentally safe project'.[179]

3.7.3.2.3 Mined Voids – Lined and Unlined. The main impacts of this form of storage relate to the production of waste rock and its use or disposal, be it at site (waste piles) or elsewhere. If local, then leaching of the waste piles can lead to contamination of surface water bodies. Dust emissions may be significant if transport is required. If water enters the storage during operations, as is the case in some fuel storages, where containment partly provided by hydrostatic pressure then produced waters, this may represent a risk in terms of pollution. This could be due to having dissolved material on its way to the storage, but also within the storage, with the potential for picking up gases from mineral breakdown, bacterial growth or contaminants from former coalmines.

In summary, conventional CAES may be of limited environmental and economic value in future energy systems, where electricity storage will increasingly be required, whereas ACAES/AACAES and ICAES may reach storage efficiencies of more than 70% for practical applications and, if reductions in costs are possible, provide a better alternative to CAES. ACAES is a promising technology for off-grid applications as it is capable of operating without the need for fossil fuels. With a demonstration plant now operating (from 2016) in the Pollegio–Loderio Tunnel ALACAES project,[165] this technology may prove to be a realistic option for support of future hybrid systems.

3.8 Safety Record of CAES and Some Potential Risks (Human and Environmental)

It should be stressed that gas storage, CAES in particular, has a good safety record. However, despite nearly 40 years of operational experience, with only two commercial CAES facilities, there has been little chance for serious problems to develop during CAES operations. Health and safety hazards associated with compression of air on this scale will be similar to those associated with natural gas or CCS, *i.e.* related to high pressures, the integrity of pipelines, storage structures and the associated electrical and mechanical equipment. However, stored compressed air in salt caverns does not have any hazardous toxic or asphyxiant properties, which is a significant advantage over some other stored energy concepts. This may not be true if porous rock storages are considered, particularly depleted gas fields, where the production of native gases with the withdrawn air might be possible.

At Huntorf, serious problems occurred during testing and commissioning of the plant, when human error led to overpressuring of the compressor unit, causing a 3.5 month shutdown and costly maintenance and repairs.[231] Also at Huntorf, after 20 years of problem-free operations, some sections of the fibreglass-reinforced plastic (FRP) production strings developed material problems, with the partial destruction of some that required replacement.[151] The potential exists for salt and other corrosive or abrasive materials from the storage cavern to enter the airshaft and be carried into the turbine, where damage to the blades and associated equipment in a CAES plant might occur.[232]

Elsewhere, a water-compensated, 11 000 m³ rock cavern air-storage system, which operated for over 30 years at Trollhättan in Sweden as part of a vehicle test facility, developed a serious problem with major consequences at the surface.[233] Around 1960, the system was overcharged and a large air bubble entered the water shaft, blowing the fluid up to the surface and destroying the building on top of the shaft.

Unlined rock caverns storing compressed air as part of a power-generation plant operate in Norway as air cushion surge chambers attached to hydropower plants. They have suffered leakages.[234] Problems of containment have also occurred in hydraulic compressed air storages in Finland. At the Pyhäsalmi Mine, extensive shotcreting (spraying of concrete at high speed as a construction and strengthening technique) and coating with an elastic bitumen was required to achieve air tightness around the void.[184] The dramatic blowout event at the Kanopolis salt mine in Kansas, USA,[235,236] was not related to storage, but highlights the problems of increased air pressures underground without proper identification and sealing of old shafts.

Depleted natural gas reservoirs present a possible safety issue resulting from residual hydrocarbons remaining in the depleted formation. Additionally, some salt beds and storage caverns produce gas, as is well documented in the Strategic Petroleum Reserve (SPR) in the USA.[237] Oxygen does not occur in the subsurface in the volumes required for major ignition and explosion. However, compressed air storage would provide sufficient oxygen and therefore in the presence of hydrocarbons in a depleted reservoir, the potential for ignition exists, with possible sources including heat of compression, piezoelectricity, static from the buildup of charge on dust particles caused by filling/emptying of the underground storage facility, lightning or friction and fracturing arising from earthquakes.[238] Mitigation and safety measures for storage in depleted reservoirs were discussed.

Acknowledgements

The authors are grateful to David A. Jones (Natural Resources Wales) and Bob Schulte (Sandia National Laboratories) for discussions and information on aspects of the environment and legislation. We also thank Drs A. Bloodworth, J. Mankelow, A. Hughes and Wei He and Professor Ron Hester, all of whose comments improved earlier versions of the text. D.J.E. and G.F. publish by permission of the Director, British Geological Survey. D.J.E. acknowledges funding from the EPSRC Grand Challenge for Grid Storage IMAGES project. G.C. thanks M. O'Brien for helpful discussions on PHS.

References

1. EC, 2013. 2020 climate and energy package. European Commission (EC) Climate Change and Targets policy documentation. https://ec. europa.eu/clima/policies/strategies/2020_en.
2. IRENA, 2017. Electricity Storage and Renewables: Costs and Markets to 2030. International Renewable Energy Agency, Abu Dhabi. International Renewable Energy Agency (IRENA) Report, October 2017, 132 pp.
3. BEIS, Energy Trends, December 2017, Department for Business, Energy & Industrial Strategy (BEIS), 117 pp. https://assets.publishing.service. gov.uk/government/uploads/system/uploads/attachment_data/file/669 750/Energy_Trends_December_2017.pdf.
4. Carbon Brief website article, https://www.carbonbrief.org/uk-low-carbon-generated-more-than-fossil-fuels-in-2017, (accessed April 2018).
5. UK Government energy statistics homepage, https://www.gov.uk/ government/collections/renewables-statistics, (accessed April 2018).
6. National Infrastructure Commission, Smart Power. Infrastructure Commission Report, 2016, 88 pp. https//assets.publishing.service.gov. uk/government/uploads/system/uploads/attachment_data/file/505218/ IC_Energy_Report_web.pdf.
7. United States Department of Energy Global Energy Storage Database, http://www.energystorageexchange.org/, (accessed Dec 2017).
8. The Benefits of Pumped Storage Hydro to the UK, Scottish Renewables, Glasgow, 2016.
9. C. Bullough, C. Gatzen, C. Jakiel, M. Koller, A. Nowi and S. Zunft, 2004. Advanced Adiabatic Compressed Air Energy Storage for the Integration

ofWind Energy. Proceedings of the European Wind Energy Conference (EWEC), London, UK, 22–25 November, 2004, 8 pp.

10. ERP, 2011. The future role for energy storage in the UK: Main Report. June 2011 Energy Research Partnership (ERP) Technology Report, 48 pp. http://erpuk.org/wp-content/uploads/2014/10/52990-ERP-Energy-Storage-Report-v3.pdf.

11. F. Díaz-González, A. Sumper, O. Gomis-Bellmunt and R. Villafáfila-Robles, A review of energy storage technologies for wind power applications, *Renewable Sustainable Energy Rev.*, 2012, **16**, 2154–2171.

12. G. Salgi and H. Lund, System behaviour of compressed-air energy-storage in Denmark with a high penetration of renewable energy sources, *Appl. Energy*, 2008, **85**, 182–189.

13. E. Fertig and J. Apt, Economics of compressed air energy storage to integrate wind power: A case study in ERCOT, *Energy Policy*, 2011, **39**, 2330–2342.

14. H. Chen, T. N. Cong, W. Yang, C. Li, Y. Tan and Y. Ding, Progress in electrical energy storage system: a critical review, *Prog. Nat. Sci.*, 2009, **19**, 291–312.

15. P. Denholm and G. L. Kulcinski, 2003. Net energy balance and greenhouse gas emissions from renewable energy storage systems. Energy Centre of Wisconsin Report, June 2003, ECW Report Number 223-1, 55 pp. http://fti.neep.wisc.edu/pdf/fdm1261.pdf.

16. P. Denholm and G. L. Kulcinski, Life cycle energy requirements and greenhouse gas emissions from large scale energy storage systems, *Energy Convers. Manage.*, 2004, **45**, 2153–2172.

17. P. Denholm and T. Holloway, Improved Accounting of Emissions from Utility Energy Storage System Operation, *Environ. Sci. Technol.*, 2005, **39**, 9016–9022.

18. C. Oldenburg and L. Pan, Porous Media Compressed-Air Energy Storage (PM-CAES): Theory and Simulation of the Coupled Wellbore–Reservoir System, *Transp. Porous Media*, 2012, **97**(2), LBNL-6529E, 201–221.

19. H. Ibrahim, A. Ilinca and J. Perron, Energy storage systems—characteristics and comparisons, *Renewable Sustainable Energy Rev.*, 2008, **12**(5), 1221–1250.

20. R. B. Schainker, N. Nakhamkin, P. Kulkarni and T. Key, 2008. New Utility Scale CAES Technology: Performance and Benefits (Including CO2 Benefits). Electric Power Research Institute (EPRI) paper, EPRI 3420 Hillview Avenue, Palo Alto, California 94304, 6 pp. http://www.espcinc.com/library/EPRI_Paper_on_CAES_Technology.pdf.

21. A. Pimm, S. D. Garvey and M. de Jong, Design and testing of Energy Bags for underwater compressed air energy storage, *Energy*, 2011, **66**, 496–508.

22. P. Y. Li, E. Loth, T. W. Simon, J. D. Van de Ven and S. E. Crane, 2011. Compressed Air Energy Storage for Offshore Wind Turbines. IFPE Staff Paper, 7 pp.

23. M. Saadat and P. Y. Li, 2012. Modeling and Control of a Novel Compressed Air Energy Storage System for Offshore Wind Turbine. 2012

American Control Conference Fairmont Queen Elizabeth, Montréal, Canada, June 27–June 29, 2012, 6 pp.

24. A. Assis, J. P. Paul, A. E. Joseph and P. G. Scholar, Energy storage system for floating wind turbines, *Int. J. Adv. Comput. Electron. Technol.*, 2015, **2**, 2394–3416.

25. S. D. Garvey, Integrating Energy Storage with Renewable Energy Generation, *Wind Eng.*, 2015, **39**, 129–140.

26. D. A. Katsaprakakis, D. G. Christakis, E. A. Zervos, D. Papantonis and S. Voutsinas, Pumped storage systems introduction in isolated power production systems, *Renewable Energy*, 2008, **33**, 467–490.

27. S. Rehman, M. Luai, L. M. Al-Hadhrami and M. M. Alam, Pumped hydro energy storage system: A technological review, *Renewable Sustainable Energy Rev.*, 2015, **44**, 586–598.

28. Petrescu & Petrescu, 2015. Hydropower And Pumped Storage. Altenergymag, online article, 17 pp. https://www.altenergymag.com/article/2015/11/hydropower-and-pumped-storage/22104.

29. B. Stefen, Prospects for pumped-hydro storage in Germany, *Energy Policy*, 2012, **45**, 420–429.

30. PSM, 1930. A ten-mile storage battery. Popular Science Monthly (PSM), July 1930, p. 60. https://books.google.co.uk/books?id=sigDAAAAMBAJ&pg=PA60&dq=1930+plane+%22Popular&hl=en&ei=zxiVTtztJ-Pr0gGvtu2kBw&sa=X&oi=book_result&ct=result&redir_esc=y#v=onepage&q=1930%20plane%20%22Popular&f=true.

31. M. Lempriere, 2017. Dinorwig: A unique power plant in the north of Wales. Power Technology online article, 28 March 2017. https://www.power-technology.com/features/featuredinorwig-a-unique-power-plant-in-the-north-of-wales-5773187/.

32. Snowdonia Pumped Hydro: Glyn Rhonwy scheme homepage, http://www.snowdoniapumpedhydro.com/index.php/en/glyn-rhonwy, (accessed April 2018).

33. La Región 2011. La Xunta se desvincula del proyecto de Iberdrola en el Sil. La Región online article, 13/09/2011. http://www.laregion.es/articulo/ourense/xunta-desvincula-proyecto-iberdrola-sil/20110914082529164794.html.

34. R. D. Allen, T. J. Doherty, L. D. Kannberg, 1984. Underground Pumped Hydroelectric Storage. Pacific Northwest Laboratory prepared for the U.S. Department of Energy, Report Number PNL-5142, 78 pp.

35. B. Cassell, 2014. FERC issues permit for 1,500-MW pumped storage project in Ohio: Developer gets three years to study project feasibility. GenerationHub online article, 10/16/2014, 2 pp. http://generationhub.com/2014/10/16/ferc-issues-permit-for-1500-mw-pumped-storage-proj.

36. ECE, 2016. Comments of Eagle Crest Energy on the CAISO Draft 2015-2016 Transmission Plan, March 3rd, 2016. Eagle Crest Energy (ECE) response document, March 2016, 4 pp. https://www.caiso.com/Documents/EagleCrestEnergyCommentsDraft20152016TransmissionPlan.pdf.

37. US DoI, 2013. Columbia Basin Project. US Department of the Interior (US DoI), Bureau of reclamation. Reclamation: Managing Water in the West.
38. JCLD, 2001. Seawater pumped-storage power plant Japan Commission on Large Dams. Japan Commission on Large Dams (JCLD) online article. https://web.archive.org/web/20020708041342/http://www.jcold.or.jp/Eng/Seawater/Seawater.htm.
39. Japanupdate 2016. Experimental power plant in Kunigami dismantled. Japan Update online article, http://www.japanupdate.com/2016/07/experimental-power-plant-in-kunigami-dismantled/.
40. X. Luo, J. Wang, M. Dooner and J. Clarke, Overview of current development in electrical energy storage technologies and the application potential in power system operation, *Appl. Energy*, 2015, **137**, 511–536.
41. Generation of electricity and district heating, energy storage and energy carrier generation and conversion: technology data for energy plants. Danish Energy Agency, Denmark, 2012.
42. B. Zakeri and S. Syri, Electrical energy storage systems: a comparative life cycle cost analysis, *Renewable Sustainable Energy Rev.*, 2015, **42**, 569–596.
43. Renewable energy technologies: cost analysis series, vol. 1 Hydropower, International Renewable Energy Agency, 2012.
44. E. Barbour, I. A. Grant-Wilson, J. Radcliffe, Y. Ding and Y. Li, *Renewable Sustainable Energy Rev.*, 2016, **61**, 421.
45. Engie: understanding the capacity market article. http://business.engie.co.uk/wp-content/uploads/2016/07/capacitymarketguide.pdf, (accessed April 2018).
46. Carbon Brief: Understanding the Capacity Market article, https://www.carbonbrief.org/guest-post-understanding-governments-capacity-market, (accessed April 2018).
47. OFGEM: Cap and Floor brochure, https://www.ofgem.gov.uk/system/files/docs/2016/05/cap_and_floor_brochure.pdf, (accessed April 2018).
48. First Hydro website: Dinorwig station, https://www.fhc.co.uk/en/power-stations/dinorwig-power-station/, (accessed April 2018).
49. First Hydro website: Ffestiniog station https://www.fhc.co.uk/en/power-stations/ffestiniog-power-station/, (accessed April 2018).
50. Cruachan station homepage, https://www.visitcruachan.co.uk/, (accessed April 2018).
51. SSE Foyers station homepage, http://sse.com/whatwedo/ourprojectsandassets/renewables/foyers/, (accessed April 2018).
52. M. Gimeno-Gutiérrez, R. Lacal-Arántegui, Assessment of the European potential for pumped hydropower energy storage - A GIS-based assessment of pumped hydropower storage potential, Joint Research Centre Institute for Energy and Transport, European Commission, 2013.
53. O. Edberg, C. Naish, *Energy Storage and Management Study*, AEA Report for Scottish Government, 2010.

54. Scottish and Southern Energy: Coire Glas pumped hydroelectric storage scheme, http://sse.com/whatwedo/ourprojectsandassets/renewables/CoireGlas/, (accessed April 2018).

55. Glenmuckloch Pumped Storage Hydro Ltd - EIA Non-Technical Summary, https://www.buccleuch.com/wp-content/uploads/2015/12/Glenmuckloch-Non-Technical-Summary_2015_low-resolution.pdf, (accessed April 2018).

56. Scottish Power: Cruachan pumped hydroelectric storage scheme, https://www.visitcruachan.co.uk/, (accessed April 2018).

57. Renewable Energy World news article, http://www.renewableenergyworld.com/articles/2010/10/sse-to-develop-sloy-pumped-storage-hydroplant.html, (accessed April 2018).

58. BBC News Article, http://www.bbc.co.uk/news/uk-scotland-highlands-islands-43712533, (accessed April 2018).

59. VOA News Article, https://www.voanews.com/a/tunnel-collapse-closes-ethiopias-new-hydropower-project-84054397/159768.html, (accessed April 2018).

60. Thanhnien News Article, http://www.thanhniennews.com/society/vietnam-tunnel-collapse-highlights-safety-management-of-hydropower-projects-36607.html, (accessed April 2018).

61. Energy Live News Article, https://www.energylivenews.com/2013/08/13/four-die-in-chinese-hydropower-tunnel-collapse/, (accessed April 2018).

62. Article: http://blogs.worldwatch.org/revolt/pump-up-that-seawater-a-remix-to-pumped-storage-hydro/, (accessed April 2018).

63. Hydropower Good Practices: Environmental Mitigation Measures and Benefits Case study 01-01: Biological Diversity - Okinawa Seawater Pumped Storage Power Plant, Japan, International Energy Agency, 2006.

64. News Article, https://www.energy-storage.news/news/energyaustralia-ponders-worlds-largest-seawater-pumped-hydro-energy-storage, (accessed April 2018).

65. News Article, https://www.hydroworld.com/articles/2015/01/ferc-receives-permit-application-for-seawater-powered-hawaii-pumped-storage.html, (accessed April 2016).

66. E. McLean and D. Kearney, *Energy Procedia*, 2014, **46**, 152.

67. European Commission Environmental Impact Assessment homepage, http://ec.europa.eu/environment/eia/eia-legalcontext.htm, (accessed April 2018).

68. Legislation.gov.uk website – Town & Town and Country Planning (Environmental Impact Assessment) Regulations 1999 (as amended), https://www.legislation.gov.uk/uksi/1999/293/contents/made, (accessed April 2018).

69. UK Government guidance on Environmental Impact Assessment (England), https://www.gov.uk/guidance/environmental-impact-assessment, (accessed April 2018).

70. Scottish Government guidance on Environmental Impact Assessment, http://www.gov.scot/Topics/Built-Environment/planning/Roles/

Scottish-Government/Enviromental-Assessment/EIA, (accessed April 2018).

71. Welsh Government guidance on Environmental Impact Assessment, http://gov.wales/topics/planning/developcontrol/environmental-impact-assessment/?lang=en, (accessed April 2018).

72. Northern Ireland Department of Agriculture, Environment and Rural Affairs, guidance on Environmental Impact Assessment, https://www.daera-ni.gov.uk/articles/environmental-impact-assessment-eia, (accessed April 2018).

73. Environmental Impact Assessment Guide to Delivering Quality Development, Institute for Environmental Management and Assessment, Lincoln, 2016.

74. The Town and Country Planning (Environmental Impact Assessment) Regulations 2017, Statutory Instruments 2017 No. 571 Town and Country Planning.

75. UK Government guidance: Water Management - Abstract or impound water, https://www.gov.uk/guidance/water-management-abstract-or-impound-water, (accessed April 2018).

76. Natural Resources Wales Information about applying for an abstraction or impoundment licence, http://naturalresourceswales.gov.uk/permits-and-permissions/water-abstraction-and-impoundment/?lang=en, (accessed April 2018).

77. Scottish Environment Protection Agency: Water regulations in Scotland, https://www.sepa.org.uk/regulations/water/, (accessed April 2018).

78. Northern Ireland Department of Agriculture, Environment and Rural Affairs, https://www.daera-ni.gov.uk/articles/abstraction-and-impoundment-licensing-requirements, (accessed April 2018).

79. European Commission Water Framework Directive homepage, http://ec.europa.eu/environment/water/water-framework/index_en.html, (accessed April 2018).

80. Legislation.gov.uk website - The Water Environment (Water Framework Directive) (England and Wales) Regulations 2003, http://www.legislation.gov.uk/uksi/2003/3242/contents/made, (accessed April 2018).

81. Legislation.gov.uk website - Water Environment and Water Services (Scotland) Act 2003, https://www.legislation.gov.uk/asp/2003/3/contents, (accessed April 2018).

82. Legislation.gov.uk website - The Water Environment (Water Framework Directive) Regulations (Northern Ireland) 2003 (as amended), http://www.legislation.gov.uk/nisr/2003/544/contents/made, (accessed April 2018).

83. Guidance document on Article 6(4) of the 'Habitats Directive' 92/43/EEC, European Commission, 2007.

84. European Commission Invasive Alien Species homepage, http://ec.europa.eu/environment/nature/invasivealien/index_en.htm, (accessed April 2018).

85. D. Bradley, D. Cadman and N. J. Milner, *Ecological indicators of the effects of abstraction and optimisation of flow releases from reservoirs WFD21d Final Report*, Scotland and Northern Ireland Forum for Environmental Research (Sniffer), 2012.

86. B. D. Smith, P. S. Maitland and S. M. Pennock, *Biol. Conserv.*, 1987, **39**, 291.

87. B. D. Smith, P. S. Maitland, M. R. Young and M. J. Carr, *Monogr. Biol.*, 1981, **44**, 155.

88. B. D. Smith, *The ecology of Cruachan pumped-storage reservoir in relation to Loch Awe, Scotland*, Institute for Terrestrial Ecology Report to NSHEB, Penicuik, Midlothian, 1980.

89. C. Mc Parland and O. Barratt, Hydromorphological Literature Reviews for Lakes - Science report: SC060043/SR1, Environment Agency, Bristol, 2009.

90. Common Standards Monitoring Guidance for freshwater lakes, Joint Nature Conservation Committee, Peterborough, 2015.

91. The Water Environment (Controlled Activities) (Scotland) Regulations 2011 (as amended) A Practical Guide Version 8.2, Scottish Environment Protection Agency, 2018.

92. R. Boak and D. Johnston, Hydrogeological impact appraisal for dewatering abstractions Science Report – SC040020/SR1, Environment Agency, Bristol, 2007.

93. UK Government guidance: Reservoirs – owner and operator requirements (England), https://www.gov.uk/guidance/reservoirs-owner-and-operator-requirements, (accessed April 2018).

94. Scottish Environment Protection Agency: Reservoir Safety in Scotland homepage, https://www.sepa.org.uk/regulations/water/reservoirs/, (accessed April 2018).

95. Natural Resources Wales: Reservoir safety homepage, http://www.naturalresources.wales/ReservoirSafety/?lang=en, (accessed April 2018).

96. Northern Ireland Department for Infrastructure: Regulating reservoir safety in Northern Ireland homepage, https://www.infrastructure-ni.gov.uk/articles/what-reservoirs-bill-northern-ireland, (accessed April 2018).

97. A. M. Folkard, *North West Geogr.*, 2008, **8**, 42.

98. H. Austin, D. Bradley, I. Stewart-Russon, and N. Milner, Literature review of the influence of large impoundments on downstream temperature, water quality and ecology, with reference to the Water Framework Directive, Scottish Environment Protection Agency, 2015.

99. G. E. Petts, *Progr. Phys. Geogr.*, 1986, **10**, 492.

100. F. Bunea, G. D. Ciocan, G. Oprina, G. Băran and C. A. Băbuțanu, *Environ. Eng. Manage. J.*, 2010, **9**, 1459.

101. F. Bunea, D. M. Bucur, G. E. Dumitran and G. D. Ciocan, *Ecol. Water Qual. – Water Treat. Reuse*, 2012, 391.

102. Joint Nature Conservation Committee: Protected sites homepage, http://jncc.defra.gov.uk/default.aspx?page=4, (accessed April 2018).

103. European Commission Habitats Directive homepage, http://ec.europa. eu/environment/nature/legislation/habitatsdirective/index_en.htm, (accessed April 2018).

104. Legislation.gov.uk website - Conservation (Natural Habitats) Regu-
    ·lations 1994 (as amended), http://www.legislation.gov.uk/uksi/1994/ 2716/contents/made, (accessed April 2018).

105. European Commission Habitats Directive homepage, http://ec.europa.eu/ environment/nature/legislation/birdsdirective/index_en.htm, (accessed April 2018).

106. Ramsar homepage, https://www.ramsar.org/, (accessed April 2018).

107. Legislation.gov.uk website - Wildlife and Countryside Act 1981, https:// www.legislation.gov.uk/ukpga/1981/69, (accessed April 2018).

108. Joint Nature Conservation Committee: Wildlife and Countryside Act 1981, http://jncc.defra.gov.uk/page-1377, (accessed April 2018).

109. Legislation.gov.uk website - The Countryside and Rights of Way Act 2000, https://www.legislation.gov.uk/ukpga/2000/37/contents, (accessed April 2018).

110. Legislation.gov.uk website - Nature Conservation (Scotland) Act 2004, https://www.legislation.gov.uk/asp/2004/6/contents, (accessed April 2018).

111. Legislation.gov.uk website - Wildlife and Natural Environment (Scotland) Act 2010, http://www.legislation.gov.uk/asp/2011/6/contents/ enacted, (accessed April 2018).

112. Legislation.gov.uk website - Nature Conservation and Amenity Lands (Northern Ireland) 1985, https://www.legislation.gov.uk/nisi/1985/170/ contents, (accessed April 2018).

113. Legislation.gov.uk website - Environment (Northern Ireland) Order 2002, https://www.legislation.gov.uk/nisi/2002/3153/contents, (accessed April 2018).

114. Joint Nature Conservation Committee: UK Biodiversity Action Plan, http://jncc.defra.gov.uk/ukbap, (accessed April 2018).

115. Legislation.gov.uk website - Natural Environment and Rural Com-
    munities (NERC) Act 2006, https://www.legislation.gov.uk/ukpga/2006/ 16/contents, (accessed April 2018).

116. Legislation.gov.uk website – Environment (Wales) Act 2016, http://www. legislation.gov.uk/anaw/2016/3/contents/enacted, (accessed April 2018).

117. Welsh Government - Environment (Wales) Act 2016, http://gov.wales/ topics/environmentcountryside/consmanagement/natural-resources-man agement/environment-act/?lang=en, (accessed April 2018).

118. Legislation.gov.uk website - Wildlife and Natural Environment Act (Northern Ireland) 2011, https://www.legislation.gov.uk/nia/2011/15/ contents, (accessed April 2018).

119. Joint Nature Conservation Committee: Country Biodiversity Strategies, http://jncc.defra.gov.uk/default.aspx?page=5701, (accessed April 2018).

120. Welsh Government Natural Resources Policy homepage, http://gov.wales/ topics/environmentcountryside/consmanagement/natural-resources-manag ement/natural-resources-policy/?lang=en, (accessed April 2018).

121. European Law and Publications webpage: ECJ Case C-127/02 (Waddenzee), https://eur-lex.europa.eu/legal-content/EN/TXT/?uri=CELEX%3A62002CJ0127, (accessed April 2018).

122. Guidance document on Article 6(4) of the Habitats Directive 92/43/EEC, European Commission, 2007.

123. Guidelines for Ecological Impact Assessment in the UK and Ireland: Terrestrial, Freshwater and Coastal, Chartered Institute of Ecology and Environmental Management, Winchester, 2nd edn, 2016.

124. Literature review and analysis of the effectiveness of mitigation measures to address environmental impacts of linear transport infrastructure on protected species and habitats, Report NECR132, Natural England, Peterborough, 2013.

125. I. F. G. McLean, *A Habitats Translocation Policy for Britain*, Joint Nature Conservation Committee, Peterborough, 2003.

126. Interim Advice Note 183/14 Environmental Management Plans, Highways England, 2014.

127. NetRegs: Environmental guidance for Scotland and Northern Ireland for the Construction sector, http://www.netregs.org.uk/business-sectors/construction/, (accessed April 2018).

128. Control of Water Pollution from Construction Sites - Guidance for consultants and contractors *C532*, CIRIA, London 2001.

129. Environmental Permitting Guidance - Core guidance for the Environmental Permitting (England and Wales) Regulations 2010, DEFRA, London, 2013.

130. *Guide to hydropower construction best practice Version 2*, Joint publication by Scottish Environment Protection Agency, Scottish Natural Heritage and Scottish Renewables, 2015.

131. P. Stone and J. Shanahan, *Sediment Matters – A Practical Guide to Sediment and Its Impact in UK Rivers*, Environment Agency, Bristol, 2011.

132. Engineering in the Water Environment Good Practice Guide - Temporary Construction Methods, Scottish Environment Protection Agency, 2009.

133. NetRegs: Environmental guidance for Scotland and Northern Ireland on oil storage, http://www.netregs.org.uk/environmental-topics/materials-fuels-and-equipment/oil-storage/, (accessed April 2018).

134. UK Government guidance on oil storage regulations for business, https://www.gov.uk/guidance/storing-oil-at-a-home-or-business, (accessed April 2018).

135. UK Government guidance on waste legislation and regulations, https://www.gov.uk/guidance/waste-legislation-and-regulations, (accessed April 2018).

136. Natural Resources Wales guidance on waste management, https://naturalresources.wales/guidance-and-advice/environmental-topics/waste-management/?lang=en, (accessed April 2018).

137. NetRegs: Environmental guidance for Scotland and Northern Ireland guidance on waste, http://www.netregs.org.uk/environmental-topics/waste/, (accessed April 2018).

138. *Good practice during wind farm construction Version 3*, Joint Publication by Scottish Renewables, Scottish Natural Heritage, Scottish Environment Protection Agency, Forestry Commission Scotland, Historic Environment Scotland, 2015.

139. GB Non-Native Species Secretariat homepage, http://www.nonnativespecies.org/home/index.cfm, (accessed April 2018).

140. *Biosecurity Guidance*, Forestry Commission, 2012.

141. Scottish Natural Heritage Peatland action, https://www.nature.scot/climate-change/taking-action/carbon-management/restoring-scotlands-peatlands/peatland-action-2018-2019, (accessed April 2018).

142. Developments on peatland: guidance on the assessment of peat volumes, reuse of excavated peat and the minimisation of waste, Joint publication by Scottish Renewables and Scottish Environment Protection Agency, 2012.

143. General Electric, 1976. Economic and Technical Feasibility Study of Compressed Air Storage. General Electric final report prepared for the US Energy Research and Development Administration Office of Conservation. Report number COO-2559-1, March 1976, 399 pp.

144. R. F. Legget, 75-year old underground compressed air plant still in use in Canada, *Underground Space*, 1979, **4**, 29–31.

145. N. Tollinsky, 2013. Prof aims to revive Ragged Chute technology. In: Technology, online article, August 13, 2013. http://www.sudburyminingsolutions.com/prof-aims-to-revive-ragged-chute-technology.html.

146. F. W. Gay, 1948. Means for storing fluids for power generation, 2,433,896, 1948.

147. D. J. Evans, A review of underground fuel storage problems and putting risk into perspective with other areas of the energy supply chain, in *Underground Gas Storage: Worldwide Experiences and Future Development in the UK and Europe*, ed. D. J. Evans and R. A. Chadwick, Geological Society of London Special Publication, 2009, vol. 313, pp. 173–216.

148. R. D. Allen, T. J. Doherty and L. D. Kannberg, 1985. Summary of selected compressed air energy storage sites. Report prepared by Pacific Northwest Laboratory for Battelle and the U.S. Department of Energy under contract DE-AC06-76RLO 1830. Report number PNL-5091, 112 pp.

149. D. W. Boyd, O. E. Buckley and C. E. Clark Jr, Assessment of Market Potential of Compressed Air Energy Storage Systems, *J. Energy*, 1983, 7, 549–556.

150. M. Budt, D. Wolf, R. Span and J. Yan, Compressed air energy storage – an option for medium to large scale electrical energy storage, *Energy Procedia*, 2016, **88**, 698–702.

151. F. Crotogino, K.-U. Mohmeyer and D. R. Scharf, 2001. Huntorf CAES: More than 20 years of successful operation. Spring 2001 Meeting Orlando, Florida, USA.

152. P. Radgen, 2008. 30 years compressed air energy storage plant Huntorf – experiences and outlook. In: 3rd International Renewable Energy Storage Conference, Berlin, p. 18.

153. PowerSouth, 2017. Compressed Air Energy Storage. McIntosh Power-plant, McIntosh, Alabama. PowerSouth Energy Cooperative Brochure, available online, 3 pp. http://www.powersouth.com/wp-content/uploads/2017/07/CAES-Brochure-FINAL.pdf.

154. S. Zavattoni, M. Barbato, L. Geissbühler, A. Haselbacher, G. Zanganeh and A. Steinfeld, 2014. CFD modeling and experimental validation of the TES system exploited in the Pollegio AA-CAES pilot plant. Swiss Competence Centers for Energy Research, Heat and Electricity Storage – 5 Symposium, 1 p. http://repository.supsi.ch/7878/1/SCCER-5th-symposium-PSI-A1.pdf.

155. J. Deign, 2017. ALACAES seeks CAES partners. Online article, March 22, 2017, 4 pp. http://energystoragereport.info/alacaes-seeks-partners-caes-quest/#sthash.sf9E7kfC.dpbs.

156. M. Budt, D. Wolf, R. Span and J. Yan, A review on compressed air energy storage: Basic principles, past milestones and recent developments, *Appl. Energy*, 2016, **170**, 250–268.

157. EPRI, 2003. EPRI-DOE Handbook ofEnergy Storage for Transmission and Distribution Applications: Electricity Energy Storage Technology Options. Electric Power Research Institute (EPRI), Palo Alto, CA, and the U.S. Department of Energy (DoE), Washington, DC. EPRI Report, 516 pp.

158. S. van der Linden, *Wind Power: Integrating Wind Turbine Generators (WTG's) with Energy Storage*, in *Wind Power*, ed. S. M. Muyeen, 2010, Ch. 18, pp. 415–436. http://www.intechopen.com/books/wind-power/wind-power-integrating-wind-turbine-generators-wtg-s-withenergy-storage.

159. R. D. Allen, T. J. Doherty and R. L. Thorns, 1982. Geotechnical Factors and Guidelines for Storage of Compressed Air in Solution Mined Salt Cavities. Report prepared by Pacific Northwest Laboratory for Battelle and the U.S. Department of Energy under contract DE-AC06-76RLO 1830. Report number PNL-4242, 105 pp.

160. R. D. Allen, T. J. Doherty, R. L. Erikson and L. E. Wiles, 1983. Factors Affecting Storage of Compressed Air in Porous Rock Reservoirs. Report prepared by Pacific Northwest Laboratory for Battelle and the U.S. Department of Energy under contract DE-AC06-76RLO 1830. Report number PNL-4707, 157 pp.

161. EPRI, 2011. Compressed Air Energy Storage Demonstration Newsletter. Electric Power Research Institute (EPRI), Palo Alto, CA, October 2011, 10 pp.

162. EPRI, 2013. CAES Projects: Development/Progress Status. Compressed Air Energy Storage Demonstration Newsletter. Electric Power Research Institute (EPRI), Palo Alto, CA, July 2013, 8 pp.

163. EPRI, 2014. Molten Salt Provides Thermal Energy Storage for ACAES. Compressed Air Energy Storage Demonstration Newsletter. Electric Power Research Institute (EPRI), Palo Alto, CA, September 2014, 8 pp.

164. J. Wang, L. Kunpeng, L. Ma, J. Wang, M. Dooner, S. Miao, J. Li and D. Wang, Overview of compressed air energy storage and technology development, *Energies*, 2017, **10**, 991–1013.
165. S. Zunft, V. Dreißigacker and M. Krüger, 2016. Electricity storage with adiabatic compressed air energy storage: Results of the BMWi-project ADELE-ING. International ETG Congress, 28–29 Nov. 2017, Bonn, Germany.
166. D. Marcus, 2011. Fuel-Free Geologic Compressed Air Energy Storage From Renewable Power. General Compression Task #1 Deliverable Report for New Technology Implementation Grant Program 582-11-13126-3225. October 24, 2011, 9 pp. https://www.tceq.texas.gov/assets/public/implementation/air/terp/ntig/prog_rpts/GC_Task1.pdf.
167. R. J. Lyman, 2014. Fuel-Free Geologic Compressed Air Energy Storage From Renewable Power. General Compression Task #7 Deliverable Report for New Technology Implementation Grant Program 582-11-13126-3225. March 25, 2014, Resubmitted April 30, 2014, 20 pp. http://tceq.com/assets/public/implementation/air/terp/ntig/prog_rpts/2014_GC_Task7.pdf.
168. T. Wilson and U. Turaga, 2016. Oil Majors Pursue Energy Storage. ADI Analytics online article, July 29, 2016. http://adi-analytics.com/2016/07/29/oil-majors-pursue-energy-storage-technology/.
169. B. Cárdenas, A. J. Pimm, B. M. Kantharaj, C. Simpson, J. A. Garvey and S. D. Garvey, Lowering the cost of large-scale energy storage: High temperature adiabatic compressed air energy storage, *Propul. Power Res.*, 2015, **6**, 126–133.
170. A. Gillhaus, F. Crotogino, D. Albes and L. van Sambeek, 2006. Compilation and evaluation of bedded salt deposit and bedded salt cavern characteristics important to successful cavern sealing and abandonment. SMRI Research Report No. 2006-2-SMRI, Clarks Summit (PA), USA, 2006, 118 pp.
171. F. Crotogino, G.-S. Schneider and D. J. Evans, Renewable energy storage in geological formations, *Proc. – Inst. Mech. Eng. Part A: J. Power Energy*, 2017, **232**, 100–114.
172. EPRI, 2010. Electricity Energy Storage Technology Options: A White Paper Primer on Applications, Costs, and Benefits. Electric Power Research Institute (EPRI), Technical Update, December 2010, 170 pp.
173. P. Quast and F. Crotogino, Initial Experience with the Compressed-Air Energy Storage (CAES) – Project of Nordwestdeutsche Kraftwerke AG (NWK) at Huntorf/ West Germany, *Erdöl-Erdgas-Zeitschrift*, 1979, **95**, 310–314.
174. S. Serata and B. Mehta, 1993. Design and Stability of Salt Caverns for Compressed Air Energy Storage (CAES). Elsevier Science Publishers B.V., Amsterdam Seventh Symposium on Salt, vol. I, pp. 395–402.
175. S. Aherne and J. Kelly, 2013. Project CAES Larne – An exploration program for compressed air energy storage in Northern Ireland.

Solution Mining Research Institute (SMRI) Fall 2013 Technical Conference, Avignon, France, 30 September–1 October 2013, 11 pp.

176. C. Haughey, 2015, Larne CAES: A Project Update. In Gaelectric Energy Storage: The Missing Link. Gaelectric, 2015, Company newsletter available online. http://www.gaelectric.ie/wpx/wp-content/uploads/2015/09/Gaelectric-Supplement-June-2015.pdf.

177. U. S. DoE, 2014. Final Environmental Assessment for the Pacific Gas and Electric Company (PG&E) Compressed Air Energy Storage (CAES) compression testing phase project, San Joaquin County, California. U.S. Department of Energy (US DoE), National Energy Technology Laboratory, Report DOE/EA-1752, 74 pp.

178. A. Ter-Gazarian, Compressed air energy storage. Energy Storage for Power Systems. Peter Peregrinus Ltd., on behalf of the Institution of Electrical Engineers, London, UK. Redwood Books, Trowbridge Wiltshire, 1994, ch. 7, 197 pp. https://books.google.co.uk/books?id=5VMotgCfJmACandpg=PA118andlpg=PA118anddq=compressed+air+sesta+italyandsource=blandots=XLXOU7exRXandsig=Q3_46bb17tTmKc61ObwnJ8sHKHQandhl=enandsa=Xandei=vPSrVIe5B6Le7AbAloGwDwandved=0CCoQ6AEwAQ#v=onepageandq=compressed%20air%20sesta%20italyandf=false.

179. R. H. Schulte, N. Critelli Jr., K. Holst and G. Huff, 2012. Lessons from Iowa: Development of a 270 Megawatt Compressed Air Energy Storage Project in Midwest Independent System Operator: A Study for the DOE Energy Storage Systems Program. Sandia Report, January 2012, Report number SAND2012-0388, 97 pp.

180. Hydrodynamics, 2011. Iowa Stored Energy Plant Agency Compressed-Air Energy Storage Project: Final Project Report Dallas Center Mt. Simon Structure CAES System Performance Analysis. Report prepared for: Iowa Storage Energy Plant Agency Des Moines, Iowa, July 22, 2011, 52 pp.

181. K. Piessens, and M. Dusar, 2003. CO_2 sequestration in abandoned coalmines. Proceedings of the International Coal bed Methane Symposium, May 5–9, Tuscaloosa, Alabama. Paper No. 346.

182. K. Schultz, 1998. Gas storage at the abandoned Leyden coal mine near Denver, Colorado. EPA report, contract 68-W5-0018, prepared by Raven Ridge Resources Inc., 12 p.

183. W. S. Keith, 2011. Surrebuttal testimony before the Public Service Commission of Missouri, Case NO. ER-2011-0004. April 2011, 23 pp. https://www.efis.psc.mo.gov/mpsc/commoncomponents/viewdocument.asp?DocId=935587378.

184. R. Matikainen, Structure and operation of underground hydraulic compressed air reservoirs in Finland, in Storage in Excavated Rock Caverns: Rockstore 77, ed. M. Bergman, Proceedings of the First International Symposium, Stockholm, 5–8 September 1977, Pergamon Press, 1977, vol. 3, pp. 705–710.

185. E. Grøv, Geological requirements and challenges for underground hydrocarbon storage, Underground Constructions for the Norwegian Oil and Gas Industry, Norwegian Tunnelling Society Publication No. 16, 2007, ch. 4.2, pp. 27–33.

186. K. Nakagawa, T. Shidahara and T. Ohyama, Field tests on air tightness of rock for compressed air storage, Proceedings 7th International Society for Rock Mechanics (ISRM) Congress, A. A. Balkema, 1991, vol. l, pp. 135–138.

187. T. Shidahara, T. Oyama and K. Nakagawa, The hydrogeology of granitic rocks in deep boreholes used for compressed air storage, *Eng. Geol.*, 1993, **35**, 125–135.

188. T. Shidahara, T. Oyama, K. Nakagawa, K. Kaneko and A. Nozaki, Geotechnical evaluation of a conglomerate for compressed air energy storage: the influence of the sedimentary cycle and filling minerals in the rock matrix, *Eng. Geol.*, 2000, **56**, 207–213.

189. J. Rutqvist, H. M. Kim, D. W. Ryu, J. H. Synn and W. K. Song, Modeling of coupled thermodynamic and geomechanical performance of underground compressed air energy storage in lined rock caverns, *Int. J. Rock Mech. Min. Sci.*, 2012, **52**, 71–81.

190. M. Hayashi, Rock mechanics of compressed air energy storage and super magnetic energy storage in Japan, in *Rock Mechanics in Japan*, Japanese Committee for ISRM, Tokyo, 1991, pp. 50–57.

191. T. Ishihata, Underground compressed air storage facility for CAES-G/T power plant utilizing an airtight lining, *Int. Soc. Rock Mech.*, 1997, **5**, 17–21.

192. H. M. Kim, J. Rutqvist, D. W. Ryu, B. H. Choi, C. Sunwoo and W. K. Song, Exploring the concept of compressed air energy storage (CAES) in lined rock caverns at shallow depth: a modeling study of air tightness and energy balance, *Appl. Energy*, 2012, **92**, 653–667.

193. P. Perazzelli and G. Anagnostou, Design issues for compressed air energy storage in sealed underground cavities, *J. Rock Mech. Geotech. Eng.*, 2016, **8**, 314–328.

194. O. Kruck, F. Crotogino, R. Prelicz and T. Rudolph, 2013. Overview on all Known Underground Storage Technologies for Hydrogen. In: Assessment of the potential, the actors and relevant business cases for large scale and seasonal storage of renewable electricity by hydrogen underground storage in Europe. HyUnder Report for Deliverable No. 31, 94 pp.

195. C. Braester and J. Bear, Some hydrodynamic aspects of compressed-air energy storage in aquifers, *J. Hydrol.*, 1984, **73**, 201–225.

196. E. Passaris and R. Noden, 2011. Geomechanical Parametric Studies for a Propane Storage Cavern in North Tees Salt Field in the UK. Solution Mining Research Institute (SMRI) Fall 2011 Technical Conference, 3–4 October 2011, York, United Kingdom, 10 pp.

197. J. Kepplinger, F. Crotogino, S. Donadei and M. Wohlers, 2011. Present Trends in Compressed Air Energy and Hydrogen Storage in Germany.

SMRI Fall 2011 Technical Conference 3–4 October 2011 York, United Kingdom, 13 pp.

198. T. J. Doherty, 1981. Air Storage Systems. In Proceedings of International Conference on Seasonal Thermal Energy Storage and Compressed Air Energy Storage, Vol. 2, CONF-811066 Technical Information Service, Springfield, Virginia, 455–468.

199. R. A. Edwards, A. J. J. Goode and J. G. Rees, 1987. Preliminary Geological Assessment of compressed air energy storage (CAES) sites. British Geological Survey commercial in confidence report, WZ/87/04.

200. S. T. Horseman, S. Holloway, P. J. Hooker and J. G. Rees, 1992. Geological Assessment of Potential Sites for Compressed Air Storage. British Geological Survey commercial in confidence report, WE/92/6C.

201. Gaelectric, 2015. Project-CAES Larne, NI Section 3: Planning and Development Context. Gaelectric Environmental Statement report, December 2015, 24 pp.

202. Hyder, 2011. Environmental Impact Assessment Scoping Report. Halite Energy Group. Preesall Underground Gas Storage Facility DCO document, 128 pp.

203. EA, 2016. Onshore Oil and Gas Sector Guidance. Environment Agency (EA), Version 1, 17 August 2016, 64 pp.

204. EA, 2017. The Environment Agency's approach to groundwater protection. Environment Agency (EA), March 2017, Version 1.0, 55 pp.

205. HSE, 2016. Natural Gas Salt Cavity Storage – Guidance to Inspectors on Borehole and Cavern Design, Cavern Leaching and Operation of the Borehole and Cavern. Health and Safety Executive (HSE) document: SPC/ENFORCEMENT/185, 14 pp. http://www.hse.gov.uk/foi/internalops/hid_circs/enforcement/spc185.pdf.

206. M. Beaudin, H. Zareipour, A. Schellenberglabe and W. Rosehart, Energy storage for mitigating the variability of renewable electricity sources: An updated review, *Energy Sustainable Dev.*, 2010, **14**, 302–314.

207. A. Abdon, X. Zhang, D. Parra, M. K. Patel, C. Bauer and J. Worlitschek, Techno-economic and environmental assessment of stationary electricity storage technologies for different time scales, *Energy*, 2017, **139**, 1173–1187.

208. F. R. Eldridge, *Wind Energy Conversion Systems Using Compressed Air Storage*, McLean, Virginia, 1976.

209. P. Denholm, Improving the technical, environmental and social performance of wind energy systems using biomass-based energy storage, *Renewable Energy*, 2006, **31**, 1355–1370.

210. P. Denholm and R. Sioshansi, The value of compressed air energy storage with wind in transmission-constrained electric power systems, *Energy Policy*, 2009, **37**, 3149–3158.

211. L. Oliveira, M. Messagie, J. Mertens, H. Laget, T. Coosemans and J. Van Mierlo, Environmental performance of electricity storage systems for grid applications, a life cycle approach, *Energy Convers. Manage.*, 2015, **101**, 326–335.

212. E. A. Bouman, M. M. Øberg and E. G. Hertwich, Environmental impacts of balancing offshore wind power with compressed air energy storage (CAES), *Energy*, 2016, **95**, 91–98.

213. D. O. Akinyele and R. K. Rayudu, Review of energy storage technologies for sustainable power networks, *Sustainable Energy Technol. Assess.*, 2014, **8**, 74–91.

214. S. Sundararagavan and E. Baker, Evaluating energy storage technologies for wind power integration, *Sol. Energy*, 2012, **86**, 2707–2717.

215. S. Succar, and R. Williams, 2008. Compressed air energy storage: theory, resources, and applications for wind power. Technical report Princeton Environmental Institute, 81 pp.

216. EPRI, 2012. EPRI Report Analyzes CAES Plant Reference Design and Costs. Compressed Air Energy Storage Demonstration Newsletter. Electric Power Research Institute (EPRI), Palo Alto, CA, January 2012, 9 pp.

217. A. A. Akhil, G. H. Aileen, B. Currier, B. C. Kaun, D. M. Rastler, S. B. Chen, A. L. Cotter, D. T. Bradshaw and W. D. Gauntlett, 2015. DOE/EPRI Electricity Storage Handbook in Collaboration with NRECA. Sandia National Laboratories, Sandia Report, SAND2015-XXX, 347 pp.

218. LAZARD, 2015. Lazard's Levelized Cost of Storage Analysis - Version 1.0; 2015, 30 pp.

219. V. Jülch, Comparison of electricity storage options using levelized cost of storage (LCOS) method, *Appl. Energy*, 2016, **183**, 1594–1606.

220. T. Weiss, J. Meyer, M. Plenz and D. Schulz, Dynamische Berechnung der Stromgestehungskosten von Energiespeichern für die Energiesystemmodellierung und -einsatzplanung, *Z. Energiewirtsch.*, 2016, **40**, 1–14.

221. M. Obi, S. M. Jensen, J. B. Ferris and R. B. Bass, Calculation of levelized costs of electricity for various electrical energy storage systems, *Renewable Sustainable Energy Rev.*, 2017, **67**, 908–920.

222. G. Locatelli, E. Palerma and M. Mancini, Assessing the economics of large Energy Storage Plants with an optimisation methodology, *Energy*, 2015, **83**, 15–28.

223. H. Lund and G. Salgi, The role of compressed air energy storage (caes) in future sustainable energy systems, *Energy Convers. Manage.*, 2009, **50**(5), 1172–1179.

224. L. Gagnon, C. Bélanger and Y. Uchiyama, Life-cycle assessment of electricity generation options: The status of research in year 2001, *Energy Policy*, 2002, **30**, 1267–1278.

225. D. Langham, 2007. The proposed Isle of Portland natural gas storage facility and Mappowder to Portland gas pipeline project: 2. Project Description, Text, 191 pp.

226. F. Dean, Salt cavity storage, *Gas Eng. Manage.*, 1978, 291–304.

227. E. Passaris, A. Dornan and M. O'Brien, 2011. King Street Energy gas storage scheme in the UK. Solution Mining Research Institute (SMRI) Fall 2011 Technical Conference, York, United Kingdom, 3–4 October 2011, 12 pp.

228. E. A. Simpson, 2007. Town & Country Planning Act 1990 Planning (hazardous substances) Act 1990 Lancashire County Council: Appeals by Canatxx Gas Storage Limited. Report by the Planning Inspectorate to the Secretary of State for Communities and Local Government. Report date30 March2007, 318 pp.

229. S. Succar and R. H. Williams, 2008. Compressed Air Energy Storage: Theory, Resources, and Applications for Wind Power. Report prepared by Princeton Environmental Institute, Princeton University, 8 April 2008, 81 pp.

230. EPRI, 2002. Compressed-Air Energy Storage Preliminary Design and Site Development Program in an Aquifer. Electric Power Research Institute (EPRI), EM-2351, November 1982.

231. H. C. Herbst, H. Hoffeins and Z. S. Stys, 1979. Huntorf 290 MW Air Storage System Energy Transfer (ASSET) Plant Design, Construction and Commissioning. In: Proceedings of the 1978 Compressed Air Energy Storage Symposium, Vol. 1, 1–17.

232. W. F. Adolfson, J. S. Mahan, E. M. Schmid and K. D. Weinstein, 1979. Geologic issues related to underground pumped hydroelectric and compressed air storage. 14th Intersociety Energy Conversion Engineering Conference, 452–454.

233. J. Pellin, 1980. Actual C.A.E.S. technology in Europe. In: Energy Storage: a vital element inMankind's quest for survival and progress. 1st International Assembly: Papers, 472–484.

234. D. C. Goodall, B. Aberg and T. L. Brekke, Fundamentals of Gas Containment in Unlined Rock Caverns, *Rock Mech.*, 1988, **21**, 235–258.

235. P. Bérest, B. Brouard and B. Feuga, 2003. Dry mine abandonment, Solution Mining Research Institute Technical Conference Paper, Fall 2003 Conference, 5–8 October, Chester, United Kingdom, 28 pp.

236. L. van Sambeek, 2009. Natural compressed air storage: a catastrophe at a Kansas salt mine. Proceedings 9th International Symposium on Salt, Beijing, Zuoliang Shaed, Vol. 1, 621–632.

237. B. Ehgartner, J. Neal and T. Hinkebein, 1998. Gas releases from salt. Sandia National Laboratories, Albuquerque, New Mexica, Sandia Report June 1998, report number SAND98-1354, 42 pp.

238. M. C. Grubelich, S. J. Bauer and P. W. Cooper, 2011. Potential Hazards of Compressed Air Energy Storage in Depleted Natural Gas Reservoirs. Sandia Report, number SAND2011-5930, 23 pp.

Electrochemical Energy Storage

D. NOEL BUCKLEY,* COLM O'DWYER, NATHAN QUILL AND
ROBERT P. LYNCH

ABSTRACT

Electrochemical energy storage systems have the potential to make a
major contribution to the implementation of sustainable energy. This
chapter describes the basic principles of electrochemical energy storage and discusses three important types of system: rechargeable batteries, fuel cells and flow batteries. A rechargeable battery consists of
one or more electrochemical cells in series. Electrical energy from an
external electrical source is stored in the battery during charging and
can then be used to supply energy to an external load during discharging. Two rechargeable battery systems are discussed in some
detail: the lead–acid system, which has been in use for over 150 years,
and the much more recent lithium system; sodium–sulfur and nickel–
metal hydride systems are also briefly discussed. A fuel cell is an
electrochemical cell in which the reactants supplying the energy are
not stored in the cell itself but rather are continuously supplied to the
electrodes from an external source. A common example is a hydrogen–
oxygen fuel cell: in that case, the hydrogen and oxygen can be generated by electrolysing water and so the combination of the fuel cell and
electrolyser is effectively a storage system for electrochemical energy.
Both high- and low-temperature fuel cells are described and several
examples are discussed in each case. A flow battery is similar to a
conventional rechargeable battery in that it can be repeatedly charged
and discharged. However, the energy storage material is dissolved in
the electrolyte as a liquid and so can be stored in external tanks.

---

*Corresponding author.

Issues in Environmental Science and Technology No. 46
Energy Storage Options and Their Environmental Impact
Edited by R.E. Hester and R.M. Harrison
© The Royal Society of Chemistry 2019
Published by the Royal Society of Chemistry, www.rsc.org

Various types of flow batteries are available or under development. Three of the more important examples are discussed in some detail: the all-vanadium flow battery, the zinc–bromine hybrid flow battery and the all-iron slurry flow battery. Some other examples are also briefly mentioned. The choice of electrochemical storage system is highly dependent on the specific requirements of the project that is being considered, the associated upfront capital and lifetime expenditure costs and end-of-life, environmental and safety considerations.

1   Introduction

Because many sustainable energy systems are intermittent, the extent to which they can be used on the electricity grid is limited unless suitable storage is available. Electrochemical systems are attractive options for such storage and so have the potential to make a major contribution to the implementation of sustainable energy. In this chapter, we describe the basic principles and discuss three important types of electrochemical energy storage system.

Electrochemical energy storage is achieved by the use of electrochemical cells.[1,2] Fundamentally, any electrochemical cell involves the passage of an electric current from an electronically conducting medium to an ionically conducting medium and back to the electronic medium. Hence it consists of two electronic–ionic interfaces (at the two electrodes) separated by an ionic medium (the electrolyte). Passage of a (direct) current from an electronic to an ionic medium necessarily involves a chemical reaction at the interface. Thus, at the cathode, the electrode at which electrons enter the cell, a chemical reduction reaction takes place; at the anode, the electrode at which electrons leave the cell, an oxidation reaction takes place.

1.1   Electrolytic and Voltaic Cells

We can think of an electrochemical cell as consisting of two half-cells. Each half-cell consists of an electrode–electrolyte interface and a so-called redox couple – corresponding to the oxidised and reduced species in the reaction that occurs at the electrode. We can divide electrochemical cells into two categories: electrolytic cells and voltaic cells. In an electrolytic cell, an externally applied potential drives a current through the cell and causes chemical reactions at the electrodes. In a voltaic cell, chemical reactions occur spontaneously in the cell and produce a potential difference between the electrodes that drives a current through the external circuit.

A simple example of an electrolytic cell is a water electrolysis cell consisting of two platinum electrodes immersed in a sulfuric acid solution (the electrolyte). When a suitable potential is applied between the electrodes, hydrogen ions are converted to hydrogen gas (reduced form) at the negative electrode:

$$2H^+ + 2e^- \rightarrow H_2 \tag{1}$$

while water is converted to oxygen gas at the positive electrode:

$$H_2O \rightarrow \tfrac{1}{2}O_2 + 2H^+ + 2e^- \tag{2}$$

Hence the overall reaction occurring in the cell is the sum of these two half-cell reactions:

$$H_2O \rightarrow \tfrac{1}{2}O_2 + H_2 \tag{3}$$

This reaction has, of course, a positive free energy of reaction $\Delta G°$ and this energy is supplied by the external source of potential (*i.e.* the power supply). Thus electrolytic cells convert electrical energy into chemical energy, in this case stored in hydrogen and oxygen.

The corresponding example of a voltaic cell also consists of two platinum electrodes in a sulfuric acid solution. However, in this case oxygen gas is supplied to one electrode and hydrogen gas to the other. In such a cell, a potential difference spontaneously develops between the electrodes: the oxygen electrode becomes positive with respect to the hydrogen electrode. This potential difference can drive current through the external circuit and so supply energy to a load. The source of this energy is the reaction of hydrogen and oxygen to form water [*i.e.* the reverse of reaction (3)]. Current flow across the electrode–electrolyte interfaces involves the reverse of the half-cell reactions (1) and (2). Hence a voltaic cell converts chemical energy in the cell (in this case stored in hydrogen and oxygen, which react when current flows) to electrochemical energy in the external circuit.

Comparing these examples of an electrolytic cell and a voltaic cell, we see that the difference between the two consists essentially of the relative concentrations of the reactants and products (*i.e.* of the oxidised and reduced forms of the redox couples at the electrodes). Initially in the electrolysis cell there is no hydrogen and oxygen present; these are formed on passing a current: electrons into the negative electrode and out of the positive electrode. In the voltaic cell, in contrast, hydrogen and oxygen are initially present and these react to cause a current to flow: electrons out of the negative electrode and into the positive electrode. Note that the directions of current flow are opposite in the two cases. Hence the reactants (hydrogen and oxygen) used during voltaic operation of the cell may be regenerated by reversing the current flow and operating the cell in electrolytic mode.

1.2 Batteries, Fuel Cells and Flow Batteries

A battery[3-5] consists of one or more voltaic cells in series. When a battery is discharged, the chemical energy stored in it is converted to electrical energy in the external circuit. In principle, the battery reactions can be reversed and the original chemical constituents regenerated by applying an external voltage and so reversing the current flow, *i.e.* by running it as an electrolytic cell. This process is known as charging. A battery that can be charged in this way is known as a secondary battery: electrical energy can be stored in the battery during charging and then used to supply energy to an external load during discharging. Such a battery may, in principle, be used indefinitely. On the other hand, a primary (non-rechargeable) battery may be used only once: it stores the chemical energy present in its constituents when it was fabricated and when this energy is exhausted on discharge it cannot be recharged. Although all batteries are, in principle, capable of being recharged, the chemistry and construction of a primary battery make recharging impractical. An example of a primary battery is the alkaline battery commonly used to power consumer electronic devices.

Since we are interested in reusable energy storage devices, primary batteries are not relevant. We will discuss in some detail two secondary (or rechargeable) battery systems: the lead–acid system, which has been in use for over 150 years, and the much more recent lithium system; we will also briefly discuss sodium–sulfur and nickel–metal hydride systems. In these conventional batteries, energy is stored principally on solid material in the electrodes. We will also discuss two other types of electrochemical energy systems: fuel cells and flow batteries.

A fuel cell is a voltaic cell in which the reactants supplying the energy are not stored in the cell itself but rather are continuously supplied to the electrodes from an external source. A common example is a hydrogen–oxygen fuel cell, the operating principles of which have already been described (hydrogen–oxygen voltaic cell) and details of which will be discussed later. A fuel cell[6] is essentially a primary cell in that the energy-storing materials (fuel and oxidant) are used only once. However, since the fuel and oxidant are supplied from an outside source, the cell can be used continuously. Furthermore, in the case of a hydrogen–oxygen fuel cell, the fuel and oxidant can be generated electrolytically from water and so the combination of the fuel cell and electrolyser is effectively a storage system for electrochemical energy.

A flow battery is similar to a conventional secondary battery in that it can be repeatedly charged and discharged. However, the energy storage material is dissolved in the electrolyte as a liquid and so can be stored in external tanks. In that regard, it resembles a fuel cell. Flow batteries may, in fact, be regarded as regenerative fuel cells. The fuel and oxidant are supplied to the cells from external tanks during discharge; during charge the fuel and oxidant are regenerated in the cell and pumped into the tanks for storage. Various types of flow batteries are available or under development. We will discuss in some detail three of the more important examples: the all-vanadium flow battery, the zinc–bromine hybrid flow battery and the all-iron slurry flow battery. We will also briefly mention some other examples.

2   Lead–Acid Batteries

First developed in 1859, the lead–acid battery (LAB) was the first ever rechargeable battery system. Compared with many other types of battery, LABs have a relatively low energy density, due primarily to their use of large amounts of lead. However, lead is also a relatively inexpensive and abundant metal. This low material cost, combined with a large power density, has made LABs attractive for a range of energy storage applications, including uninterruptible power supplies (UPSs), electric vehicles, electricity grid backup[7] and car batteries. The last application is where the majority of LABs end up, where the high surge currents that they can provide make them useful for starting, lighting and ignition (SLI).

Over 80% of global lead production is motivated by its use in LABs.[8] This fraction has been increasing over time owing to recent efforts to curb the use of lead wherever possible. However, the LAB share of the overall battery market has been falling in recent years as it has been supplanted by lithium-ion batteries for many applications. Compared with lithium-ion batteries, LABs have a lower energy density, a shorter cycle lifetime and a shallower depth of discharge. However, their low cost, safety, simplicity, recyclability, low rate of self-discharge and ability to deliver large surge currents still enable them to outperform lithium-ion batteries in many applications.[9]

2.1 Fundamental Aspects of Lead–Acid Batteries

A LAB typically consists of a lead anode and a lead dioxide cathode immersed in ~30% sulfuric acid.[10–12] At the anode, the sulfuric acid and lead will react to form $PbSO_4$ at the electrode surface. The oxidation of the metallic lead (Pb^0) electrode to Pb^{2+} in $PbSO_4$ results in the release of two electrons to the anode:
Anode reaction:

$$Pb(s) + H_2SO_4(aq.) \rightarrow PbSO_4(s) + 2H^+ + 2e^- \tag{4}$$

At the cathode, the reaction of the lead oxide and sulfuric acid will also form $PbSO_4$ at the electrode surface. In this case, the lead in the PbO_2 (Pb^{4+}) is reduced to Pb^{2+} in $PbSO_4$ and takes two electrons from the cathode:
Cathode reaction:

$$PbO_2(s) + H_2SO_4(aq.) + 2H^+ + 2e^- \rightarrow PbSO_4(s) + 2H_2O(l) \tag{5}$$

The overall cell reaction is therefore
Overall reaction:

$$Pb(s) + PbO_2(s) + 2H_2SO_4(aq.) \rightarrow 2PbSO_4(s) + 2H_2O(l) \tag{6}$$

This reaction produces ~2 V between the electrodes (depending on battery type and discharge current). The chemical reactions in the cell can be made to go in reverse by application of a larger (typically <3 V) opposing potential, and in this way the battery can be restored to its original condition (recharged).

Typically, LABs are not discharged below ~50% state-of-charge (SoC). This is because of both the irreversible formation of crystalline sulfate deposits on the electrodes (see Section 2.2.2) and the dilution of the electrolyte by consumption of H^+ ions and the production of water in the discharge re-action. This gradual dilution decreases the electrolyte conductivity. Below a certain conductivity, the ohmic losses in the electrolyte become so great that the battery cannot be operated in an efficient manner (*i.e.* the voltage losses become too great and the energy used to charge the battery can no longer be recovered efficiently).[11] Overdischarging can also cause unwanted lead pre-cipitation in some cell designs (particularly where the space between the anode and cathode is small), leading to an electrical short in the cell. In response to this failure mechanism, an electrode separator is usually in-cluded between the anode and cathode. The separator can be made of a variety of materials, including rubber, wood, glass fibre mat and a range of polymers. The separator must prevent physical contact between the anode and cathode while also facilitating electrolyte transport and ionic con-duction. It must also be robust enough to survive the acidic environment and last for the lifetime of the cell. While it is necessary to prevent a short-circuit, the separator generally increases the electrical resistance of the cell, lowering the voltage efficiency. In the case of batteries designed for deep discharge, the thickness of the electrodes, the space between the electrodes and the space at the base of the battery can all be increased so as to protect against short-circuits caused both by precipitates on the electrodes and the accumulation of precipitated material at the base of the cell.[10]

Overcharging a LAB results in electrolysis of the water in the electrolyte. Some sealed cell designs contain an overpressure vent for the release of H_2 and O_2 built up during the charging process. This electrolysis also results in the gradual depletion of water from the electrolyte and, in some cell designs, the water level may need to be topped up.[11] However, if the rate of formation of these gases is slow, *e.g.* during 'trickle charging', they can diffuse to the opposing electrode, *i.e.* hydrogen can diffuse to the positive electrode where it is oxidised and oxygen can diffuse to the negative electrodes where it is reduced. The overall effect is that they recombine to form water. The option to 'trickle charge' (slightly overcharge) the electrodes of the battery allows all electrodes of a battery to be fully charged and rebalanced at the end of a charge cycle, without the use of any complicated monitoring or control of the system; the system is simply charged at a constant potential of ∼2.3 V per cell. This characteristic greatly increases the reliability and simplicity of LAB systems. Indeed, some LABs are used in standby conditions where they are rarely cycled. These batteries can have a very long service life if they are continuously charged at a 'float' potential (of between 2.25 and 2.3 V per cell) that maintains the electrodes and keeps them at a high state of charge without the loss of significant amounts of water.

The SoC can be determined by measuring the open-circuit voltage of the cell. However, this method can be inaccurate owing to the relatively small changes in potential that occur throughout most of the charge/discharge process and the sensitivity of this voltage to temperature and composition, the latter of which changes as the battery is cycled. Since the discharging process involves a decrease in the sulfuric acid concentration, the linear variation of specific gravity with SoC can be used to determine the SoC of a LAB.[12] This is often routinely done where access to the electrolyte is easy.

2.2 Electrodes

2.2.1 Electrode Design. At the heart of both LAB electrodes is a lead alloy grid (common additives include Sb, Sn and Ca).[10–12] This grid acts as a conductive backbone on which the active materials are deposited. The grid gives structural stability to the electrode in addition to assuring an even current distribution throughout the electrode. Lead alloys are chosen because of their resistance to oxidation at high potentials and their stability within the electrolyte. However, the lead alloys are relatively dense and have a negative influence on the energy density of the battery, typically accounting for as much as 50% of the weight of the electrode.[11]

To form both anode and cathode, the grid is covered with a lead oxide paste (typically a mixture of a lead oxide powder, water and sulfuric acid). This paste is then dried in air to form a stiff, porous lead oxide electrode. The next step is to convert the lead oxide paste on each grid electrochemically into appropriate active materials for anode and cathode. This 'forming' process is essentially a slow charge of each electrode in a sulfuric acid bath or, in the case of flooded batteries (see Section 2.3), this initial charge may be carried out when installing the battery: after addition of sulfuric acid and before the first use of the battery. The current passed between the two electrodes slowly changes the paste to a spongy metallic lead on the negative electrode and PbO_2 on the positive electrode.[12]

The spongy lead of the anode is a highly porous form of metallic lead that has a high surface area. The PbO_2 that is formed on the cathode is also highly porous. Often small amounts of Sb or Ca are also included.[10,11] Pb/Sb electrodes are cheaper, have greater mechanical strength and are capable of deep discharge. However, they also increase the rate of gassing reactions and have a shorter lifetime owing to the gradual corrosion of the Sb from the anode and its subsequent deposition on the cathode. Pb/Ca electrodes are intermediate in cost between Pb and Pb/Sb electrodes. The addition of Ca adds strength to the electrode but also reduces gassing. However, Pb/Ca electrodes cannot be deeply discharged. Pb/Sb/Ca electrodes can combine the advantages of both additives, but at a higher cost.

2.2.2 Sulfation. During discharge, the lead sulfate that is initially formed on each electrode exists in an amorphous state. In this state, it will readily convert back to lead or lead oxide. However, if the battery remains in a discharged state, or is recharged incompletely, some of this sulfate can convert to a non-conductive, crystalline form.[13] This crystalline sulfate will no longer dissolve on charging, effectively decreasing the capacity of the battery by limiting the amount of active material. These deposits may also passivate a portion of the electrode surface, blocking reaction sites on the electrodes. Furthermore, the crystalline deposits can eventually expand by the Ostwald ripening process,[14] potentially causing catastrophic damage to the cell. Sulfation can be an issue when batteries remain unused for long periods of time or if the battery is never fully re-charged. Regularly cycled batteries are sometimes slightly overcharged at regular intervals to minimise the sulfation effect.

2.2.3 Effect of Carbon. Some modern LAB applications, such as storing energy from regenerative braking in electric vehicles, require high-rate partial-state-of-charge (HRPSoC) operation, *i.e.* the battery must be capable of rapidly charging and discharging for short bursts while rarely, if ever, being fully charged. Under these operating conditions, sulfation is a major problem owing to the gradual buildup of crystalline sulfate deposits and the lack of a full recovery charge. This problem is particularly bad on the anode (negative electrode), owing to its lower effective surface area. As a result, traditional LAB designs deteriorate rapidly under HRPSoC conditions.[15]

It has been demonstrated, however, that the addition of a small amount of certain forms of carbon to the anode can result in a significant decrease in sulfation effects, prolonging the life of the battery and maintaining greater capacity over time.[16] This carbon can be integrated in many ways, *e.g.* by incorporating carbon into the spongy lead, depositing lead on a carbon backbone or coating the lead electrode surface with carbon.[15] The carbon additive influences the electrode performance in a number of ways. First, it can act as a capacitive buffer, capable of accepting charge in excess of what the Faradaic processes can accommodate. This charge can then be gradually consumed by the reduction of more stable sulfates from the electrode sur-face. Second, during charging, lead can be deposited on the carbon surface, increasing the reaction area. Both of these effects are enhanced by forms of carbon with a relatively high surface area. Finally, the carbon can provide steric resistance to the crystallisation of $PbSO_4$ and maintain channels that allow improved wetting of the electrode.

The addition of carbon must be carefully considered, however, owing to an increase in the rate of H_2 evolution at these electrodes.[15] Carbon can lower the overpotential for hydrogen evolution, and any increase in surface area will increase the rate of this reaction. Therefore, the amount and type of carbon must be carefully chosen to optimise capacitance, active area and the kinetics of the hydrogen evolution reaction.

2.3 Cell Designs

2.3.1 Flooded. This is the simplest design and implies that both elec-trodes are simply immersed in a pool of the liquid electrolyte.[10–12] There will typically be a separator between anode and cathode to prevent short-circuits. These batteries must be installed and operated with a specific orientation owing to the liquid electrolyte, which can spill if the battery is not correctly oriented or if the battery casing is punctured. They provide the lowest cost per amp hour of any LAB design. These cells also require the most maintenance, including charge equalising, water balancing and terminal cleaning. However, they also allow easy access to the electrolyte, permitting enhanced monitoring and assessment of the state of the system. Furthermore, careful design of this type of cell can facilitate very high power outputs for short durations (*e.g.* SLI batteries in cars), multiple discharges to below 50% SoC (*e.g.* deep-discharge batteries for use in energy storage) or long-lifetime systems designed for the reliable backup of important systems (*e.g.* stand-by batteries used for backing up critical systems).

2.3.2 Absorbed Glass Mat. In absorbed glass mat (AGM) batteries, the liquid electrolyte and separator are replaced by a glass-fibre mat that is soaked in the electrolyte.[11,17] The primary reason for this is to increase the rate of gas transport between the electrodes, allowing the H_2 and O_2 to diffuse to, and react with, their opposing electrodes. This keeps the water within the cell, thereby reducing maintenance requirements. It also allows the cell to be completely sealed with the obvious advantage that the cell will not leak if punctured and can be used in any orientation. As discussed previously, calcium is alloyed with the electrodes to reduce the amount of gas produced. However, gas buildup can still be a problem under certain conditions. As a result, AGM LAB designs usually include a one-way pressure valve, designed to release gas to the atmosphere if a certain pressure is exceeded in the cell. These designs are sometimes known as valve-regulated lead–acid (VRLA) batteries or sealed LABs.

The AGM design also prevents the gradual stratification of the electrolyte that can occur in flooded designs, where the heavier sulfuric acid settles at the bottom of the cell. The water at the top of the cell can then freeze in low-temperature conditions. The concentration of the sulfuric acid at the bottom of the cell also results in most of the current flowing through the bottom of the electrolyte and the reactions centring on the bottom of the electrodes. This accelerates degradation of this region of the electrode, leading to premature failure of the battery. For these reasons, AGM LABs are ideally suited to low-temperature conditions and conditions of infrequent use. However, the overall lifetime of AGM cells is often shorter than those of typical flooded designs. This is because the acid within the mat is relatively concentrated in order to minimise water loss. The gradual loss of water as the cell ages will cause the acid to become even more concentrated over time. The increased acid concentration accelerates electrode corrosion, leading to shorter cell lifetimes.

2.3.3 Gelled. A 'gel' electrolyte can be obtained by mixing a silica gelling agent into the electrolyte.[11,17] This converts the electrolyte to a high-viscosity paste that confers many of the advantages of AGM cells such as the ability to seal the battery completely and use it in any orientation. However, the ionic mobility of the gel is considerably lower than that of the liquid electrolyte, so these batteries are unsuitable for applications that require large surge currents (*e.g.* a starting battery in a car). The main advantages of this design are the low maintenance requirements and the ability of the gel electrolyte to withstand a wider range of temperatures than a liquid electrolyte. These batteries are also sometimes referred to as sealed or VRLA batteries.

2.4   Cycle Depth

The design of a LAB can be tailored to allow for either rapid, high-power discharge or low-power discharge over a longer period of time. The batteries typically used to start a car engine are of the former type and, therefore, are not designed for deep discharge. The high power is achieved by using a large number of very thin electrodes. This increases the reaction surface area, allowing for rapid discharge. These thin electrodes are also more susceptible to damage owing to the mechanical stresses in the charging/discharging process. Limiting the cycle depth (and therefore minimising the range of stresses on the electrodes) of these batteries is essential to ensuring a reasonable electrode lifetime.[18]

Alternatively, batteries used for UPSs or battery electric vehicles (BEVs), known as deep-cycle batteries, are designed to be less susceptible to degradation due to mechanical stress. These batteries have relatively thick electrodes and, as a result, have a relatively small reaction surface area and lower maximum power output. However, the thicker electrodes are more robust, allowing for a much greater depth of discharge and a longer cycle life in general.

2.5   Environmental Aspects

The simplicity of a LAB contributes to its relatively low environmental footprint. Apart from the casing, the main components are lead electrodes and sulfuric acid. Although lead and many of its compounds are potentially harmful to life over long exposure times,[19] most of the lead in all LABs is recycled, the recycling rate being well over 90% in most developed countries.[20] For example, it is reported to be >99% in the USA and EU.[21,22] Recovery of lead from LABs is easier and requires less energy than extracting lead from ore, often making recycled lead an attractive starting material.[20,23] However, owing to the ubiquity of LABs, there are still tens of thousands of tonnes of lead that are not recycled and end up in landfills or worse. In addition, even though most of the lead in the batteries is recycled, the mining and manufacturing processes required to produce the batteries also produce a lot of waste lead and require a significant amount of energy. In the case of mining, in particular, concentrated lead may make its way into the local environment, leading to long-term contamination of the ecosystem. When all of these factors are considered, the use of LABs can have a considerably greater impact on the environment than, for example, a vanadium flow battery system.[24]

As with other rechargeable battery systems, LABs can be charged by renewable energy sources and used to provide backup power with no harmful emissions due to the operation of the battery. However, their current limited cycle life and low energy density mean that they are outcompeted in many such applications by other battery technologies that can offer significantly longer cycle life (flow batteries) or higher energy density (lithium-ion batteries). Recent advances in LABs have increased their cycle life greatly and, along with their simplicity of operation and recycling, this will result in LABs continuing to have the most significant share of the battery market for the foreseeable future.

3 Lithium and Lithium-ion Batteries

The first rechargeable lithium batteries were developed by Whittingham at Exxon Corporation[25] in the mid-1970s. These lithium batteries used lithium metal as the anode and TiS_2 as the cathode; the electrolyte was a solution of $LiClO_4$ in propylene carbonate. On discharge, Li^+ ions were formed at the anode and intercalated into the layered crystal structure of the TiS_2 cathode. The lithium-ion battery was invented by Basu and co-workers[26] at Bell Laboratories in the late 1970s, following on his earlier work[27] at the University of Pennsylvania. In these batteries, the lithium anode was replaced with graphite into which Li^+ ions intercalated on charging; the cathode was $NbSe_3$ into which Li^+ ions intercalated on discharging. In 1980, Goodenough and co-workers[28] and Godshall *et al.*,[29] working independently, demonstrated that $LiCoO_2$ could be used as a cathode material and this made practical batteries possible. The first commercial lithium-ion battery was released by Sony and Asahi Kasei in 1991.

3.1 Basic Theory, Structure and Operation

Lithium and lithium-ion batteries consist of a sandwich of two materials with an electrolyte central filling, in which electrical energy can be stored as chemical potential energy.[30] The two materials are the positive (cathode) and negative (anode) electrodes, and the electrolyte promotes the movement of Li^+ ions from one side to the other during charging or discharging, while keeping the cathode and anode from short-circuiting. During charge, a Li^+ ion inserts into the layered crystal structure of the negative electrode while a corresponding Li^+ ion is released from the positive electrode; this process is reversed during discharge. The uptake and release of the Li^+ cations are balanced by electrons flowing through the external circuit, as shown in Figure 1. During charge and discharge, Li^+ ions pass from one electrode to the other through the electrolyte, which is ionically conductive, while the corresponding electronic current (one electron per Li^+ ion) is passed through the outer circuit. Conduction of Li^+ ions through the electrolyte and

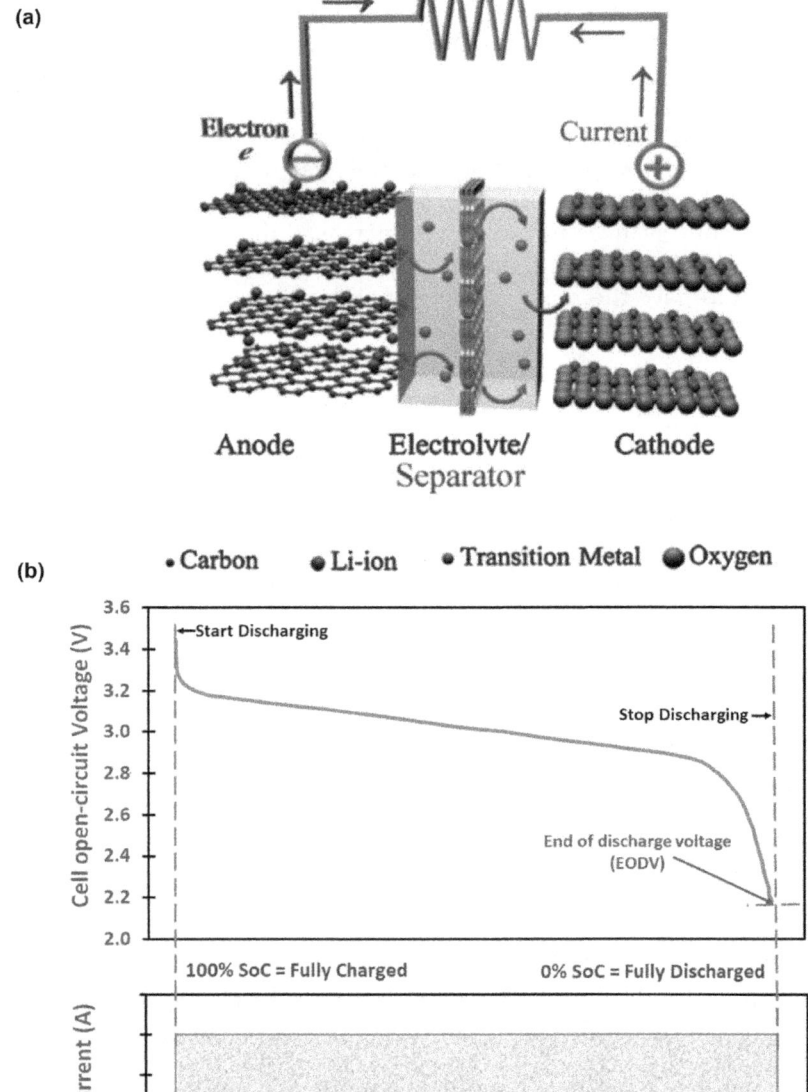

Figure 1   (a) Schematic of the electrochemical process in a lithium-ion battery. Reproduced from ref. 31 with permission from Elsevier, Copyright 2011. (b) Schematic illustration of a typical lithium-ion battery discharge curve showing the cell voltage as a function of capacity/time.

their insertion in or removal from the physical material of the cathode or anode are generally sluggish compared with the high electronic conductivity in the external circuit. To compensate for this conductivity difference as much as possible, large-area cathode and anode materials are used, each attached to a metallic current collector (backing electrode), and these are separated by a very thin region of electrolyte.

The cathode materials are most often a transition metal compound, typically an oxide (usually $LiMO_2$, where M = transition metal) for which the chemical potential of intercalated lithium is very low. In a lithium-ion battery, the anode is a material, typically graphitic carbon, in which intercalated lithium has a chemical potential close to that of elemental lithium, whereas in a lithium battery the anode is metallic lithium. The electrochemical reactions take place in a lithium-ion battery are summarised as follows:

Cathode reaction:

$$LiMO_2 \rightleftharpoons Li_{1-x}MO_2 + xLi^+ + xe^- \tag{7}$$

Anode reaction:

$$xLi^+ + xe^- + xC_6 \rightleftharpoons xLiC_6 \tag{8}$$

Overall reaction:

$$LiMO_2 + C_6 \rightleftharpoons Li_{1-x}MO_2 + Li_xC_6 \tag{9}$$

3.2 Materials

The $LiCoO_2$/carbon (LCO) and $Li(CoNiAl)O_2$/carbon (NCA) cells are the dominant lithium-ion battery types for consumer electronics. Lithiated NMC materials, $Li(NiMnCo)O_2$, and high-voltage alternatives have also been commercialised, and these materials and stoichiometric analogues are part of the predominant chemistries for lithium-ion technology.[5,32–35] A variety of material systems are summarised in Figure 2.

To improve existing rechargeable lithium-ion battery systems, efforts are under way to screen many new structures and compositions of new materials.[36–38] Lowering the cost, increasing safety for more demanding applications and ensuring environmental compatibility through the life cycle[39] of the batteries' construction, use and recycling will be paramount as the volume of lithium-ion battery use increases, particularly in use for large-scale storage.

3.3 Electrolytes

The electrolyte should be stable with respect to oxidation at the positive electrode and reduction at the negative electrode. Only non-aqueous electrolytes have such a stability window for lithium or lithium-ion batteries. Ideally, the electrolyte (solid or liquid) should have a high Li^+ ion conductivity over a practical temperature range (between −40 and 60 °C) and

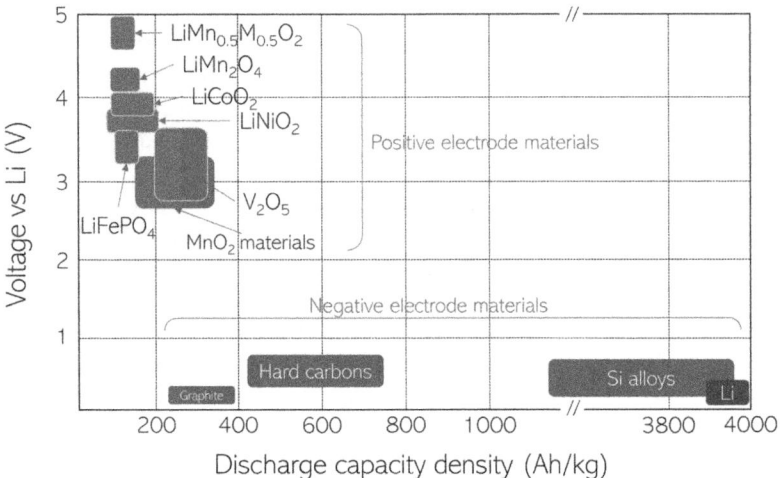

Figure 2 Electric potential *versus* charge-capacity density for some positive and negative electrode materials in lithium-ion batteries.

allow a thermodynamically stable open-circuit voltage $V_{oc} \geq 4$ V. Typical electrolytes consist of lithium salts such as $LiPF_6$ or $LiBF_4$ in an organic solvent such as ethylene carbonate, dimethyl carbonate or diethyl carbonate. Typical conductivities are in the region of 10 mS cm^{-1} at room temperature.

A solid electrolyte interphase (SEI) layer is typically formed on the surface of the anode by reductive decomposition of the salt-containing organic electrolyte at potentials (*versus* Li/Li$^+$) of ~0.7 V and below. It forms continuously for many carbonaceous electrode surfaces over a few cycles. This layer provides a surface that limits decomposition of the electrolyte during long-term cycling. Hence the SEI should be insoluble in the electrolyte and be electronically insulating, while maintaining a very low cationic impedance.[40,41] Another function of the SEI is to aid in the removal of the solvation shell around Li$^+$ ions to avoid solvent co-intercalation, a problem that is particularly evident for layered graphitic anode material in some solvents such as propylene carbonate. For the current lithium-ion battery technology using carbonaceous electrodes, the SEI is critical.[42] Alternative electrolytes, including ionic liquids and solid-state electrolytes, change the type of the SEIs and the nature of some SEIs is different for lithium-ion battery cells that use anodes that intercalate Li$^+$ at voltages >1 V.

A widely held assumption that solid-state ionically conducting ceramic or solid-state electrolytes could never compete with flammable organic electrolytes was based on cationic conductivity considerations. This is no longer the case, as glass-based electrolytes offer a cationic conductivity of ~10 mS cm^{-1} at room temperature and are stable against electrochemical reduction on contact with lithium metal, giving them a large voltage window for operation.[43]

3.4 Separators

Modern separators are typically made of microporous polyethylene (PE) that is moistened with the electrolyte and allows relatively unimpeded ion flow from anode to cathode *via* the electrolyte, while also forming a physical barrier between the anode and cathode to prevent short-circuiting. Separators are physically different from the electrolyte, if the electrolyte is liquid or a very low-viscosity polymer. PE-based separators are, in effect, a type of fuse, which can melt at high temperature and block electronic and ionic transfer, effectively turning off the cell in cases of thermal runaway or excessive heating. More recent modifications to this technology include tri-layer separators, using polypropylene outer coatings to a central PE porous membrane. As these polymers have two different melting temperatures, more effective control of ion flow cut-off is possible. Research into separator development is currently focused on a ceramic coating as one of several modifications to PE-based thin membranes, to offer better safety for higher voltage (>4 V) lithium-ion cells.

3.5 Sustainability of Lithium-ion Batteries

With an ever-growing need for lithium-ion batteries, greener materials and production methods and entire life-cycle assessment of lithium-ion battery production will become more important.[44] Sustainability issues exist for the materials that are used and it is essential that material abundance and ecologically benign methods for sustainable synthesis, extraction and refining are developed that are consistent with the future projected battery demand, and with protection of human health.[45] In short, a greener and more sustainable approach to lithium-ion storage systems from a materials perspective is needed.[46,47]

Elemental abundance for future lithium-ion battery demand is critically important. The increase in lithium-ion battery demand will likely require parallel developments in second use and recycling capabilities that are eco-friendly to maximise the use of energy storage raw materials. Cobalt is a critical raw material, and one-fifth of the world's cobalt resources are found in Democratic Republic of Congo, where mining and extraction methods are under scrutiny from a humanitarian perspective. Predicting alternative materials from computational investigations for lithium-ion batteries is gaining traction,[48] and this approach is not only becoming more powerful at developing understanding and predictive capability for materials, but may also provide options that use earth-abundant elements. Larcher and Tarascon[47] suggested the open phosphorus cycle and the elements constituting most biomass as a naturally accumulating source of elements for energy storage systems such as lithium-ion batteries.[47] Figure 3 highlights the mismatch between the main elements from biomass and those used in today's lithium-ion batteries. The sustainable elements (H, O, S, P, Na, K, Ca, Mg, Mn, Fe, Ti) are some of the predominant elements in biomass[49] and

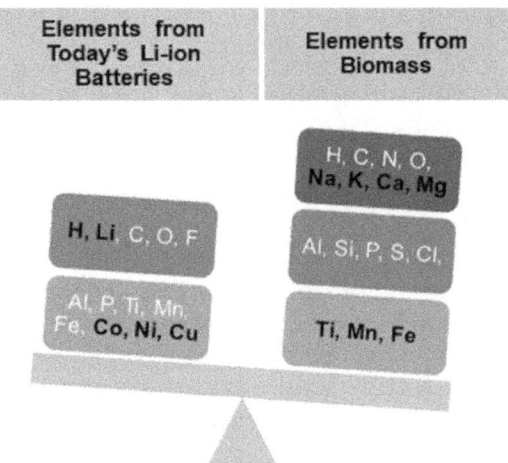

Figure 3   The mismatch between the main elements from biomass and those used in today's lithium-ion batteries. Elements highlighted in bold indicate possible alternatives for more sustainable battery materials to improve sustainability, *e.g.* replacing Ni or Co with Fe in oxides or phosphates.

recycle through natural processes, whereas our current elements base for lithium-ion batteries, including Li, Co, Ni, Cu and Fe, is limited in relative sustainability. With the right choice of cation, metal phosphates are an abundant form of cathode material ($LiFePO_4$ is an example of a sustainable cathode material in today's lithium-ion batteries), with Ti-based oxides[50] being an option for non-metallic or non-carbonaceous anodes.

Recycling current lithium-ion batteries may also be an important source of critical raw materials. For portable electronics, it is reported that cobalt and lithium constitute roughly 5–15 and 2–7 wt%, respectively, of lithium-ion batteries in their current form.[51] The recovery of thousands of tonnes of Co and Li is plausible, given the total number of cells produced annually, a number that is increasing considerably. As a representative example, 1 tonne of lithium can be refined from 250 tonnes of mineral or 750 tonnes of brine, but the same quantity can be recycled from just 30 tonnes of used batteries. Closing the loop of slow formation and fast usage for energy storage and also energy generation, as had been the case for fossil fuels and raw materials, will be crucial for the long-term viability of clean energy storage that is sustainable.

4   Other Battery Chemistries

4.1   Sodium–Sulfur Batteries

Sodium–sulfur batteries were first developed in the 1960s by Kummer and Weber at the Ford Motor Company.[52] A sodium–sulfur battery consists of a negative electrode of molten sodium metal and a positive electrode of molten sulfur separated by a solid electrolyte of sodium β-alumina.[53] It operates in the temperature range 300–350 °C, at which temperatures the solid electrolyte is a good Na^+ ion conductor. The battery has a high voltage (~2 V). The positive (sulfur) electrode uses a carbon (typically carbon felt) current collector. The system uses inexpensive and abundant materials. It has a relatively high energy density (150–750 Wh kg^{-1}) and a long cycle life (~4000 cycles). Large systems (up to 10 MW, 60 MWh) are commercially available and are suitable for large-scale storage. However, because of the reactivity of molten sodium and sulfur, safety is a concern.

4.2 Nickel–Metal Hydride Batteries

The nickel–metal hydride (NMH) battery was originally developed at the Battelle Geneva Research Centre (1967) and later by the Ovonic Battery Company (1990s).[54] It uses a nickel hydroxide/oxyhydride positive electrode and a metal hydride negative electrode with an alkaline electrolyte. The cell voltage is 1.2 V. The positive electrode is the same as that used in the older nickel–cadmium battery; the negative electrode may be considered to be a hydrogen electrode (similar to a fuel cell) where the hydrogen is stored in the form of a metal hydride. A variety of intermetallic compounds can be used for the metal hydride electrode. Nickel is often used with multiple rare-earth metals, but titanium, vanadium and zirconium are used in some combinations. NMH batteries have a high power density (discharge rate) and relatively long cycle life. They have principally found application in consumer electronics and hybrid electric vehicles. The materials of the battery are relatively environmentally friendly.

5 Fuel Cells

A fuel cell (FC) combines hydrogen and oxygen in an electrochemical reactor to generate electricity. Although FCs were first demonstrated in the nineteenth century, commercial interest in the development of FC technology did not begin in earnest until the middle of the twentieth century when NASA, in collaboration with General Electric, developed and used FC technology in Project Gemini. Since then, a wide range of different FC technologies have been developed, all based on the same fundamental principles. Here, the fundamentals are reviewed first, followed by a brief description of each of the main varieties of FC. Finally, the environmental and economic impacts of FC technology are discussed.

An FC can be thought of as a compromise between a battery and a combustion engine. In an FC, fuel (typically, pure hydrogen gas is the fuel, but methane, natural gas, methanol and other hydrocarbons are also sometimes used) is combined with oxygen, and this is accompanied by a release of energy. In a similar manner to an internal combustion engine, the fuel and oxygen are stored externally and are supplied to the reaction site on demand. However, in an engine, a direct reaction between the fuel and oxygen

(the fuel is oxidised by the oxygen) is induced by high temperatures and pressures. In an FC, the fuel is independently oxidised at the cell anode and the oxygen is independently reduced at the cell cathode.
Anode reaction:

$$H_2 \rightarrow 2H^+ + 2e^- \tag{10}$$

Cathode reaction:

$$\tfrac{1}{2}O_2 + 2e^- \rightarrow O^{2-} \tag{11}$$

Overall reaction:

$$H_2 + \tfrac{1}{2}O_2 \rightarrow H_2O \tag{12}$$

The direct conversion of chemical potential energy to electricity in an FC is not limited by the Second Law of Thermodynamics as it is in a steam turbine, where the prime mover is the expansive thrust of steam from a boiler heated by the burning of the fuel (or engine. where the prime mover is the expansive thrust of hot gases that are the product of the explosive burning of the fuel). Therefore, theoretically, FCs offer much greater efficiency in the transformation of chemical potential energy to electricity.

As in any electrochemical cell, the electrons liberated in the oxidation reactions at the anode flow through an external circuit to facilitate reduction reactions at the cathode. The anode and cathode typically are separated by a solid, ionically conducting phase. The purpose of this separator is to separate physically the fuel and oxygen, while also allowing ions (typically either H^+ or O^{2-}) generated from the oxidation or reduction reactions to pass through and combine to form H_2O. The reactions occur in a catalytic layer (CL), typically containing a noble metal catalyst, where a triple phase boundary (TPB) exists between the reactant gases, the electronically conducting electrode and the ionically conducting electrolyte. Although each FC will produce only a small voltage (typically <1 V), they can be connected in series (much like batteries) to obtain the desired output voltage. There are many different types of FC but they all operate on these basic principles. In the next section, FCs are described in more detail with specific reference to the simplest type, the proton exchange membrane fuel cell.

5.1   Low-temperature Fuel Cells

5.1.1   Proton Exchange Membrane Fuel Cells.   Similarly to all fuel cell designs, a proton exchange membrane fuel cell (PEMFC) consists of a proton exchange membrane (*i.e.* a solid that is a good conductor of H^+) sandwiched between the anode and cathode, as shown schematically in Figure 4. These are collectively referred to as the membrane electrode assembly (MEA). Typically, perfluorinated sulfonic acid membranes, such as Nafion, are used.[55,56] The anode and cathode typically are made of electronically conductive porous carbon-based materials.[56] The fuel and oxygen are made to flow through the anode and cathode, respectively. At the anode, H_2 is oxidised, with the resulting H^+ ions being conducted through the membrane and the electrons being conducted through the electrode to the external circuit. At the cathode, O_2 is reduced to OH^-, consuming electrons from the external circuit and combining with hydrogen ions generated at the anode, which have passed through the membrane, to form H_2O. Along with the electrical work done in the external circuit, the FC reactions also generate considerable heat. This can increase the operating temperature, reducing activation/polarisation losses.

The temperature must be kept below ~80 °C for the membrane to remain hydrated.[55,57] The necessity to keep the membrane hydrated requires a heat-management system and a gas humidification facility, which add considerable complexity and inefficiency to the system.

Owing to the sluggish kinetics of the reactions at each electrode (the reduction of O_2 in particular), a catalyst is needed to achieve a sufficient reaction rate at the relatively low temperatures in PEMFCs. This catalyst is typically platinum nanoparticles, which are deposited on larger support particles, typically made of carbon.[58] The two greatest issues with the PEMFC system are the cost (of both the platinum catalyst and the membrane) and the long-term durability.[56,59] A number of variations on this basic system have also been studied and are detailed in the following sections.

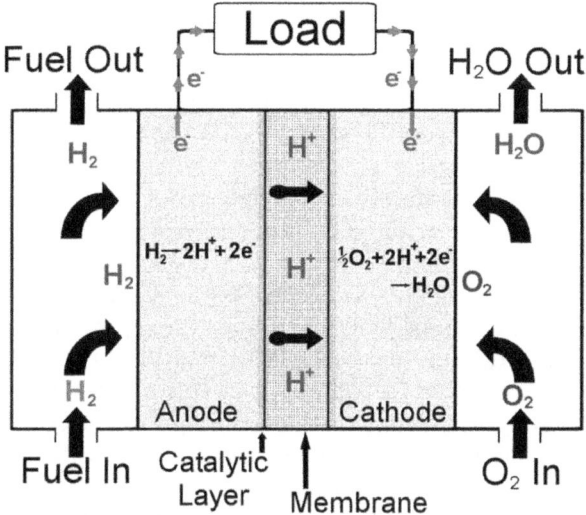

Figure 4 Schematic of a typical PEMFC design. The fuel (H_2) is pumped through the anode and is converted to H^+ ions and electrons, which travel through the membrane and external circuit, respectively. Oxygen is pumped through the cathode, is reduced by electrons from the external circuit and combines with H^+ ions from the membrane to form H_2O. The oxidation and reduction reactions occur in the Pt-seeded catalytic layer, where a triple phase boundary (TPB) exists between the electronically conducting electrodes, the ionically conducting membrane and the gaseous reactants.

5.1.2 Phosphoric Acid Fuel Cells. A phosphoric acid fuel cell (PAFC) is similar to a PEMFC except that the ionically conducting electrolyte is phosphoric acid instead of a proton-exchange membrane. The acid is typically contained within a Teflon-bonded silicon carbide matrix and facilitates the transport of protons while separating the gas streams at the anode and cathode. Removing the membrane from the FC design decreases the production cost and allows the operating temperature to be increased to between 150 and 200 °C. The higher temperature increases the rates of the reactions at each electrode but a significant amount of platinum catalyst is still required to achieve satisfactory reaction rates.[60] However, the PAFC is more resistant to CO poisoning than the PEMFC, potentially allowing a less clean fuel to be used.

The higher temperature of operation results in a relatively low thermodynamic efficiency (37–42% – slightly greater than that of a thermal power plant) unless the energy from this heat is also extracted in a CHP system (85% efficient).[61] The highly acidic nature of the phosphoric acid electrolyte, combined with the higher operating temperature, leads to more severe material degradation issues. Material selection is limited by the requirement for high corrosion resistance. PAFCs generally have a low power-to-weight ratio compared with other FC designs and, as a result, find more use in stationary applications. However, in most of these applications, it has been supplanted by the high-temperature molten carbonate fuel cell, which is discussed later.

5.1.3 Alkaline Fuel Cells. There are some significant advantages to running a fuel cell under alkaline conditions. A primary advantage is that the rate of the oxygen reduction reaction is much faster in an alkaline environment than in an acidic environment.[61] This allows the use of non-noble metal catalysts, while still achieving satisfactory reaction rates.[62,63] Alkaline electrolytes also tend to be less corrosive, increasing the range of available materials.

The electrolyte can be either concentrated aqueous KOH solution [an alkaline fuel cell (AFC)] or a polymeric anion-exchange membrane [an alkaline anion-exchange membrane fuel cell (AAEMFC)].[61,64] In both cases, it is the OH^- ion that is transported through the electrolyte. Oxygen at the cathode is reduced to OH^- which is then conducted through the electrolyte to combine with the H^+ that is produced by oxidation of hydrogen at the anode. AFCs can be operated in excess of 200 °C but AAEMFCs are typically limited to <90 °C.

A problem with AFCs is the relatively low conductivity of the OH^- ion compared with the H^+ ion. This requires highly concentrated KOH solutions to be used to ensure a sufficient flux of OH^- ions.[61] A similar problem exists for AAEMFCs, where a balance must be struck between increasing the membrane conductivity by increasing the ion-exchange capacity of the polymer and the decrease in the mechanical stability of the membrane that this entails. One of the greatest problems with traditional AFC technology is its susceptibility to CO_2 poisoning, *i.e.* the precipitation of metal carbonates within the electrolyte.[61] This limits the effectiveness of the technology owing to the costly process of purifying the incoming gas streams. AAEMFC technology is generally immune to this problem and allows for more compact, lightweight designs. However, the cost of these membranes, their relatively low ionic conductivity and their lack of stability at high pH are still barriers to the widespread commercial deployment of AAEMFCs.[61,64,65]

5.2 High-temperature Fuel Cells

5.2.1 Solid Oxide Fuel Cells.

In solid oxide fuel cells (SOFCs), the ionically conducting electrolyte is a solid metal oxide. Oxides with a cubic fluorite structure such as yttria-stabilised zirconia are common, but recent studies have focused on perovskites such as $La_{1-x}Sr_xGa_{1-y}Mg_yO_3$ owing to their high conductivity and material compatibility.[66] These oxides allow the conduction of O^{2-} ions through oxygen vacancies within the material. The electrodes are typically composed of electronically conducting ceramics. Composite electrodes, where an ionically conducting metal oxide is combined with an electronically conducting metal, have also been studied.[66] These extend the region of the TPB, allowing the gases to react anywhere on the electrode surface. The solid ceramic design allows a wide range of cell designs to be adopted.[67] More importantly, it allows the SOFC to run at very high temperatures, typically between 800 and 1000 °C. This is necessary to increase the O^{2-} conductivity within the electrolyte to acceptable levels.

The high operating temperature allows the use of non-noble metal catalysts, which offers a significant cost reduction compared with low-temperature FCs.[68] Another advantage of operating at such high temperatures is that a wide range of fuels can potentially be used, including natural gas and even coal gas, as the high temperatures and presence of H_2O in the cells allow these gases to be reformed *in situ*.[67,69] SOFCs generally have high tolerance for contaminants in the gas stream but are still susceptible to electrode degradation from adsorbed sulfur. The high operating temperature also has disadvantages, which include complex thermal management issues, increased rate of material degradation, enhancement of the rates of undesirable reactions and heightened material compatibility issues, such as differences in thermal expansion coefficient leading to mechanical stresses and cracking.[65] The high operating temperature also leads to a slow start-up time as the cell needs to warm up.

At present, the greatest research issue is simply finding appropriate materials that can operate for long periods of time at such high temperatures. The high temperatures and slow start-up times make SOFCs unsuitable for transportation applications, but they have potential for use in stationary CHP plants where the overall efficiency can be as high as 70%.[67,70]

5.2.2 Molten Carbonate Fuel Cells.

Molten carbonate fuel cells (MCFCs) are another class of high-temperature FC. The electrolyte in an MCFC is an alkaline metal carbonate (*e.g.* K_2CO_3) suspended in a porous matrix of

γ-lithium aluminate.[71] The metal carbonate is solid at room temperature but is a paste at typical MCFC operating temperatures of ~650 °C. At the anode, hydrogen from the fuel is oxidised and combined with CO_3^{2-} from the electrolyte to form H_2O and CO_2 [eqn (13)]. At the cathode, O_2 is reduced and combined with CO_2 to form the CO_3^{2-} ion [eqn (14)], which is transported through the electrolyte.

Anode reaction:

$$H_2 + CO_3^{2-} \rightarrow H_2O + CO_2 + 2e^- \tag{13}$$

Cathode reaction:

$$\tfrac{1}{2}O_2 + CO_2 + 2e^- \rightarrow CO_3^{2-} \tag{14}$$

Overall reaction:

$$H_2 + \tfrac{1}{2}O_2 + CO_{2,\text{cathode}} \rightarrow H_2O + CO_{2,\text{anode}} \tag{15}$$

From an environmental perspective, the production of CO_2 at the anode is undesirable but, since it can be recycled to the cathode and consumed there, it can theoretically be a carbon-neutral process. However, since an excess of CO_2 is desired to achieve an efficient reaction, CO_2 must be provided by another source. Owing to the high temperature of operation, hydrogen reforming can be performed *in situ* and therefore, if natural gas or other hydrocarbons are used as fuel, the CO_2 produced in the reforming process can be used.[72] The excess CO_2 generated in the cell can, in principle, be collected and stored, potentially allowing for operation with zero carbon emissions.[73] MCFCs themselves can also be used as a CO_2 separator or concentrator.[74]

Many of the advantages of MCFCs are due to the high operating temperature and thus are similar to the advantages of SOFCs. They include the use of non-noble metal catalysts, internal gas reforming and the high resistance to fuel contaminants of the materials used.[75] The disadvantages are again the lack of durability[76] of the cell materials due to high operating temperatures and contact with the corrosive electrolyte. As in the case of SOFCs, MCFCs are also prone to long start-up times.

The high temperatures, long start-up times and relatively high efficiencies[77] that are achieved in a CHP MCFC system make them ideal for stationary generation applications and they have received more attention for this type of application than any other FC technology. Field demonstrations of multi-megawatt units fuelled by a range of gases have been running for many years.[60]

5.3   Fuel Cells for Energy Storage

In principle, FCs can be made to run in reverse, *i.e.* to take in water and electrochemically generate H_2 at the cathode and O_2 at the anode.[78] Such systems are referred to as unitised regenerative fuel cells (URFCs) and are most commonly based on PEMFCs, but AFCs and SOFCs are also suitable.

URFCs can potentially compete with rechargeable batteries for energy storage. They have advantages such as high energy density, deep cycle ability, low self-discharge and inherently decoupled power and energy capacity in a similar manner to flow batteries (see later). Unfortunately, additional catalyst materials such as RuO_2 or IrO_2 are required for the oxygen evolution reaction,[79,80] complicating the design and increasing the cost.[81] Furthermore, they are subject to rapid degradation due to the high potentials used in the regenerative mode.[78,79] They are also very expensive and they generally show a lower round-trip efficiency than competing battery technologies.[79,81]

Hydrogen can, of course, also be produced by splitting water in a traditional electrolysis cell. However, the electrolysis of water is a relatively inefficient process.[82] The electrical energy used to produce this hydrogen for FC applications could also be used to charge batteries, which generally have much higher round-trip efficiency (*i.e.* a larger fraction of the energy used to charge a battery can be recovered than from an FC fuelled by hydrogen produced by electrolysis). Nevertheless, the use of off-peak electricity from renewable sources for water electrolysis may be an attractive way of generating hydrogen for fuel cell-powered transport.

Thermochemical water splitting has been shown to be technically viable but its economic viability has not been demonstrated.[82] There also has been considerable research into using specific microbes either to split water using sunlight or to break down various biomass feedstocks into hydrogen, but these processes are expensive and slow and have not been demonstrated at a large scale.[81,82]

5.4 Environmental Issues with Hydrogen Production and Distribution

If FCs are considered in the context of energy storage, then the storage medium is likely to be hydrogen. The most common method of producing hydrogen (95% of global production) is the steam reformation process in which steam and hydrocarbons (*e.g.* methane) are combined at high temperatures in the presence of a catalyst. Although this process results in less CO_2 emission than combustion of the original methane, it is not carbon neutral. Hydrogen can also be produced by the gasification of biomass or coal, but this too produces harmful emissions.[82] If clean, renewable sources are used to produce the electrical energy for this electrolysis, then both pure O_2 and H_2 can be produced without any harmful emissions. Carbon capture and sequestration technologies[83] are also advancing and may eliminate many of the harmful emissions from the steam reformation or coal gasification processes, thus allowing hydrogen to be produced in a clean and efficient manner.

As has already been discussed, stationary FCs can reach very high efficiencies when used in CHP generation systems. However, the cost of, and associated emissions from, producing the hydrogen fuel offset this efficiency such that the overall energy efficiency of generating the fuel and using it in the fuel cell is less than the efficiency obtained by charging a battery directly with a renewable energy source. However, FCs are potentially more suitable for transport applications than batteries, owing to their relatively large energy density, and under certain circumstances using electrical energy relatively inefficiently to produce fuel for FCs may be justified. However, H_2 is typically pressurised to hundreds of atmospheres in an energy-intensive process for storage. When the weight of the high-pressure vessels is considered, the energy density may no longer be suitable for transport applications. There is currently much research into developing storage vessels of lower weight from fibre-composite materials and high-strength polymers,[84] storage in metal hydrides[84-86] and storage in carbon nanotubes and other nanostructures.[84,85,87]

The large-scale distribution of H_2 is also problematic.[88,89] Existing pipelines for natural gas, *etc.*, would require significant modification to deal with the physical properties of hydrogen, *i.e.* its reactivity and the increased likelihood of leaks. The capital and energy cost of developing this infrastructure may outweigh the advantages that the low emissions from FCs provide.[88,89]

6 Flow Batteries

A flow battery[90-92] is a type of rechargeable battery. It was first proposed by Thaller at NASA in the 1970s.[93] In a flow battery, the electroactive species are dissolved in the electrolytes. The electrochemical cell is comprised of two half-cells, each with an electrode, separated by an ion-exchange membrane. As with any battery, charging involves passing a current through the cell, thereby converting electrical energy to chemical energy; in the case of a flow battery, this energy is stored in the electrolyte. Discharging involves allowing current to flow in the opposite direction, converting the chemical energy stored in the electrolyte back to electrical energy. The electrolyte in each half-cell is pumped through it from an external storage tank. Thus the chemical energy of the battery is stored in the tanks. Because of this, the energy capacity of the battery can be scaled separately from the power capacity, giving design flexibility not available in conventional batteries. The energy storage capacity of the battery can be increased by increasing the volume of the electrolyte, *i.e.* by increasing the size of the storage tanks, and the power available from the battery can be increased by increasing the size and number of electrochemical cells in a cell stack.

A schematic of a flow battery consisting of four cells in series is shown in Figure 5. The size (area) of each cell is chosen to give the desired current capability and the number of cells in the stack is chosen to give the desired battery voltage. The cell size and number together determine the power capability of the battery. Each cell in the stack has a positive and a negative half-cell. Electrolyte from one reservoir is pumped through all positive

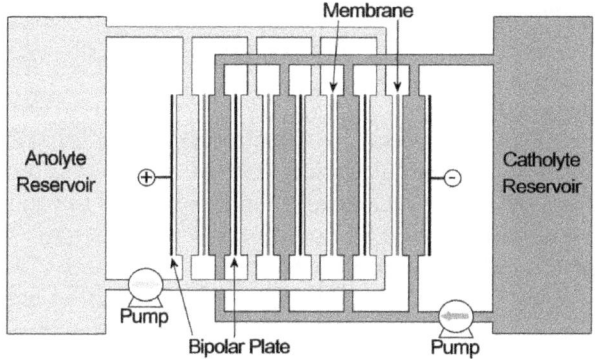

Figure 5 A redox flow battery system.

half-cells, plumbed in parallel. Similarly, electrolyte from another reservoir is pumped through all negative half-cells.

There are three main types of flow battery: traditional redox flow battery, hybrid flow battery and slurry flow battery. In all three types, the liquid electrolyte is pumped from external storage tanks through the electrochemical cells. However, in a traditional redox flow battery, all electroactive species remain dissolved in the electrolyte, whereas in a hybrid flow battery one or more of the electroactive species is solid and is deposited as a solid layer on an electrode. The energy capacity of a hybrid flow battery is thus limited by the amount of solid material that can be deposited in the cell. In a redox flow battery, on the other hand, the energy is limited only by the volume of the electrolyte, and this can be increased by increasing the size of the external storage tanks. A slurry flow battery uses the same type of battery chemistry as a hybrid flow battery, where one or more of the electroactive species is solid. However, it overcomes the limitations of hybrid flow batteries by depositing the solid on conductive particles suspended in the electrolyte as a slurry. The three types are described in the following sections, each with the aid of a common example.

6.1 Traditional Redox Flow Batteries: The All-vanadium Flow Battery

The use of vanadium in batteries was first proposed by Pissoort in 1933.[94] However, Skyllas-Kazacos and colleagues at the University of New South Wales, Australia, were the first to demonstrate an all-vanadium flow battery (VFB).[95] It is based on vanadium(II)/vanadium(III) and vanadium(IV)/vanadium(V) as the redox couples in the negative and positive half-cells, respectively. The vanadium species are dissolved in sulfuric acid; typically, the acid concentration is 3 M and the vanadium ions are at a concentration of 1.5 M. The electrode reactions may be represented as

$$VO^{2+} + H_2O = VO_2^+ + 2H^+ + e^- \tag{16}$$

at the positive electrode and

$$V^{3+} + e^- = V^{2+} \qquad (17)$$

at the negative electrode.

In commercial batteries, electrolyte is often supplied initially as a solution of vanadium with an average oxidation state of 3.5 (an equimolar mixture of V^{3+} and VO^{2+} in 3 M H_2SO_4). This can then be oxidised to vanadium(v) at the positive electrode and reduced to vanadium(ii) at the negative electrode in a 1.5-electron process during the first charge of the battery. During the subsequent discharge, a one-electron process is regarded as a full discharge, leaving the vanadium in the positive reservoir in an oxidation state of +4 and the vanadium in the negative reservoir in an oxidation state of +3. The ion-exchange membrane is usually Nafion, and the electrodes are usually carbon felt in contact with a current-collector plate made from a carbon–polymer composite.

The standard potential of the vanadium(ii)/vanadium(iii) redox couple (negative half-cell) in 3 M H_2SO_4 solution is reported to be −0.291 V and that of the vanadium(iv)/vanadium(v) couple (positive half-cell) 0.991 V *versus* the standard hydrogen electrode (SHE).[96] Hence the standard potential of the cell is ~1.3 V. The measured open-circuit potential of actual flow cells is typically 1.26–1.7 V, depending on the state of charge. Thermodynamically, since the standard potential of the negative half-cell reaction is below the standard potential of H_2 evolution, H_2 should evolve in the negative half-cell. Fortunately, however, the rate of the H_2 evolution reaction is slow and the V^{2+}/V^{3+} redox reaction occurs preferentially.

Voltage losses occur in the cell during charge and discharge due to overpotential at the electrodes and ohmic losses in the electrodes, the electrolytes and the membrane. The overpotential losses appear to be greatest at the negative electrode. This is attributed to slow kinetics at this electrode compared with the positive electrode, where the kinetics are generally faster.[97–99] Typical values of overpotential measured in our laboratory are 160 mV at the negative electrode and 20 mV at the positive electrode at a current density of 40 mA cm^{-2}.

The membrane acts as a physical separator to prevent cross-mixing of the electrolyte in both half-cells while allowing the migration of protons to complete the electrical circuit. The membrane therefore should suppress diffusion of vanadium ions from one half-cell to the other but should not hinder transport of protons. A high area resistivity of the membrane can cause a large internal ohmic (iR) drop and therefore a large undesirable internal energy loss. Ohmic losses across the membrane cause low voltage efficiencies. The area resistivity of Nafion 117 membrane is ~1 Ω cm^2, which would give a voltage drop of ~40 mV at 40 mA cm^{-2}.

Coulombic efficiencies are typically 95–99%. Losses are attributed to self-discharge caused by transport of vanadium ions through the membrane and also by side reactions, including H_2 evolution. Energy efficiencies are typically 70–80%.

The VFB is currently probably the most technologically advanced redox flow battery and is commercially available from several manufacturers. It has been used in many prototype renewable energy projects[100] and there is considerable ongoing research on understanding and improving the system.[101–103] Other redox flow batteries include the bromine–polysulfide (Regenesys) system,[104] a 12 MW prototype of which was built in 2002 but was not commercialised, and the iron–chromium,[105] uranium[106] and vanadium–bromine systems.[107]

6.2 Hybrid Flow Batteries: The Zinc–Bromine Flow Battery

In a hybrid flow battery (sometimes called a semi-flow battery), at least one of the four species involved in the redox process is a solid that deposits on the electrode. An example of a commercially available hybrid flow battery is zinc–bromine.[90,108] Other examples of hybrid flow batteries include the zinc–cerium[90,109] and lead–acid flow batteries.[90,110]

In a zinc–bromine flow battery in the fully discharged state, the electrolyte in both the positive and negative half-cells consists of aqueous $ZnBr_2$. On charging, bromide ions are oxidised to bromine on the positive side of the cell and zinc ions are reduced to metallic zinc on the negative side. As in a traditional redox flow battery, the positive and negative electrolytes are stored in separate tanks and circulated through the respective half-cells, which are separated by a membrane (ion-exchange or microporous). On the positive side, the system is similar to a traditional flow battery since both species (Br_2 and Br^-) are soluble in the electrolyte. However, on the negative side, the reduced species, elemental zinc, is not dissolved in the electrolyte but rather deposits on the electrode. Hence the negative side of the cell operates differently from a traditional flow battery and resembles somewhat a conventional battery, where the active species are present as solids on the electrodes. As a consequence of this, the energy storage capacity of a hybrid flow battery is not determined by tank size alone but also depends on the amount of solid deposit that the electrode can accommodate. Thus the energy storage capacity of a hybrid flow battery is not completely independent of the power capability as it is in a traditional flow battery.

Hybrid flow batteries allow the use of redox couples where one or more of the active species is a solid (such as an elemental metal) but at the expense of limiting the energy storage capacity. This limitation may be overcome by the use of a slurry flow battery.

6.3 Slurry Flow Batteries: The All-iron Flow Battery

An example of a slurry-based flow battery is the all-iron system. The all-iron redox flow battery was first described by Hruska and Savinell[111] in 1981. The chemistry consists of the Fe^{2+}/Fe^{3+} couple on the positive side and Fe^{2+}/Fe on the negative side. The electrolyte is an aqueous mixture of $FeCl_2$ and $FeCl_3$ on the positive side and aqueous $FeCl_2$ on the negative side, separated by a microporous membrane. The positive and negative electrolytes are stored in separate tanks. On charging, Fe^{2+} is converted to Fe^{3+} on the positive side and both the oxidised and reduced species are thus soluble in the electrolyte. On the negative side, Fe^{2+} is reduced to elemental iron on charging, which is not, of course, soluble in the electrolyte.

Hence the all-iron system may be implemented as a hybrid flow battery where the iron deposits on the negative electrode during charging. However, it may also be implemented as a slurry flow battery.[112] In this implementation, a fine dispersion of carbon particles is suspended in the negative electrolyte. This slurry electrolyte is circulated from the negative tank through the negative side of the cell where the carbon particles can make electronic contact with the cell electrode. Thus, on charging, elemental iron is deposited on the carbon particles, which are then circulated back to the tank. During discharging, the slurry containing iron-coated carbon particles is circulated through the cell, where it is oxidised to Fe^{2+}, which dissolves in the electrolyte. Thus, the active species on the negative side are all contained in the slurry electrolyte, the oxidised species Fe^{2+} in solution and reduced species Fe in the form of a coating on the suspended particles.

6.4   Other Flow Battery Systems

There is currently research on a wide range of flow battery designs and materials.[113] Research includes work on novel electrode materials, structures and treatment; use of nanostructured materials to enhance electrocatalytic activity; novel electrolyte flow arrangements; membranes, *etc.* Many alternative electrolytes and redox couples are being investigated for next-generation systems, including non-aqueous and ionic liquid electrolytes, organic redox species, use of ligands to modify the potential of transition metal couples, use of suspended nanoparticle systems, *etc.* Such systems will be important for improved performance and lower costs if flow battery technology captures a significant portion of the energy storage market.

7   Summary and Conclusions

Electrochemical energy storage systems have the potential to make a major contribution to the implementation of sustainable energy. Because many sustainable energy systems are intermittent, the extent to which they can be used on the electricity grid is limited unless suitable storage is available. Both conventional batteries and flow batteries provide attractive options for such storage. Fuel cells are also an attractive option for the efficient use of hydrogen that has been generated from sustainable sources, particularly for electric vehicles.

The choice of battery chemistry and design for a particular project is highly dependent on the specific requirements of the project that is being considered, the upfront capital and lifetime expenditure costs associated with the chosen battery type and the end-of-life, environmental and safety considerations. The characteristics of some of the batteries that we have discussed are summarised in Table 1.

Table 1 Characteristics of various battery types.

Type	Subtype	Specific power/ $W\,kg^{-1}$	Power density/ $W\,L^{-1}$	Specific energy/ $Wh\,kg^{-1}$	Energy density/ $Wh\,L^{-1}$	Battery price/€ kWh^{-1}	Lifetime/years	Cycles (for depth of discharge, DoD)
Lead–acid	Lead–acid (SLI)	180	360	30	60	120	10	3500 (10% DoD)
	Lead–acid (deep discharge)					200		500 (50% DoD)
	Hybrid lead–acid					200		>10 000 (10% DoD)
Li ion	LiFePO$_4$	300	820	180	490	350	15	2200 (80% DoD)
Ni	NiMH	800	1600	100	200	500	15	1000
NaS		200		150–750	150	200	5–15	4000 (80% DoD)
Flow battery	Vanadium flow battery	Power and energy are decoupled		25	30	220 (900 per kW)	20	Unlimited

For applications where it is predicted that the system will be called upon on average twice per month, then traditional lead–acid batteries, which have the greatest market share of worldwide battery sales and are estimated to be 99% recycled in the EU and USA, may be considered a very good option. Such batteries can be designed to have standby operation lifetimes that exceed 20 years, cycle life of greater than 500 discharges to a depth of discharge of 50%, relatively low battery cost and low balance of plant costs.

However, if the application requires continuous charging and discharging or operation at low states of charge, where sulfation of electrodes can occur, then the latest designs of lead–acid battery that use carbon in their electrodes, or a different battery design, such as vanadium flow batteries, could be used since both battery types greatly increase the number of available charge–discharge cycles (>10 000) without greatly increasing the system costs or resulting in additional end-of-life, environmental and safety considerations.

Currently, lithium-ion battery systems are gaining a significant market share because of their ability to charge and discharge multiple times and their high energy density. However, unless the application requires a high power density, *i.e.* unless there are significant restrictions on available space for the installation of the energy storage system, lower energy density solutions such as advanced lead–acid and flow batteries have greater long-term feasibility when end-of-life considerations are taken into account. Furthermore, the case for lithium-ion batteries is hampered by the need for individual cell monitoring and for close thermal control, which are not issues in many other battery systems.

References

1. A. J. Bard and L. R. Faulkner, *Electrochemical Methods: Fundamentals and Applications*, Wiley, New York, 2nd edn, 2000.
2. J. Newman, K. E. Thomas-Alyea, *Electrochemical Methods*, Wiley, New York, 3rd edn, 2004.
3. V. S. Bagotsky, A. M. Skundin and Y. M. Volfkovich, *Electrochemical Power Sources: Batteries, Fuel Cells, and Supercapacitors*, Wiley, New York, 2015.
4. C. Vincent and B. Scrosati, *Modern Batteries*, Elsevier Science, Oxford, 2nd edn, 1997.
5. R. A. Huggins, *Advanced Batteries: Materials Science Aspects*, Springer, New York, 2009.
6. R. O'Hayre and S. W. Cha, *Fuel Cell Fundamentals*, Wiley, 2016.
7. J. O'Donnell, C. Lenihan, N. Quill, D. N. Buckley, E. Pican and R. P. Lynch, *Frequency Stabilisation of Island Electricity Grid: A New Application for Lead-Acid Batteries, 10th International Conference on Lead-Acid Batteries (LABAT 2017)*, Bulgaria, 2017.
8. T. Xi, W. Yufeng, G. Yu and Z. Tieyong, *Waste Manage. Res.*, 2015, **33**, 986–994.

9. I. Hadjipaschalis, A. Poullikkas and V. Efthimiou, *Renewable Sustainable Energy Rev.*, 2009, **13**, 1513–1522.
10. H. Bode, R. J. Brodd and K. V. Kordesch, *Lead-Acid Batteries*, Wiley Interscience, New York, 1977.
11. D. Pavlov, *Lead-Acid Batteries: Science and Technology*, Elsevier, Oxford, 2011.
12. O. A. White, *The Automobile Storage Battery Its Care and Repair*, American Bureau of Engineering, Chicago, 1922.
13. H. A. Catherino, F. F. Feres and F. Trinidad, *J. Power Sources*, 2004, **129**, 113–120.
14. J. Albers and E. Meissner, in *Lead-Acid Batteries for Future Automobiles*, ed. J. Garche, E. Karden, P. T. Moseley and D. A. J. Rand, Elsevier, Oxford, 2017, ch. 6, p. 201.
15. K. Peters, D. A. J. Rand and P. T. Moseley, in *Lead-Acid Batteries for Future Automobiles*, ed. J. Garche, E. Karden, P. T. Moseley and D. A. J. Rand, Elsevier, Oxford, 2017, ch. 7, pp. 217–232.
16. P. T. Moseley, R. F. Nelson and A. F. Hollenkamp, *J. Power Sources*, 2006, **157**, 3–10.
17. D. W. H. Lambert, P. H. J. Greenwood and M. C. Reed, *J. Power Sources*, 2002, **107**, 173–179.
18. P. Ruetschi, *J. Power Sources*, 2004, **127**, 33–44.
19. M. G. Mayer and D. N. Wilson, *J. Power Sources*, 1998, **73**, 17–22.
20. F. Ahmed, *J. Power Sources*, 1996, **59**, 107–111.
21. National Recycling Rate Study 2012–2016, Battery Council International, http://batterycouncil.org/resource/resmgr/Recycling_Rate/BCI_201212-17_FinalRecycling.pdf. Accessed March 2018.
22. The Availability of Automotive Lead-based Batteries for Recycling in the EU, A joint industry analysis of EU collection and recycling rates 2010–2012, http://eurobat.org/brochures-reports. Accessed March 2018.
23. W. Zhang, J. Yang, X. Wu, Y. Hu, W. Yu, J. Wang, J. Dong, M. Li, S. Liang, J. Hu and R. V. Kumar, *Renewable Sustainable Energy Rev.*, 2016, **61**, 108–122.
24. C. J. Rydh, *J. Power Sources*, 1999, **80**, 21–29.
25. M. S. Whittingham, *Science*, 1976, **192**, 1126.
26. S. Basu, C. Zeller, P. J. Flanders, C. D. Fuerst, W. D. Johnson and J. E. Fischer, Synthesis and properties of lithium-graphite intercalation compounds, *Mater. Sci. Eng.*, 1979, **38**(3), 275–283.
27. US Patent 4304825, Basu, Samar, "Rechargeable battery", issued 8 December 1981, assigned to Bell Telephone Laboratories.
28. K. Mizushima, P. C. Jones, P. J. Wiseman and J. B. Goodenough, *Mater. Res. Bull.*, 1980, **15**, 783.
29. N. A. Godshall, I. D. Raistrick and R. A. Huggins, *Mater. Res. Bull.*, 1980, **15**, 561.
30. G. A. Nazri and G. Pistoia, *Lithium Batteries: Science and Technology*, Kluwer Academic/Plenum, 2004.

31. M.-K. Song, S. Park, F. M. Alamgir, J. Cho and M. Liu, *Mater. Sci. Eng., R*, 2011, **72**, 203.
32. Y. Takei, *NISTEP Sci. Technol. Trends Q. Rev.*, 2010, **36**, 40–52.
33. J. M. Tarascon and M. Armand, *Nature*, 2001, **414**, 359.
34. J. B. Goodenough and K.-S. Park, The Li-ion rechargeable battery: A perspective, *J. Am. Chem. Soc.*, 2013, **135**, 1167–1176.
35. M. D. Bhatt and C. O'Dwyer, *Phys. Chem. Chem. Phys.*, 2015, **17**, 4799–4844.
36. M. Osiak, H. Geaney, E. Armstrong and C. O'Dwyer, *J. Mater. Chem.*, 2014, **A2**, 9433.
37. D. McNulty, H. Geaney, D. Buckley and C. O'Dwyer, *Nano Energy*, 2018, **43**, 11.
38. W. McSweeney, H. Geaney and C. O'Dwyer, *Nano Res.*, 2015, **8**, 1395.
39. M. Zackrisson, L. Avellán and J. Orlenius, *J. Cleaner Prod.*, 2010, **18**, 1519.
40. N. Takenaka, Y. Suzuki, H. Sakai and M. Nagaoka, *J. Phys. Chem. C*, 2014, **118**, 10874.
41. X.-B. Cheng, R. Zhang, C.-Z. Zhao, F. Wei, J.-G. Zhang and Q. Zhang, *Adv. Sci.*, 2016, **3**, 1500213.
42. M. Winter, *Z. Phys. Chem.*, 2009, **223**, 1395.
43. M. Tatsumisago and A. Hayashi, *Solid State Ionics*, 2012, **225**, 342.
44. D. A. Notter, M. Gauch, R. Widmer, P. Wäger, A. Stamp, R. Zah and H.-J. Althaus, *Environ. Sci. Technol.*, 2010, **44**, 6550.
45. D. H. P. Kang, M. Chen and O. A. Ogunseitan, *Environ. Sci. Technol.*, 2013, **47**, 5495.
46. J. M. Tarascon, *Philos. Trans. R. Soc.*, 2010, **A368**, 3227.
47. D. Larcher and J. M. Tarascon, *Nat. Chem.*, 2014, **7**, 19.
48. A. Jain, S. P. Ong, G. Hautier, W. Chen, W. D. Richards, S. Dacek, S. Cholia, D. Gunter, D. Skinner, G. Ceder and K. A. Persson, *APL Mater.*, 2013, **1**, 011002.
49. S. V. Vassilev, D. Baxter, L. K. Andersen and C. G. Vassileva, *Fuel*, 2010, **89**, 913.
50. D. McNulty, E. Carroll and C. O'Dwyer, *Adv. Energy Mater.*, 2017, **7**, 1602291.
51. J. Xu, H. R. Thomas, R. W. Francis, K. R. Lum, J. Wang and B. Liang, *J. Power Sources*, 2008, **177**, 512.
52. J. T. Kummer and N. Weber, *SAE Trans.*, 1968, **76**, 1003–1028.
53. Ku D. Kumar, S. K. Rajouria, S. B. Kuhar and D. K. Kanchan, *Solid State Ionics*, 2017, **312**, 8–16.
54. S. R. Ovshinsky and R. Young, *U. S. Pat.*, 6413670 (B1), 1998.
55. W. Dai, H. Wang, X.-Z. Yuan, J. J. Martin, D. Yang, J. Qiao and J. Ma, *Int. J. Hydrogen Energy*, 2009, **34**, 9461–9478.
56. R. Borup, J. Meyers, B. Pivovar, Y. S. Kim, R. Mukundan, N. Garland, D. Myers, M. Wilson, F. Garzon, D. Wood, P. Zelenay, K. More, K. Stroh, T. Zawodzinski, J. Boncella, J. E. McGrath, M. Inaba, K. Miyatake,

M. Hori, K. Ota, Z. Ogumi, S. Miyata, A. Nishikata, Z. Siroma, Y. Uchimoto, K. Yasuda, K.-i. Kimijima and N. Iwashita, *Chem. Rev.*, 2007, **107**, 3904–3951.

57. T. Ous and C. Arcoumanis, *J. Power Sources*, 2013, **240**, 558–582.
58. X. Yu and S. Ye, *J. Power Sources*, 2007, **172**, 145–154.
59. S. Zhang, X.-Z. Yuan, J. N. C. Hin, H. Wang, K. A. Friedrich and M. Schulze, *J. Power Sources*, 2009, **194**, 588–600.
60. X. Zhu and B. Huang, *Electrochemical Technologies for Energy Storage and Conversion*, ed. R. Liu, L. Zhang, X. Sun, H. Liu and J. Zhang, Wiley VCH, Chichester, 2012, p. 730.
61. A. International Energy, *Hydrogen and Fuel Cells: Review of National R&D Programs*, 2004.
62. G. Merle, M. Wessling and K. Nijmeijer, *J. Membr. Sci.*, 2011, **377**, 1–35.
63. K. Asazawa, K. Yamada, H. Tanaka, A. Oka, M. Taniguchi and T. Kobayashi, *Angew. Chem., Int. Ed.*, 2007, **46**, 8024–8027.
64. M. Gong, D.-Y. Wang, C.-C. Chen, B.-J. Hwang and H. Dai, *Nano Res.*, 2016, **9**, 28–46.
65. J. Cheng, G. He and F. Zhang, *Int. J. Hydrogen Energy.*, 2015, **40**, 7348–7360.
66. N. Mahato, A. Banerjee, A. Gupta, S. Omar and K. Balani, *Prog. Mater. Sci.*, 2015, **72**, 141–337.
67. A. Buonomano, F. Calise, M. D. d'Accadia, A. Palombo and M. Vicidomini, *Appl. Energy*, 2015, **156**, 32–85.
68. A. B. Stambouli and E. Traversa, *Renewable Sustainable Energy Rev.*, 2002, **6**, 433–455.
69. T. M. Gür, *Prog. Energy Combust. Sci.*, 2016, **54**, 1–64.
70. A. Choudhury, H. Chandra and A. Arora, *Renewable Sustainable Energy Rev.*, 2013, **20**, 430–442.
71. J.-E. Kim, J. Han, S. P. Yoon, S. W. Nam, T.-H. Lim and H. Kim, *Curr. Appl. Phys.*, 2010, **10**, S73–S76.
72. X. Zhang, H. Liu, M. Ni and J. Chen, *Renewable Energy*, 2015, **80**, 407–414.
73. S. Campanari, P. Chiesa, G. Manzolini and S. Bedogni, *Appl. Energy*, 2014, **130**, 562–573.
74. J.-H. Wee, *Renewable Sustainable Energy Rev.*, 2014, **32**, 178–191.
75. N. Di Giulio, E. Audasso, B. Bosio, J. Han and S. J. McPhail, *Int. J. Hydrogen Energy*, 2015, **40**, 6430–6439.
76. E. Antolini, *Appl. Energy*, 2011, **88**, 4274–4293.
77. X. Zhang, J. Guo and J. Chen, *Int. J. Hydrogen Energy*, 2012, **37**, 8664–8671.
78. F. Barbir, *Sol. Energy*, 2005, **78**, 661–669.
79. Y. Wang, D. Y. C. Leung, J. Xuan and H. Wang, *Renewable Sustainable Energy Rev.*, 2016, **65**, 961–977.
80. T. Sadhasivam, K. Dhanabalan, S.-H. Roh, T.-H. Kim, K.-W. Park, S. Jung, M. D. Kurkuri and H.-Y. Jung, *Int. J. Hydrogen Energy*, 2017, **42**, 4415–4433.

81. M. Gabbasa, K. Sopian, A. Fudholi and N. Asim, *Int. J. Hydrogen Energy*, 2014, **39**, 17765–17778.
82. I. Dincer and C. Acar, *Int. J. Hydrogen Energy*, 2015, **40**, 11094–11111.
83. N. Z. Muradov and T. N. Veziroğlu, *Int. J. Hydrogen Energy*, 2005, **30**, 225–237.
84. H. T. Hwang and A. Varma, *Curr. Opin. Chem. Eng.*, 2014, **5**, 42–48.
85. A. Züttel, *Mater. Today*, 2003, **6**, 24–33.
86. Q.-L. Zhu and Q. Xu, *Energy Environ. Sci.*, 2015, **8**, 478–512.
87. E. S. Cho, A. M. Ruminski, S. Aloni, Y.-S. Liu, J. Guo and J. J. Urban, *Nat. Commun.*, 2016, **7**, 10804.
88. M. Ball and M. Weeda, *Int. J. Hydrogen Energy*, 2015, **40**, 7903–7919.
89. J. Alazemi and J. Andrews, *Renewable Sustainable Energy Rev.*, 2015, **48**, 483–499.
90. A. Z. Weber, M. M. Mench, J. P. Meyers, P. N. Ross, J. T. Gostick and Q. Liu, *J. Appl. Electrochem.*, 2011, **41**, 1137.
91. M. Skyllas-Kazacos, M. H. Chakrabarti, S. A. Hajimolana, F. S. Mjalli and M. Saleem, *J. Electrochem. Soc.*, 2011, **158**, R55.
92. M. J. Watt-Smith, P. Ridley, R. G. A. Wills, A. A. Shah and F. C. Walsh, *J. Chem. Technol. Biotechnol.*, 2013, **88**, 126.
93. L. H. Thaller, *Electrically Rechargeable Redox Flow Cells*. 1974: NASA TM X-71540.
94. P. A. Pissoort, *FR Pat.*, 754065, 1933.
95. M. Skyllas-Kazacos, M. Rychcik, R. G. Robins, A. G. Fane and M. A. Green, *J. Electrochem. Soc.*, 1986, **133**, 1057.
96. M. Gattrell, J. Park, B. MacDougall, J. Apte, S. McCarthy and C. W. Wu, *J. Electrochem. Soc.*, 2004, **151**, A123.
97. A. Bourke, M. A. Miller, R. P. Lynch, X. Gao, J. Landon, J. S. Wainright, R. F. Savinell and D. N. Buckley, *J. Electrochem. Soc.*, 2016, **163**, A5097.
98. M. A. Miller, A. Bourke, N. Quill, J. S. Wainright, R. P. Lynch, D. N. Buckley and R. F. Savinell, *J. Electrochem. Soc.*, 2016, **163**, A2095.
99. A. Bourke, M. A. Miller, R. P. Lynch, J. S. Wainright, R. F. Savinell and D. N. Buckley, *J. Electrochem. Soc.*, 2015, **162**, A1547.
100. H. Binder, *et al.*, Characterization of Vanadium Flow Battery, Report Risø-R-1753(EN), ISBN 978-87-550-3853-0. 2010, Risø National Laboratory for Sustainable Energy Technical University of Denmark.
101. C. Petchsingh, N. Quill, J. T. Joyce, D. Ní Eidhin, D. Oboroceanu, C. Lenihan, X. Gao, R. P. Lynch and D. N. Buckley, *J. Electrochem. Soc.*, 2016, **163**, A5068.
102. D. Oboroceanu, N. Quill, C. Lenihan, D. Ní Eidhin, S. P. Albu, R. P. Lynch and D. N. Buckley, *J. Electrochem. Soc.*, 2016, **163**, A2919.
103. D. Oboroceanu, N. Quill, C. Lenihan, D. Ní Eidhin, S. P. Albu, R. P. Lynch and D. N. Buckley, *J. Electrochem. Soc.*, 2017, **164**, A2101.
104. H. Zhou, H. Zhang, P. Zhao and B. Yi, *Electrochim. Acta*, 2006, **51**, 6304.
105. C. H. Bae, E. P. L. Roberts and R. A. W. Dryfe, *Electrochim. Acta*, 2002, **48**, 279.

106. T. Yamamura, Y. Shiokawa, H. Yamana and H. Moriyama, *Electrochim. Acta*, 2002, **48**, 43–50.
107. M. Skyllas-Kazacos, G. Kazacos, G. Poon and H. Verseema, *Int. J. Energy Res.*, 2009, **34**, 182–189.
108. D. Linden, *Handbook of Batteries*, McGraw-Hill, Inc., 2 edn, 1995.
109. P. K. Leung, C. Ponce-de-León, C. T. J. Low, A. A. Shah and F. C. Walsh, *J. Power Sources*, 2011, **196**, 5174.
110. D. Pletcher and R. Wills, *J. Power Sources*, 2005, **149**, 96.
111. L. W. Hruska and R. F. Savinell, *J. Electrochem. Soc.*, 1981, **128**, 18.
112. T. J. Petek, N. C. Hoyt, R. F. Savinell and J. S. Wainright, *J. Power Sources*, 2015, **294**, 620–626.
113. M. L. Perry and A. Z. Weber, *J. Electrochem. Soc.*, 2016, **163**, A5064.

Electrical Storage

HAN SHAO, PADMANATHAN NARAYANASAMY, KAFIL M. RAZEEB,
ROBERT P. LYNCH AND FERNANDO M. F. RHEN*

ABSTRACT

Short-duration fluctuations of electricity supply and demand can cause instability in electricity grids and utility systems. These fluctuations are usually corrected using fast response time electrical energy storage devices that have high power and low storage capacity, hence they are never noticed by the end user. This chapter covers basic aspects of the most important electrical energy storage devices in this category, including supercapacitors and supercapatteries, superconducting magnetic energy storage devices, flywheels and synchronous condensers. Their fast response times and high efficiencies, particularly when storing energy for short durations, makes these devices very suitable for the continuous stabilisation of energy systems. Furthermore, the coupling of these devices to dc and ac electrical networks can provide natural instantaneous responses that stabilise such systems.

1 Introduction

This chapter is dedicated to electrical energy storage devices with high power and low storage capacity, covering supercapacitors and supercapatteries, superconducting magnetic energy storage (SMES) devices and flywheels, including synchronous condensers. These are ranked as auxiliary energy storage devices in terms of energy capacity as they are designed to operate in combination with high-energy capacity devices such as wind farms, pumped

---

*Corresponding author.

Issues in Environmental Science and Technology No. 46
Energy Storage Options and Their Environmental Impact
Edited by R.E. Hester and R.M. Harrison
© The Royal Society of Chemistry 2019
Published by the Royal Society of Chemistry, www.rsc.org

hydroelectric plants, electrical hydroelectric plants and thermal electrical power plants. They can provide high power for short periods of time, helping to maintain the stability of electricity grids and utility systems.

Supercapattery, combining the best properties of both batteries and supercapacitors, was first proposed in 2011.[1] The benefit of building a hybrid device is to obtain high energy from a battery-type material and couple this with the ability to deliver high power from supercapacitor-type material. Owing to the combination of two mechanisms, supercapattery devices could potentially broaden the cell voltage and extend the cell lifetime;[2,3] it is, therefore, particularly suited as an energy device for a wide range of applications such as portable devices and biosensors. This approach has afforded a significant increase in energy density, though not without sacrificing power density. A single device that combines all of these positive attributes could change the entire technological landscape of today, leading to lighter, more compact phones and electric cars that charge in minutes instead of hours. The amount of energy that can be stored in such a device depends in large part on the contact area between the electrolyte and the two electrodes: the greater the contact area, the more energy can be stored.

SMES is basically an energy inductor with zero resistance appropriately coupled to the electricity grid for storing energy. The physical characteristics of SMES allow for fast energy storage response and high power as the energy is stored in the magnetic field. The drawback of this technology currently seems to be the capital cost, which is directly linked to the fabrication of inductors from superconducting wires and operating at low temperatures.

Flywheels operate on the principle of inertia as the energy is stored in the kinetic energy of rotating parts. This technology is well established; it shows great advantages in terms of implementation. The conversion from mechanical energy to electrical energy usually requires an electromagnetic component in combination with an electrical interface to attach to the grid. Additionally, while making use of kinetic energy, synchronous condensers can provide and absorb reactive power, stabilising the load of a grid. They operate in a similar manner to motors; however, they are not intended to drive mechanical loads.

2 Supercapacitor and Supercapattery

2.1 Basics of Energy Storage Devices

The rapid consumption of conventional fossil fuels and their global environmental issues triggered the search for new energy storage/conversion devices and were highly focused in past decades, such as fuel cells, batteries, supercapacitors and their hybrid systems. Among them, the lithium-ion battery has been used commercially in electric vehicles and military and other mobile applications owing to its high energy capacity. On the other hand, the supercapacitor has been intensively researched owing to its high specific power density and long life span. Recently, there has been a surge of research and development aimed at making new types of hybrid energy storage device with high energy and power density that can bridge the gap between a conventional capacitor and the battery. These different types of energy storage devices are suitable for a variety of applications.

2.1.1 Comparison of Batteries, Supercapacitors and Supercapatteries.

Batteries are one of the most common energy storage devices in our day-to-day lives; they can provide power for electrical devices such as smartphones, flashlights and electric cars. Batteries are classified into two types: primary and secondary batteries. Primary batteries are also called single-use or disposable batteries, which are convenient but detrimental to the environment. Secondary batteries are rechargeable batteries, which are able to be charged and discharged multiple times along with reversible chemical reactions. Compared with widely used lead–acid[4] and nickel–cadmium secondary batteries,[5,6] lithium-ion (Li-ion) batteries have attracted great attention because of their high capacity, high energy density, small self-discharge and reasonable cycle stability.[7,8]

In the Li-ion battery system, Li ions move between the positive electrode and the negative electrode during charge and discharge processes. Typical charge storage mechanisms for a Li-ion battery are classified into three types: intercalation, conversion and alloying[9,10] (see Figure 1). In general, intercalation materials are supposed to provide better cyclic stability, but the theoretical capacity is low, whereas the other two mechanisms offer ultra-high capacity but the large volume change limits the coulombic efficiency and cyclic stability.[9,10] Work on improving the existing materials of Li-ion battery electrodes is in progress. Although Li-ion batteries can provide large operating voltages and high energy densities, their poor cyclic stability and low power output limit their applications so that an alternative type of energy storage device with higher power density and longer lifetime is required.

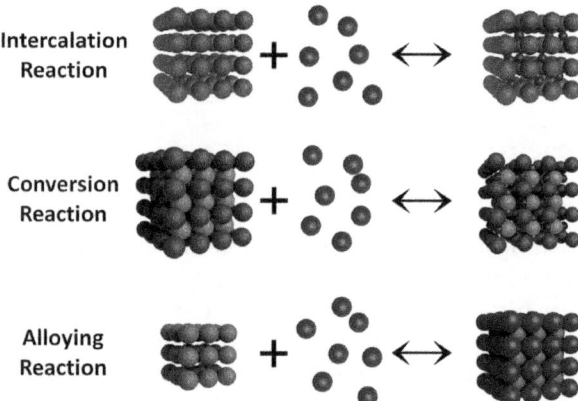

Figure 1 Schematic diagram of the different reaction mechanisms of lithium-ion batteries.

Supercapacitors, owing to their fast charge–discharge capability and high power density, give unique energy storage properties compared with a conventional energy storage system. A supercapacitor consists of positive and negative electrodes, which are electrically isolated from each other by a separator (see Figure 2). In a classic electrical double-layer capacitor (EDLC), an electrical double layer is formed at the interface between the surface of a conductor electrode and the electrolyte. Activated carbon, graphene and other carbon-based materials have been widely investigated as high-performance supercapacitor electrode materials owing to their high surface area, good stability and high conductivity. Interestingly, some metal oxide materials such as RuO_2[11,12] and MnO_2[13,14] can exhibit capacitive behaviour achieved by Faradaic reaction on the surface of the electrode by specifically adsorbed electrolyte ions. Owing to the different electron-transfer mechanisms of capacitive behaviour, these materials have been classified as pseudocapacitive materials. In fact, owing to the Faradaic process involved in reduction–oxidation (redox) reactions, these metal oxides have higher theoretical capacitances than carbon-based materials, but also suffer from the lack of cyclic stability and lower power density.[15–18] Further research is necessary to maximise both the capacitance and cyclic stability of the supercapacitor materials in the near future.

It is worth pointing out that not all catalytic materials exhibit pseudocapacitive behaviour, such as NiO[19,20] and Co_3O_4.[21] Owing to the misunderstanding in the literature of the term 'pseudocapacitance', many researchers have classified materials with the behaviour of a noticeable discharge plateau to be capacitive[22–24] and used the same equation to estimate the charge storage capacity. However, such materials demonstrate either

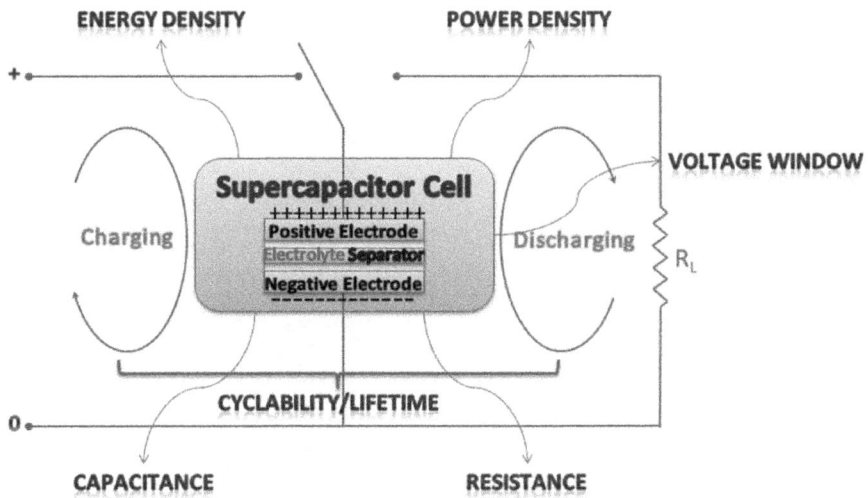

Figure 2 Schematic diagram of the supercapacitor working process and relative parameters.

diffusion-controlled or surface-controlled reversible redox reactions on the electrode surface with the behaviour of a noticeable discharge plateau. Hence they should be classified as pseudobattery-type materials, since there is no intercalation/de-intercalation process involved, and can be distinguished from ion capacitors. Therefore, the new terminology *supercapattery* was created to define the electrochemical behaviour between capacitor-like and battery-like hybrids.[1,2] Generally, a supercapattery cell consists of pseudobattery-type and supercapacitor-type electrode materials, as shown in Figure 3. Owing to the combination of two mechanisms, the supercapattery devices could obtain high energy from the battery-type material and couple this with the ability to deliver high power from supercapacitor-type material. Moreover, the supercapattery could also potentially broaden the cell voltage range and extend the cell lifetime.[2,3]

However, with the rapid development of high-performance energy storage devices, more and more novel devices have been investigated. Therefore, the definition and classification of supercapattery remain controversial. For instance, the lithium/sodium ion capacitor, which is a hybrid electrochemical energy storage device that couples a high-capacity bulk intercalation-based battery-style negative electrode (anode) and a high-rate surface adsorption-based capacitor-style positive electrode (cathode), has attracted much attention recently. The lithium/sodium ion can intercalate into the anode and counter-ions will form the double layer at the cathode (*i.e.* activated carbon) during the charge process. Although the ion capacitor exhibits a capacitance-type behaviour, which is a rectangular shape of the

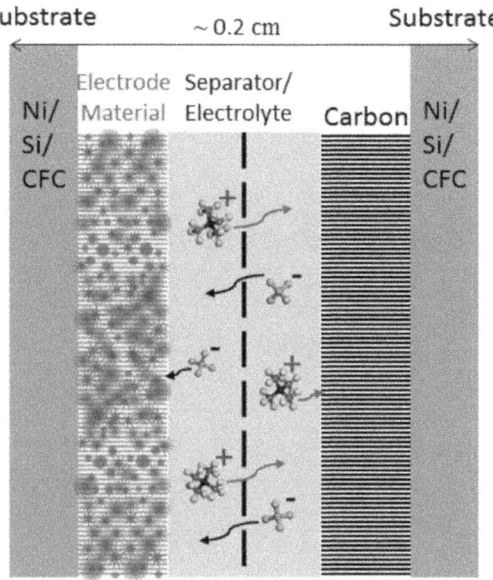

Figure 3   Schematic diagram of the supercapattery structure.

cyclic voltammogram (CV) and a linear charge–discharge curve, it is difficult to categorise the ion capacitor as a supercapacitor or supercapattery owing to the different storage mechanisms. Some researchers claim that the Li/Na ion capacitor is a hybrid device, which belongs to supercapattery,[25] because the discharging voltage reaches zero instead of V_{min} as in batteries; however, it is difficult to distinguish the devices without involving an intercalation mechanism. Furthermore, another type of symmetric device (in which both the positive and negative electrodes are made from the same material) fabricated with inexpensive pseudobattery-type materials normally exhibits a noticeable discharge plateau or quasi-linear discharge behaviour,[23] with higher energy density than supercapacitors and higher power density and much better cyclic stability than a battery. Therefore, they are worthy of further investigation and need to be classified properly.

2.1.2 Classification of Energy Storage Devices. In general, the electrode materials of energy storage devices can be classified into four different types based on their different energy storage mechanisms: capacitive materials (activated carbon, graphite, graphene, *etc.*[26–30]), pseudocapacitive materials (MnO_2, RuO_2, *etc.*[11,13,14]), battery-type materials ($LiTi_4O$, $LiFePO_4$, *etc.*[31–33]) and pseudobattery-type materials [NiO, Co_3O_4, $CoMoO_4$, $Co_3(PO_4)_2$, *etc.*[2,34–36]]. Theoretically, all of these materials can be used as either positive or negative electrodes to assemble an energy storage device according to their voltage window. Different electrolytes also determine the charge storage characteristics of an energy storage device, where each electrode exhibits a specific voltage window in a particular electrolyte. Therefore, a large number of energy storage devices can be fabricated by selecting and pairing various electrodes and electrolytes. In order to maximise the electrochemical performance of a device, researchers are now focusing on developing asymmetric cells assembled using two different electrodes with positive and negative working potential in a specific electrolyte. Compared with symmetric cells (same electrode applied), asymmetric cells can broaden the cell voltage range so that the energy density may increase accordingly. Figure 4 illustrates the cyclic voltammetry (CV) of the electrodes and the charge–discharge behaviour of the devices with different paired electrodes.

Obviously, a rechargeable battery consists of two battery-type electrodes whereas a supercapacitor consists of two capacitive-type electrodes. In addition, there are also many other choices for the selection and pairing of electrodes. Typically, ion capacitors are fabricated by pairing a battery-type electrode and a capacitive-type electrode, and supercapatteries are assembled with pseudobattery-type electrodes. However, there may still be many devices that cannot fit into any of the above terminologies; then the term supercapattery may extend to describe all the energy storage devices apart from the rechargeable battery and supercapacitor. Figure 5 presents the classification of energy storage devices with various selections and

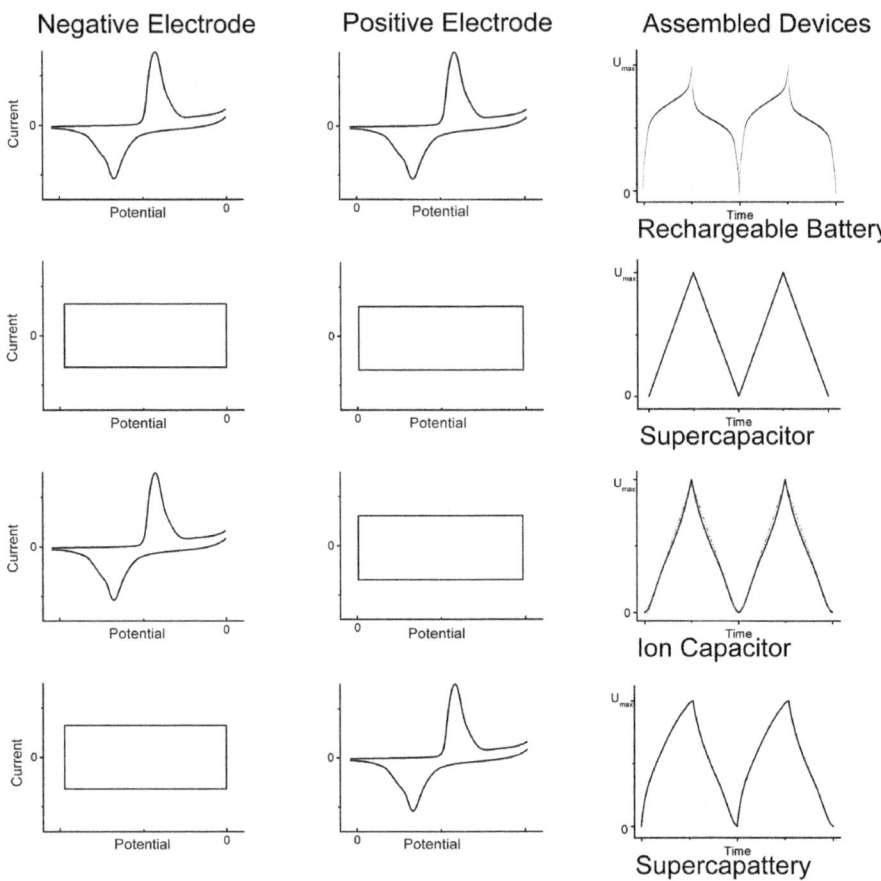

Figure 4   Schematic illustration of selecting and pairing various electrodes to assemble different types of energy storage devices.

pairings of the same or different electrodes. Further discussion on the detailed mechanisms is given in the following sections.

2.2   *Pseudobattery-type Electrode Materials*

Recent research efforts have been devoted to advancing electrode materials for energy storage devices, most often involving lithium- or sodium-based intercalation materials,[37–41] carbon-based materials,[28,42,43] metal oxides,[23,36,44,45] conducting polymers[46,47] and, in recent years, metal–organic frameworks (MOFs).[48–50] In order to achieve high electrochemical performance, optimisation of the material structure and morphology is crucial. The nanostructure morphology of the electrode materials can bring drastic variations in their electrochemical properties. Owing to the high surface-to-volume ratio of the nanostructure, the specific surface area of the electrode material can be increased and consequently enhance the electrochemical

Energy Storage Devices	Supercapacitor					Supercapattery					Battery
	EDLC		Pseudocapacitor			Supercapattery				Ion capacitor	
Electrode Materials	CM	CM	CM	PCM	PCM	CM	PCM	PBM	PBM	CM	BM
	I	II	II	I	II	II	II	I	II	II	II
	CM	CM	PCM	PCM	PCM	PBM	PBM	PBM	PBM	BM	BM

CM: capacitive material
PCM: pseudocapacitive material
BM: battery material
PBM: pseudo battery material
I: symmetric device using same electrode material
II: asymmetric device using different electrode material

Figure 5 Classification of energy storage devices with various selection and pairing of the same or different electrodes.

performance of the storage device. The nanostructure morphology of different materials may vary significantly depending on their material composition, crystal structure and manufacturing method. The various structures of different electrode nanomaterials are depicted in Figure 6.[51] Compared with traditional carbon-based materials, pseudocapacitive materials can obviously enhance the specific capacitance and energy density of the supercapacitors *via* interfacial Faradaic reactions to store energy. Previously, many book chapters and review articles have explained the various electrode materials and their suitability for different types of energy storage device fabrications. However, there is no appropriate distinction of these materials for the fabrication of supercapatteries. Hence, in this chapter, our main focus is on the pseudobattery-type electrode materials and their electrochemical performance as an electrode for supercapattery.

As discussed in Section 2.1.1, many materials, such as NiO,[19,20] Co_3O_4,[21] $Ni(OH)_2$[52] and $CoHPO_4$,[53] were presented in the literature as pseudocapacitive materials, which is technically inaccurate.[24] These materials do not exhibit capacitive behaviour of rectangular CV and linear charge–discharge curves like carbon-based materials. On the other hand, their Faradaic reactions are not dominated by intercalation/de-intercalation mechanism like a lithium/sodium ion battery, but are controlled by diffusion and absorption on the electrode surface. 'Pseudo' means almost and approaching; therefore, these materials, which have battery-like behaviour but no intercalation or huge structure changes caused by alloying and conversion, should be named pseudobattery-type materials. Metal oxides and phosphates are the most popular pseudobattery-type materials and their storage mechanisms are introduced in the following sections.

2.2.1 Metal Oxide-based Materials. It is well accepted that metal oxides are the most promising materials for energy storage devices owing to their high theoretical capacities (NiO with 359 mAh g^{-1}/2584 F g^{-1}, Co_3O_4 with 445 mAh g^{-1}/3560 F g^{-1}), controllable structures and simple fabrication methods.[54–56] Nickel oxide has been widely investigated because of its low cost, easy synthesis and environmental friendliness.[34] Figure 7a shows the CV graph of NiO in 1 M KOH electrolyte at different scan rates.[57] Two obvious redox peaks indicate the electron transfer and the valence change of the nickel during the process. The peak shift with increasing scan rate exhibits the quasi-reversible reaction of nickel oxide. The Faradaic reaction of NiO in alkaline electrolyte is

$$NiO + OH^- \rightleftharpoons 2NiOOH + e^- \tag{1}$$

The maximum capacitance (2018 F g^{-1} at 2.27 A g^{-1}) of NiO was obtained by Lu *et al.*[20] and the rate capacity is over 76% (1536 F g^{-1} at 22.7 A g^{-1}). NiO also shows stable electrochemical activity in an organic electrolyte (1 M $TEABF_4$; see Figure 7b).[34] The assembled symmetric device demonstrated a wide potential window of 2 V with a good energy density of 19.4 Wh kg^{-1}.

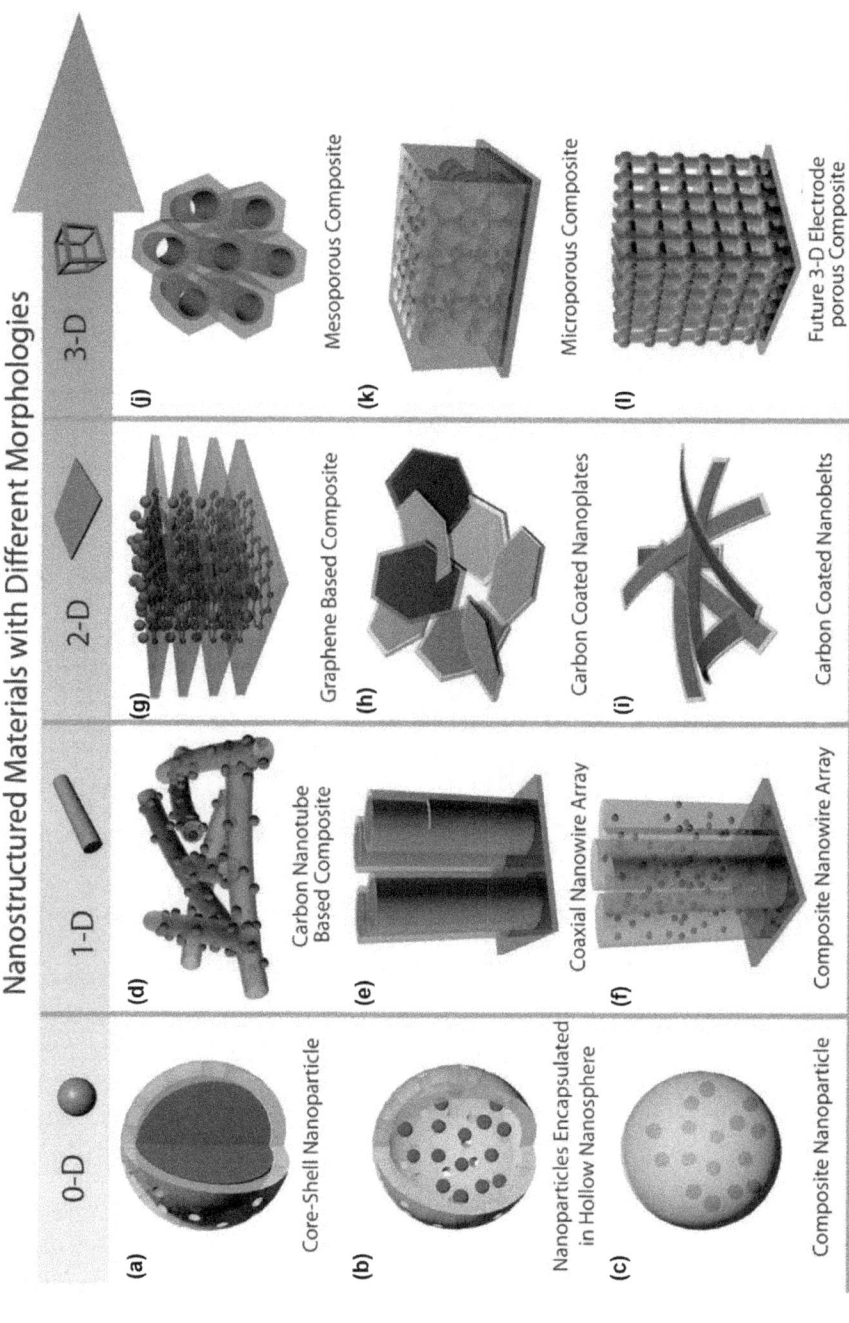

Figure 6 Schematic of heterogeneous nanostructures based on (a) 0D, (b) 1D, (c) 2D, and (d) 3D nanostructures. Reproduced from ref. 51 with permission from The Royal Society of Chemistry.

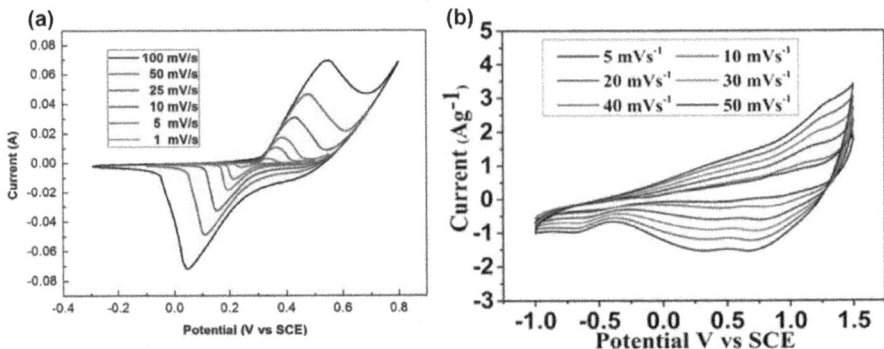

Figure 7 Cyclic voltammograms of (a) NiO film electrode at different scan rates in 1 M KOH aqueous solution. Adapted from ref. 57 with permission from The Royal Society of Chemistry. (b) NiO/CFC nanostructure at different scan rates in 1 M tetraethylammonium tetrafluoroborate/propylene carbonate (TEABF$_4$/PC) organic electrolyte.[34]

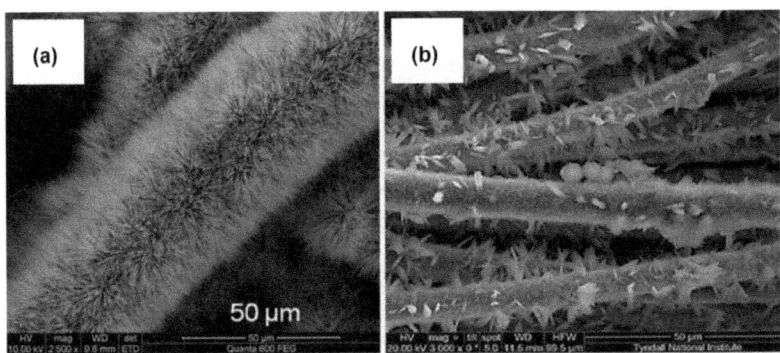

Figure 8 Different nanostructured cobalt oxide materials. (a) Cobalt oxide nanowires. Reproduced from ref. 18 with permission from American Chemical Society, Copyright 2012. (b) Cobalt oxide nanoflowers. Reproduced from ref. 35 with permission from The Royal Society of Chemistry.

Cobalt oxide has attracted much attention owing to its various nanostructure possibilities (see Figure 8)[21,35] and excellent reversible redox behaviour during the chemical reactions. Co$_3$O$_4$ nanotubes displayed a good specific capacitance of 574 F g^{-1} in 6 M KOH solution,[58] and Co$_3$O$_4$ nanowires showed a high specific capacitance of 1525 F g^{-1}.[21]

Instead of single metal oxides, binary transition metal oxides have attracted attention as active materials for energy storage devices as they possess multiple oxidation states to realise multiple redox reactions. Many binary transition metal oxides have been reported to exhibit better electrochemical properties than single-component metal oxides, such as NiFe$_2$O$_4$,[59] FeCo$_2$O$_4$,[60] NiCo$_2$O$_4$[61] and NiO–In$_2$O$_3$,[23] with high specific capacitances of

Figure 9 (a) SEM image of $FeCo_2O_4$ nanoflowers on nickel foam after calcination at 400 °C for 2 h in air. Reproduced from ref. 60 with permission from American Chemical Society, Copyright 2014. (b) SEM image of NiO–In_2O_3 microflower (3D)/nanorod (1D) hetero-architecture. Reproduced from ref. 20 with permission from The Royal Society of Chemistry. (c) SEM image of $NiCo_2O_4$ nanoflowers. Reproduced from ref. 62 with permission from The Royal Society of Chemistry.

1135.5 $F\,g^{-1}$, 2445 $mA\,h\,g^{-1}$, 506 $F\,g^{-1}$ and 766.65 $C\,g^{-1}$ (1096.8 $F\,g^{-1}$), respectively. The materials show unique performance owing to the coexistence of two different cations in highly porous nanoflake/nanoflower structures (see Figure 9).[23,60,62] The surface area of binary metal oxides may be larger than that of single metal oxides; therefore, it may enhance the overall electrochemical performance of the electrode materials.

Among different types of binary systems, $NiCo_2O_4$ is one of the most investigated materials because of its much better electrical conductivity and higher electrochemical activity compared with monometallic nickel oxide or cobalt oxide. However, owing to the powder form of this material, to fabricate the bulk electrode requires a polymer binder, which hinders the electron transport from the oxide materials to the current collector. Hence the ion transport rate in the electrode and electrolyte is limited. To resolve this issue, growing the materials over a conductive substrate becomes an attractive choice. Yang and co-workers[63] reported the fabrication of self-supported Ni–Co oxide nanowires grown on TiO_2 nanotubes. By controlling the molar ratio at 1:1, the specific capacitance of a single electrode was calculated to be 2545 $F\,g^{-1}$, and the theoretical capacitance was calculated to be 3108 $F\,g^{-1}$. A symmetric device based on this material was also investigated. Owing to the limited diffusion space in the two-electrode system, the laboratory-made cell demonstrated only 187 $F\,g^{-1}$ at a current density of 1 $A\,g^{-1}$.

Since both nickel and cobalt are heavy metals and are harmful to the environment, the abundant, cheap and environmentally friendly metal iron is an ideal candidate to replace nickel or cobalt in $NiCo_2O_4$. Owing to serious aggregation and the low specific surface area of active sites, both $NiFe_2O_4$ and $FeCo_2O_4$ were grown on some three-dimensional conductive substrates such as nickel foam or carbon fibre cloth. Yu and co-workers fabricated high-performance $NiFe_2O_4$ nanoparticles over carbon cloth with a high specific capacitance of 1032 $F\,g^{-1}$ in H_2SO_4 and 871 $F\,g^{-1}$ in KOH.[59] In addition, after 3000 cycles, the capacitance remained >91% in both electrolyte systems.

Furthermore, in the case of a poly(vinyl alcohol)–H_2SO_4 gel electrolyte-based symmetric device, a high energy density of 2.07 mWh cm^{-3} was calculated at 2 mA cm^{-2}. Even at a high current density of 10 mA cm^{-2}, the energy density remained at 0.56 mWh cm^{-3}.

Compared with carbon materials, metal oxide materials give a higher specific capacitance because of the different mechanism of chemical redox reactions. However, the surface areas of metal oxide materials are limited to 300 m^2 g^{-1}, and are mostly around 100 m^2 g^{-1} or even less. Owing to the large atom size, aggregated particles formed by the fabrication method and the small pores in the layer-by-layer structure, increasing the surface area becomes a major challenge for metal oxide materials. Recently, much effort has been focused on developing core–shell structured materials such as $MnCo_2O_4$@Ni(OH)$_2$[64] and $NiCo_2S_4$ nanosheets@$NiCo_2S_4$ nanotube.[65] These hybrid electrodes exhibit a significant increase in the surface area, which further improves the electrochemical properties of the material.

2.2.2 Metal Phosphate-based Materials.

Metal phosphate-based materials have huge potential as electrode materials for energy storage devices. Phosphorus can react with most of the elements in the Periodic Table to form various phosphides and phosphates. Many lithium metal phosphates have already been commercialised in battery manufacturing,[66,67] and nickel and cobalt phosphates have been widely investigated in the past decade as positive electrodes for hybrid devices. The phosphates present good ion conductivity and charge storage capacity due to their open structure, with large channels and cavities,[68] and good chemical stability due to the covalent P–O bonds. In general, metal phosphate electrodes are synthesised *via* simple hydrothermal processes with controllable temperature, fabrication time and other conditions. Various structures can be obtained, such as nanoparticles and nanoflakes, as can be seen in Figure 10.[2,69] Other methods, such as chemical precipitation and aqueous-based reflux, are also widely applied owing to the low cost and simplicity of the fabrication processes and the structure and morphology can be modified by an extra calcination process to obtain highly stable phosphates.[70,71]

The phosphate-based materials exhibit a pseudobattery-type behaviour similar to the oxide-based materials. The Faradaic reaction of a metal phosphate in an alkaline electrolyte is as follows:

$$M_3(PO_4)_2 + OH^- \rightleftharpoons M_3(PO_4)_2\,OH + e^- \tag{2}$$

$$M_3(PO_4)_2\,OH + OH^- \rightleftharpoons M_3(PO_4)_2\,(OH)_2 + e^- \tag{3}$$

$$M_3(PO_4)_2\,(OH)_2 + OH^- \rightleftharpoons M_3(PO_4)_2\,(OH)_3 + e^- \tag{4}$$

where M represents a transition metal such as Ni, Co, *etc*. Previously, high specific capacitances of 1876 and 1578 F g^{-1} have been achieved for nickel phosphate and cobalt phosphate.[2,72] The nickel–cobalt binary phosphates demonstrate   improved   electrochemical   performance   owing   to   the

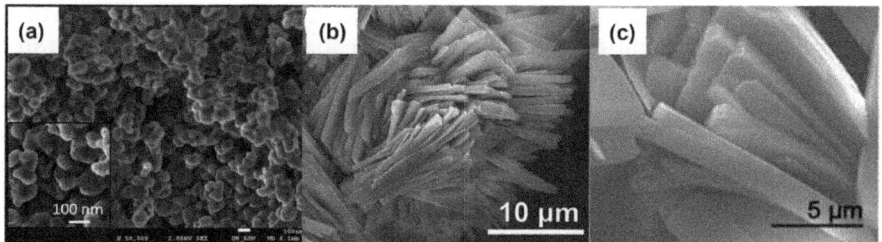

Figure 10 (a) SEM image of nickel phosphate nanoparticles. Reproduced from ref. 69 with permission from The Royal Society of Chemistry. (b) SEM image of cobalt phosphate nano/microflakes. Reproduced from ref. 22 with permission from American Chemical Society, Copyright 2016. (c) SEM image of nickel phosphate nano/microflakes. Reproduced from ref. 79 with permission from American Chemical Society, Copyright 2018.

synergistic effect of nickel–cobalt species. Thus, the $Ni_3P_2O_8$–$Co_3P_2O_8$ nano/microflower achieved 1980 Fg^{-1} in 6 M KOH solution with excellent cyclic stability of over 90% after 1000 cycles.[73] However, doubts remain in understanding the exact electron-transfer mechanism during the redox reaction, which warrants further *in situ* characterisations to understand the behaviour of the phosphates in the chemical redox process.

2.2.3 Other Pseudobattery Materials and Summary. Other materials, such as sulfides and phosphides, have also been studied in the past decades.[74,75] An Ni_3S_2-coated indium tin oxide core–shell structure exhibited an excellent capacitance of 1865 Fg^{-1} and the assembled symmetric devices delivered a maximum energy density of 1.02 $mWh\,cm^{-3}$.[74] $Ni_{1-x}Zn_xS$ multilayers showed a high capacitance of 1815 Fg^{-1} and the complete cell demonstrated a high specific energy of 38.9 $Wh\,kg^{-1}$.[75] In addition, amorphous NiP delivered a capacitance of 1597.5 Fg^{-1} at a current of 0.5 Ag^{-1}, and ~50% of the capacitance remained when the density increased to 8 Ag^{-1}. An asymmetric supercapattery was assembled using Co_2P nanoflowers as the positive electrode and graphene as the negative electrode. The device demonstrated a high specific energy of 24 $Wh\,kg^{-1}$ with 97% retention after 6000 charge–discharge cycles.[76] Interestingly, fluorides and chlorides offer high theoretical specific and volumetric capacities so they may be employed as supercapattery electrodes.[77,78] Therefore, pseudobattery-type materials have a huge potential for supercapattery applications owing to the abundant resources, high theoretical capacity, easy fabrication process, various morphology, good cyclic stability, low toxicity and environmental friendliness.

2.3 Supercapattery Performance

Different nanostructured electrode materials have been intensively investigated for application as electrodes for the fabrication of supercapattery cells.

However, to bring material studies to commercial products is a long process and, therefore, fabrication of complete supercapattery cells in the laboratory started to attract more attention in recent years. Both symmetric (using the same material) and asymmetric (pairing different materials) supercapattery cells have been widely reported. For instance, the $NiO–In_2O_3$ symmetric supercapattery showed a good specific energy of 26.24 Wh kg^{-1} with an excellent retention of 79% after 50 000 charge–discharge cycles.[23] A $Co_3(PO_4)_2$/activated carbon asymmetric cell demonstrated a high energy density of 1.17 mWh cm^{-3} (29.29 Wh kg^{-1}) and a high power density of 187.5 mW cm^{-3} (4687 W kg^{-1}).[22] An $Ni_3(PO_4)_2$/activated carbon asymmetric cell exhibited a high specific energy of 33.4 Wh kg^{-1} at a specific power of 399 W kg^{-1} and, even at a high power of 2058.7 W kg^{-1}, the energy remains 16 Wh kg^{-1}.[79] Generally, the asymmetric cells give better electrochemical properties owing to the combination of two different materials as electrodes, which can easily broaden the potential window and enhance the performance of the complete cell.

2.3.1 Selection of Electrode Materials and Electrolytes. Since one specific material normally displays either a positive or a negative working potential in a specific electrolyte, a large potential range may be obtained by combining a positive potential material and a negative potential material. Notably, supercapattery has attracted much interest because it combines two different mechanisms in one system. The benefit of building a hybrid device is to obtain high energy from the battery-type/pseudobattery-type material and couple this with the ability to deliver high power from supercapacitor-type material. Activated carbon is now the most popular electrode material for both hybrid supercapatteries and ion capacitors owing to its excellent electrochemical performance in inorganic, organic and ionic electrolytes. Interestingly, the material behaves differently in different electrolytes, which is related to the different reaction mechanism, diffusion efficiency and charge-transfer efficiency. Therefore, it is crucial to select an electrolyte that is suitable for both electrode materials. Overall, aqueous electrolytes are widely used in various energy storage devices owing to their high conductivity and high ionic concentration.[80] However, aqueous electrolytes suffer from a small working potential window, which limits the energy capability. Organic electrolytes can offer a wider working potential, but may raise serious safety issues, whereas ionic liquids have an acceptable conductivity and non-flammable nature, but are not cost-effective.[81] Therefore, the understanding and optimisation of electrode pairings and electrolyte selection need to be explored further through fundamental studies on their electrochemical mechanism.

2.3.2 Comparison of Ion Capacitors and Supercapatteries. Much effort has been placed on studying ion capacitors and hybrid supercapatteries. Both of these devices belong to the supercapattery group (as discussed in

Table 1 Comparison of ion capacitor and supercapattery.

Specification	Ion capacitor	Hybrid supercapattery
Anode/positive electrode	Ion adsorption cathode	Metal oxide/sulfate/ phosphate, *etc.*
Cathode/negative electrode	Li/Na-doped electrode	Activated carbon or other material with negative window
Electrolyte	Organic electrolyte with dissolved lithium/sodium ion salt	Inorganic aqueous solution $KOH/NaOH/H_2SO_4$
Mechanism	Intercalation/de-intercalation	Non-capacitive Faradaic redox reaction

Sections 2.1.1 and 2.1.2), but the energy storage mechanisms and electro-chemical properties are significantly different. In general, the hybrid ion capacitor couples a high-capacity bulk intercalation-based battery-style negative electrode (cathode) and a high-rate surface adsorption-based cap-acitor-style positive electrode (anode). The anode is usually an Li/Na-doped material, the cathode materials are usually carbon based and the electro-lyte is an organic electrolyte with dissolved lithium/sodium ion salts. The energy storage mechanism in an ion capacitor is dominated by an inter-calation/de-intercalation mechanism like that in a lithium/sodium ion bat-tery. However, the hybrid supercapattery is another type of storage device, which has no intercalation/de-intercalation ions in the system. In super-capattery, the energy storage mechanism is diffusion controlled through a reversible redox reaction on the electrode surface. Therefore, an ion cap-acitor exhibits several times higher energy than a hybrid supercapattery, but the latter can survive more than 10 000 charge–discharge cycles. Table 1 lists the differences between the two different supercapatteries.

2.4 Prospects and Future

The study of supercapattery has entered a high-speed development period through the development of electrode materials and complete cells. How-ever, the following aspects need to be considered during the development of a complete cell: packaging volume, packaging materials, electrolyte ma-terials, operating temperature range, environmental friendliness, large-scale production, toxicity, cost, safety and reliability. The specific energy/energy density would decrease tremendously if the whole packaging mass and volume were to be considered when comparing with the values calculated from the mass and area of the active materials and electrodes. Another issue is that previous research was carried out under ideal repeatable process conditions; for instance, charging/discharging measurements were made under a full- or half-voltage window set-up. However, it is sometimes im-possible to achieve a fully charged or discharged condition in real life.

Therefore, more investigations under realistic conditions are needed, including various partial charge–discharge, self-leakage current testing and performance over a wide range of temperatures.

3 Superconducting Magnetic Energy Storage (SMES)

3.1 Basic Aspects of SMES

Superconducting magnetic energy storage (SMES) is a technology used to store energy in a magnetic field, which can persist for a long period of time in standby mode. It works based on a non-dissipative current passing through a superconducting wire, which generates a magnetic field. In the normal state, a wire has a resistance, R, obeying Ohm's law ($V = RI$), and the power dissipated scales with the square of the current, I ($P = RI^2$). In a superconducting state, the resistance of a wire is virtually zero and does not dissipate energy, hence an electrical current can flow without power dissipation. According to basic principles, an electrical current generates a magnetic field in which energy is stored. The circuital law allows us easily to obtain the magnetic field resulting from a current-carrying wire according to the equation

$$\oint B \mathrm{d}l = \mu_0 I \tag{5}$$

where I is the electrical current, B is the magnetic flux density and μ_0 is the permeability of vacuum. As a current-carrying wire produces a magnetic field in its surroundings, the energy stored in the magnetic field is represented by

$$E = \tfrac{1}{2} L I^2 \tag{6}$$

where L is the inductance. The relationship in eqn (6) shows that at a fixed inductance the stored energy scales with the square of the current. Therefore, a high current allows the storage of large amounts of energy, which can be obtained in superconducting wires. The value of the inductance is closely linked to the geometry of the SMES, which commonly employs inductors with solenoidal or toroidal shapes. Typically, superconducting wires are designed to carry from a few hundred to thousands of amps. Therefore, it is easy to verify that for a power of 10 MW for 10 s, leading to a storage capacity of 100 MJ, an inductance of 100 H is required for a current of 1000 A. Although more complex to assemble, a toroidal shape is preferred for high energy density as the magnetic field is mostly confined to the toroidal region. The superconductivity is a collective behaviour of electrons in a solid material and is dependent on the temperature.

A SMES usually consists of three components, as illustrated schematically in Figure 11: a superconducting unit, a cryostat system and a power conversion unit. Typically, superconducting wires are made of niobium–titanium and require a temperature as low as 1.8 K to operate. The superconducting unit is the inductor in the system and can be built as a solenoid or a toroid. The cryostat is built so as to have very low heat losses and uses

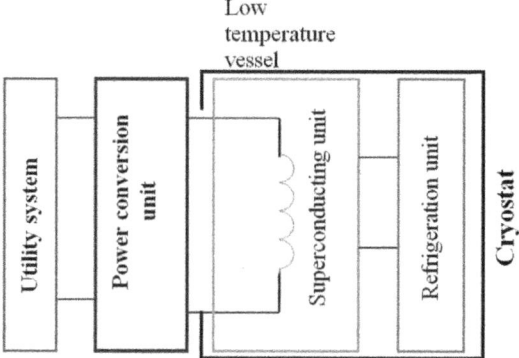

Figure 11 Schematic diagram of an SMES device showing the three elementary units connecting to the utility system.

vacuum shielding to minimise the heat transferred by conduction. The need for a low temperature means that in standby condition the energy consumed to maintain the unit operation contributes negatively to the overall efficiency of the SMES. The power unit is responsible for storing and retrieving energy from the superconducting unit and for interfacing with the utility system.

3.2 State-of-the-Art, Trends and Challenges for SMES

Some of the first ideas for SMES were proposed by Ferrier in France in 1969. The potential for applications of SMES has been investigated since 1970 and has been considered for load levelling, frequency support (spinning reserve) during loss of generation, enhancing transient and dynamic stability, dynamic voltage support, improving power quality and increasing transmission line capacity. SMES devices are mostly important for application in the field of transmission and distribution grids and are dedicated to improving the voltage stability of a utility network, uninterruptible power supply (UPS), pulse power sources for dedicated applications and flexible ac transmission (FACTS).[82] Research started in 1971 at University of Wisconsin, leading to the first SMES device. Other devices have been developed by Hitachi (1986), ISTEC (1998), Wisconsin Public Service Corporation (2000) and ACCEL Instruments GmbH (2005) and currently SMES units with over 100 MW are operational.[83] The advantages of SMES are the high-energy storage efficiency (>97%), rapid response in the milliseconds range, reliable energy output and high cycle life. However, SMES cannot replace large-scale storage units.

The high energy efficiency of SMES devices is associated with the fact that under static conditions there are virtually no losses in the superconducting unit and the relatively small losses are associated with the dynamics of storing and retrieving energy from the coils, the power conversion unit and the cryostat system. The reliability of the energy output of the device is related to the fact the energy state and internal resistance during discharge do not depend on the charge state of the SMES device. This contrasts with chemical processes in batteries, where the internal resistance is strongly dependent on the charge state; hence the electrical performance of the battery is strongly dependent on the charge state, with high internal resistance in the depleted energy state. SMES devices have high durability and operate easily in constant or fully charge–discharge modes; some units are designed to operate at high cycling rates for periods of up to 30 years. SMES has the fastest response time for a storage device and is much better than supercapacitors.[84] The greatest drawback for this technology is the high cost.

SMES devices are mostly employed in large industries where voltage stability and power quality are required. Commercial units are available with typical power ratings of 1–10 MW with storage times of a few seconds. There is also current research on large SMES devices with power ranging from 10 to 100 MW and a comparison with other technologies is illustrated in Figure 12. SMES shows the best response for known storage devices and has much greater output power than the supercapacitor.[84]

Progress has been made on the power conversion unit.[85] Two types of power converter unit are currently used: the current-source converter (CSC) and voltage-source converter (VSC). The CSC interfaces with the ac system and the charge–discharge superconducting coil. The VSC interfaces with the ac system and a dc–dc chopper to charge–discharge the superconducting coil. A number of parameters are taken into account for the design of SMES, including coil configuration, stored energy, structure of the coil and operating temperature. Usually, a compromise is necessary when considering the parameters of specific energy ($J\,kg^{-1}$), Lorentz forces, stray magnetic field and minimisation of losses for stable and cost-efficient SMES systems. The design of the superconducting inductor restricts the critical parameters such

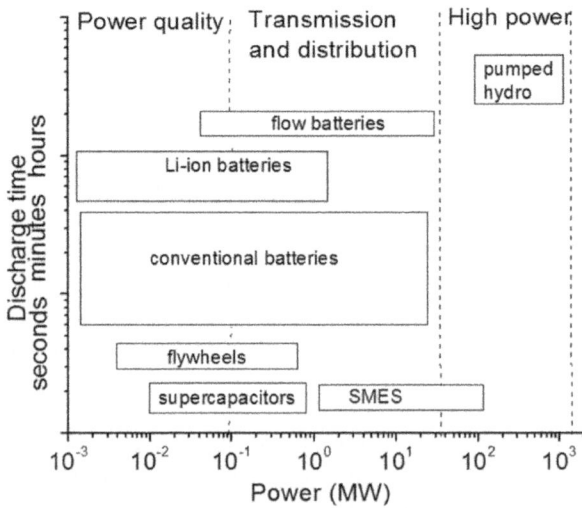

Figure 12   Comparison of discharge times for different energy storage technologies. SEMS can store the highest power for short periods of time.

as coil inductance (L), maximum operating voltage (V_{max}) and maximum current (I_{max}), which determine the maximum power/energy rating for the SMES device. The operating temperature of the SMES is usually a compromise between cost and operational requirements.

4 Flywheels, Flywheel Batteries and Synchronous Condensers

Flywheels are used to store energy in the form of rotational kinetic energy. Energy storage using flywheels is used in a range of applications, including electric vehicles, conventional vehicles, uninterruptible power supplies, power smoothing and integration of renewable energy. In the following sections, we overview the basic aspects of mechanical energy storage, flywheels and synchronous condensers, including current and future trends in this area.

4.1 Fundamental Theory of Mechanical Energy Storage

Mechanical energy can be divided into two types: potential energy and kinetic energy. Potential energy is the energy possessed by a mass due to its location with respect to other masses, for example, due to the height of a mass above the ground. Kinetic energy is the energy possessed by a mass due to its motion with respect to another mass, for example, due to the speed at which a skier travels down the incline of a hill. The study of various forms of energy led to the principle of mechanical advantage and then to the law of conservation of energy: provided that no non-conservative, *e.g.* friction, or external forces act on a system, the sum of the potential and mechanical energy is conserved.

Similar observations and study eventually led to the much wider First Law of Thermodynamics, which states that the total energy in a system may be neither created nor destroyed but instead may be transformed from one form to another. It follows that each form of energy can be thought of as a form of energy storage. For example, a mass m at the top of a hill will have a potential energy $U_E = mgh$, due to its elevated height, h, above the base of the hill (where g is the acceleration due to gravity). If this mass slides down the hill, the energy possessed as kinetic energy, $K_E = \frac{1}{2}mv^2$, at the bottom of the hill equals the potential energy the mass possessed at the top of the hill (where v is the velocity of the mass and we presume that none of the energy is converted to any other form of energy, *e.g.* heat). Furthermore, provided that no non-conservative or external forces act on a system, the mechanical energy will be conserved and, therefore, the kinetic energy may be converted back to potential energy.

There are many examples of energy stored as potential energy and kinetic energy. However, as you may have experienced from cycling on a bicycle, it is much easier to maintain your height (and therefore your potential energy) when you are sitting on your bicycle admiring the view from the top of a hill than it is to maintain velocity (and therefore kinetic energy) when you are cycling along a horizontal road at a high speed. Even though kinetic energy can be more accessible and easier to perceive, it is often the case that energy storage methods that rely on kinetic energy suffer from greater parasitic energy losses than techniques that use potential energy. It follows, when storing energy for extended periods of time, that kinetic energy storage methods often have low efficiency. However, kinetic energy can be more accessible than potential energy, it can be easier to appraise and it can be coupled to systems that are based on kinetic energy. In addition, such coupling of kinetic energy storage to systems has historically been used to improve system stability through intrinsic responses that give reliable negative feedback, which is proportional to the rate of change of velocity. Recently, there has been a rebirth in the use of this old technology for the stabilisation of electricity grids.[86]

4.1.1 *Linear Energy* Versus *Rotational Energy.*

Linear momentum and rotational momentum correspond to stored linear kinetic energy and rotational kinetic energy, respectively. For example, assuming negligible drag and friction, if a cyclist attempts to climb a steep hill and the maximum power capability of the cyclist is less than the rate of gain of potential energy, the bicycle will slow as it ascends the hill. This slowing of the bicycle is the conversion of kinetic energy to potential energy.

In such a scenario, the effect of addition of mass to the bicycle is of interest. For instance, if we consider a similar case but where the cyclist is not pedalling – *i.e.* not doing any work – all of the energy going into gaining potential energy will come from the kinetic energy of the bicycle. The rate at which the bicycle slows as it ascends the hill is independent of the mass of the cyclist. The height the bicycle reaches is dependent only on its initial speed, since both the stored linear kinetic energy, $K_E = \frac{1}{2}mv^2$, and the potential energy gained, $U_E = mgh$, are proportional to the mass, m. However, if a greater fraction of the bicycle's mass was placed in the spinning wheels, or if the diameter of the wheels was increased, the bicycle would initially have more rotational kinetic energy, $K_R = \frac{1}{2}I\omega^2$, and therefore with the same initial speed the bicycle would roll higher up the hill. (Note: I is the moment of inertia, which is $I_w = \frac{1}{2}m_w r_w^2$ for a solid disk wheel of mass m_w and radius r_w or $I_w = m_w r_w^2$ for a wheel with all its mass at its rim, and ω is the angular velocity of the wheels, *i.e.* the rate of change of angle.) There are, of course, disadvantages to having heavy or large wheels in contact with the ground on a moving vehicle and therefore most early steam traction engines, and right up to today's modern combustion engines, have an extra wheel that is driven by the engine and is not in contact with the ground.

This wheel is called a flywheel and it rotates at an almost constant rate that is determined by the average input and output torques. When the input torque is less than the output torque, the flywheel slows, converting some of its rotational energy to extra output energy that compensates for the engine power being temporarily less than the required output power. Similarly, whenever the engine's output power exceeds the power required for the task, the flywheel absorbs this energy, increasing stored rotational kinetic energy by speeding up slightly.

4.2 Basic Aspects of Flywheels

Flywheels are an old technology that has recently begun to find use in modern energy storage applications for the efficient delivery of power, rapid absorption and reuse of energy and stabilisation of electricity grids.[86–89] Flywheels, flywheel batteries and synchronous condensers all have similar physical characteristics but, depending how each system is coupled to the power supply/output, they can behave very differently. The following section provides an overview of the physics that governs this behaviour.

4.2.1 Rotational Kinetic Energy in Flywheels. Flywheels are used for a range of applications, including opposing changes in direction (*e.g.* in gyroscopes, satellite stabilisers and spinning tops), smoothing of engine output, delivery of intermittent pulses and storage of energy. The rotational kinetic energy of a flywheel can be written in terms of angular velocity ω as $K_R = \frac{1}{2}I\omega^2$ or in terms of frequency as $K_R = 2\pi^2 If^2$, since $\omega = 2\pi f$, where f is the revolutions of the flywheel per unit time. It follows that the stored energy is not proportional to the frequency (or angular velocity) but instead increases with the square of the frequency, as illustrated in Figure 13a. The rate of charge or discharge of a flywheel's energy, *i.e.* the net power transfer $P = dK_R/dt$ can be defined as

$$P = I\omega \frac{d\omega}{dt} \qquad (7)$$

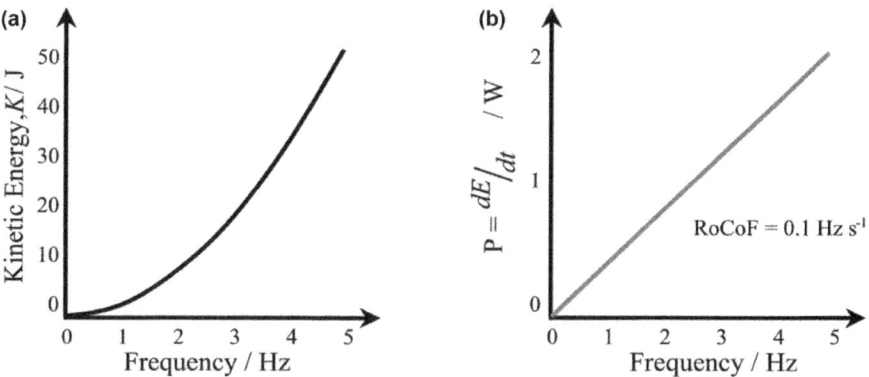

Figure 13 Illustration of the relationship between (a) rotational energy stored and (b) transfer rate of energy (at constant RoCoF of 0.1 Hz s^{-1}) to/from a flywheel of inertia of 0.1 kg m^2 *versus* frequency (in cycles per second, Hz).

where dω/dt is the rate of change of angular velocity, or as

$$P = 4\pi^2 If \frac{df}{dt} \tag{8}$$

where df/dt is the rate of change of frequency (RoCoF) of the flywheel. Therefore, the RoCoF and hence the rate of change in angular velocity of the flywheel are proportional to the rate of energy added or subtracted but inversely proportional to the frequency:

$$\frac{df}{dt} = \frac{P}{4\pi^2 If} \tag{9}$$

For example, for a specific frequency, *i.e.* a specific angular velocity, a fast energy transfer rate to/from the flywheel will result in a significant RoCoF, but the higher the frequency of the flywheel the smaller is the RoCoF, *i.e.* the smaller the rate of change of angular velocity. This response is illustrated in Figure 13b, where it can be seen that for a specific RoCoF, the rate of energy transfer is proportional to frequency and therefore the angular velocity of the flywheel.

4.2.2 Use of Flywheels with Reciprocating Engines. The main use of flywheels is for systems that require continuous motion but have a non-continuous source of power. The use of flywheels has been around for a long time (*e.g.* pottery wheels from the fourth millennium BCE[90]); however, a significant increase in the use and development of flywheels occurred at the start of the industrial revolution, coinciding with the invention of the steam engine.

In a steam engine, heat is transformed into mechanical work, resulting in reciprocating motion that is in turn transformed into rotation of a flywheel.[91] The reciprocating motion of the piston and connecting rods is provided by the expansion of pressurised steam within a cylinder. This piston is connected to a crankpin at the end of a crank arm, which acts as a lever to turn the flywheel. Therefore, as the crank arm is rotated the angle between the crank arm and the rod connecting it to the piston changes from 0 to 360° and therefore the magnitude of the torque varies. The piston undergoes two strokes during each rotation of the flywheel. At the end of each stroke of the piston, the flywheel carries the crank past the 'dead centres', giving a continuous motion to the system.

During the expansion stroke, a force on the piston from the steam pressure (see the inset of Figure 14) causes the piston to move out of the cylinder, applying a torque *via* the rods and the crank arm to the flywheel, resulting in rotation. At a flywheel angle of 0°, the torque is zero (as shown in Figure 14) since the crank arm is in line with the connecting rod. The torque is zero even though the pressure, and therefore the expansion force

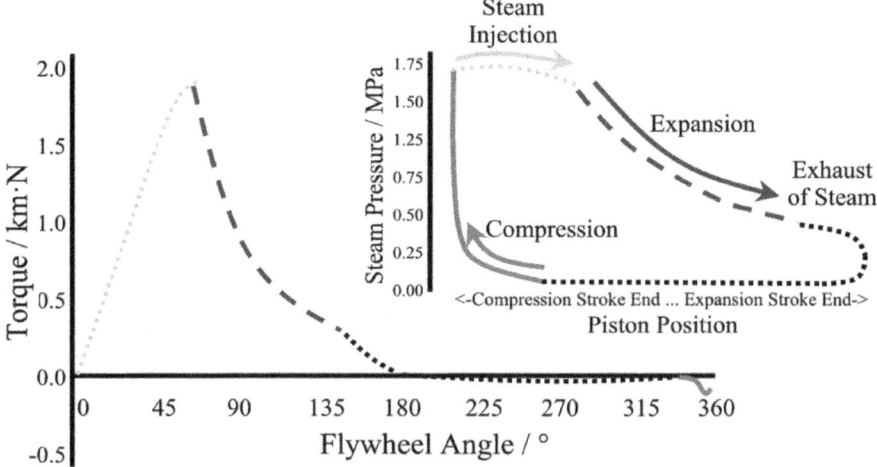

Figure 14 Flywheel torque plotted against angle during a double-piston stroke from a single-acting cylinder of a steam engine operating between 1.7 and 0.9 MPa (where positive and negative torque, respectively, refer to the expansion and compression strokes). In the inset, steam pressure above atmospheric pressure is plotted against piston position in the cylinder. Regions of steam injection, expansion, exhaust and compression are indicated in both plots.
Data calculated from ref. 91.

on the piston, are at a maximum (see the inset of Figure 14). In the same manner as torque is provided through the pedals of a bicycle, as the flywheel angle increases from 0° the torque increases and angular acceleration of the flywheel increases until maximum torque is applied at 90° (see Figure 14). Further increase in the angle beyond 90° causes the torque to decrease. In addition, no more steam is added to the cylinder, causing the pressure of the steam and therefore the torque on the flywheel to decrease rapidly as the steam expands. Nearing a flywheel angle of 180° – as the piston reaches the end of the expansion stroke – a valve is opened to release the remaining steam pressure from the cylinder.

In a single-acting cylinder, *i.e.* where the steam is introduced on only one side of the piston, there would be no driving force to push the piston back into the cylinder. However, since the piston is connected to the flywheel and the flywheel now has angular momentum, a torque will result from the force required to push the piston into the cylinder, *i.e.* from the compression stroke, slowing the rotation of the flywheel. Near the end of the compression stroke, the steam release valve is closed, trapping and compressing the remaining steam into the clearance volume of the cylinder, reducing the shock at the end of the stroke, after which a valve is opened to allow more steam into the cylinder, causing the action to be repeated.

Therefore, there are two strokes – an expansion and a compression stroke – during each flywheel rotation. During each expansion stroke the flywheel rotation is accelerated and during each compression stroke it is decelerated (as plotted in the torque *versus* angle plot of Figure 14). However, overall the flywheel accelerates during each rotation until it reaches a fairly constant rotation rate. As explained in the previous section, the change in angular velocity of a rotating mass is proportional to the energy added but inversely proportional to the angular velocity. Therefore, as the flywheel speeds up, the jolts in angular acceleration and deceleration due to the expansion and compression strokes becomes less apparent and the motion of the flywheel becomes smoother. In addition, the motion of the piston more closely resembles simple harmonic motion, reducing engine wear and tear.

The motion of the system can be made smoother by using a double-acting cylinder that allows steam to be introduced on the other side of the piston as it moves back into the cylinder – so that acceleration maxima occur twice per flywheel rotation – or by adding torque through an additional cylinder acting at 90° to the first. However, the fundamental key to the smooth operation of the engine is the flywheel. The same technology is used in modern internal combustion engines, allowing each piston in a four-stroke engine to deliver significant angular acceleration to the crankshaft (and flywheel) during only one of the four strokes, *i.e.* during half a revolution per two revolutions. That is, positive power is delivered only during the expansion stroke while the intake stroke, exhaust stroke and compression stroke consume power but, thanks to the flywheel, the engine can deliver smooth motion, quietly and free of vibration.

In the same manner that flywheels facilitate systems to ride through sudden changes in power generation, they can also facilitate systems to ride through increases or decreases in workload or power generation and they are, therefore, used for driving of reciprocating pumps, riveting machines and other systems.

As with a bicycle, to store this stabilising energy as kinetic energy the flywheel has to be kept rotating and therefore extra energy has to be generated to overcome the drag and friction of the flywheel. It follows that, although flywheels can increase the system efficiency and reduce wear and tear by facilitating smoother running of systems and delivery of extra power over short periods of time, there is an optimal size of a flywheel for any system: for any particular flywheel design, the bigger the flywheel the greater is the loss of energy added to the system and therefore the greater the overall inefficiency of the system. All rotating masses, such as wind turbines, steam turbines and water wheels, have rotational energy and therefore supply the same benefits that flywheels do. Careful design of such systems can add the benefits of flywheels without significantly increasing system inefficiency.[92] Such stored rotational energy is an intrinsic characteristic of conventional electric power generation and an essential component in the stabilisation of electricity grids.[93]

4.3 Basic Aspects of Synchronous Motors, Generators and Condensers

A synchronous condenser, sometimes called a synchronous capacitor, is a type of machine that is similar to a synchronous motor or generator. Ac motors rotate due to a force from the interaction between a magnetic field and magnetic dipole moment that rotate at the same frequency.[94] There are two types of ac motors: synchronous and induction (which are asynchronous) motors. These machines have a stator, which does not rotate, and a rotor that rotates. In induction motors, a rotating magnetic field in the stator is formed by supplying the stator windings with an alternating current, the frequency of which is an integer multiple of the frequency of rotation of the resulting magnetic field. As this magnetic field rotates relative to the rotor, electromagnetic coupling induces electric currents in the conducting material of the rotor, producing a rotor magnetic dipole moment. The stator and the rotor interact so as to minimise the distance between their north and south poles, forcing the rotation rate of the rotor to approach the rotation rate of the stator magnetic field. However, as the rotation rates approach each other, the induced currents will decrease and therefore the rotor magnetic dipole moment becomes weak. It follows that the rotation rate of the rotor will approach but never reach that of the stator magnetic field.

In a synchronous motor, the magnetic field originates from both the rotor and stator. In many designs, permanent magnets that are attached to the rotor form a magnetic field that rotates if the rotor rotates. Windings in the stator of such machines can be used to produce a rotating magnetic field in the same way as in an induction motor (by exciting the windings using an alternating current). Therefore, if the stator magnetic field is rotating at approximately the same frequency as the rotor, a force will act so as to cause the rotor to rotate at exactly the same frequency as the stator magnetic field.

In such a system, once the rotor is rotating, if the exciting current being sent to the stator windings is disconnected, the angular momentum of the rotor will cause it to continue to rotate. Therefore, the rotating magnetic field of the rotor will pass through the windings of the stator and this field will induce a current that alternates at the same frequency as the magnetic field alternates as it passes through the windings. This alternating current will initially have the same frequency as the current originally used to excite the windings. The electrical energy of this current comes from the rotational kinetic energy of the rotor and, therefore, as this kinetic energy decreases the rotor will slow and the frequency and power of the electricity will decrease. In effect, the motor is now a generator and, if a turbine were attached to the rotor, such a system could be used to generate electricity.

The alternating current passing through the windings in the stator originates from two sources, first from the electricity supplied to the windings and second from the rotating magnetic field inducing a current in the same windings. In short, such a synchronous machine is always both a motor and a generator and when connected to an electricity grid it will either be absorbing energy, so as to increase its stored magnetic and rotational energy, or be providing energy to the grid from its stored energy. Furthermore, if two of these machines are connected to the same electricity grid, increasing the rotation rate of one of the rotors will cause the frequency on the grid to increase, which will, in turn, cause the rotor of the second machine to rotate faster. The synchronous rotation of all such machines is electromagnetically locked to one another. This electromagnetic coupling is the basis of the stable operation of synchronous electricity grids.

Similarly to flywheels used with reciprocating engines, synchronous machines connected to an electricity grid act as a store of energy, smoothing out sudden changes and fluctuations in the supply and consumption of electricity. Excess electricity generation causes these machines to speed up and scarcity of generation or excess load causes them to slow, as illustrated in Figure 15. As the flywheels speed up they absorb energy and as they slow they return energy to the grid at a rate that is proportional to the RoCoF. The rotational energy stored is also dependent on the inertia of the rotors in these machines and, therefore, increasing the sum of inertia coupled to the grid decreases the RoCoF due to temporary excess or scarcity of electricity. That is, greater inertia results in greater stability. However, greater inertia also results in greater losses to friction, drag and other operational inefficiencies.[94–96]

Synchronous condensers are a type of synchronous machine specifically designed to increase grid stability through the addition of extra inertia to the grid.[95] Like many synchronous motors and generators, these machines use electromagnets, excited by direct current, instead of permanent magnets to generate the rotor magnetic field. Therefore, the degree to which the rotor magnetic field crosses the windings of the stator can be altered in real time

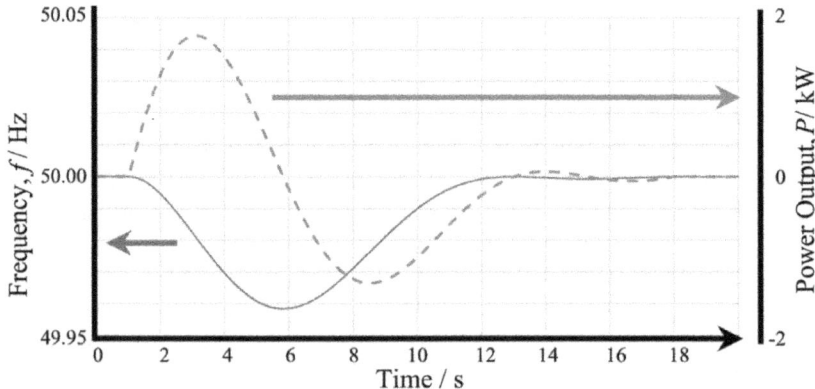

Figure 15 Calculated power output of a synchronous condenser [using eqn (8)] *versus* time during the temporary drop in frequency plotted on the same graph. Note: as the frequency drops, the synchronous condenser instantly delivers extra power to grid, *i.e.* reducing the RoCoF and, as the grid recovers, power is reabsorbed, steadying the frequency.

depending on the requirements of the system. These devices utilise not only the energy stored in the rotation of their flywheel/rotor but also the energy stored in their magnetic field to provide a range of services for stabilisation of electricity supply.

The ac current in the stator windings and the dc current in the electromagnets of the rotor sustain this rotating magnetic field. When the dc excitation current is small, the system behaves similarly to an induction motor. The magnetic field formed by the current in the stator delays the change in direction of current. That is, the machine acts as an inductive load-absorbing reactive power, resulting in a grid with lagging power. However, by increasing the dc excitation current in the rotor electromagnets, the magnetic field can be fully sustained by the rotor magnetic field, resulting in a completely resistive load and power unity. Further excitation of the dc current increases the rotor magnetic field, providing reactive power to the grid. The synchronous condenser acts as a capacitor, providing power factor correction to the grid. Its principal advantage is the ease with which the amount of correction to reactive power can be adjusted by controlling the rotor magnetic field *via* the dc current, but since this current is controlled *via* power electronics, the response rate for stabilisation of reactive power is not instantaneous. Furthermore, since synchronous condensers require their rotor to be always spinning, they have higher energy losses than capacitors and, therefore, capacitors or static synchronous condensers (STATCONs) are often used to provide power factor correction.[95,96] Nevertheless, the natural stabilisation of the grid's real power provided by synchronous condensers in response to RoCoF may be very important in the future as we move away from conventional synchronous electricity generators to grids with high penetration of energy from renewable sources, *i.e.* towards grids with reduced inertia. In addition, synchronous condensers can generate or absorb reactive power, improving the local power factor and voltage of the grid, reducing energy loss in transmission and maximising transmission capacity.

4.4 Current Trends and Challenges for Flywheels

The major advantages of flywheel energy storage systems, compared with other small- to medium-scale energy storage options, are their low wear and tear, resulting in long operational life spans and a predictable and rapid response.[97] These advantages are the result of the intrinsic properties of such systems. Unlike electrochemical batteries, flywheel batteries do not undergo structural changes during charging and discharging. Furthermore, the state of charge of these systems, unlike electrochemical batteries, can be determined directly and reliably from the rotational speed of the flywheel. Since the energy storage capacity of the flywheel does not deteriorate significantly over time and the state of charge is easily determined, the response from the flywheel can be accurately predicted and can, therefore, be faster than from many other systems. Furthermore, direct electromagnetic coupling of systems to the flywheels in synchronous condensers can provide instantaneous feedback that stabilises such systems by reducing RoCoF.

Flywheel energy storage systems can be broken down into two main classifications: low and high speed.[98] Traditionally, flywheels are operated at relatively low speeds (less than 150 Hz). Since the kinetic energy stored in a flywheel is proportional to the flywheel inertia and, therefore, to the mass of the flywheel, such systems often use steel disks. These disks provide high power density but they do not provide high energy density. Stored energy is also proportional to the square of the rotation rate of the flywheel and, therefore, increasing the rotation rate of a flywheel by 40% almost doubles the energy stored. However, increasing the rotation rate of a steel disk often leads to catastrophic failure of the disk,[99] which is disastrous to the destroyed system but also a safety issue for anyone nearby. Furthermore, for flywheels that are heavier and faster, the bearings on which they rotate experience faster wear and tear, resulting in decreased system efficiency and the requirement for H_2 cooling in large systems.

Modern flywheels use lightweight materials such as carbon fibre-reinforced plastics that have good tensile strength, allowing them to rotate at increased frequencies (*e.g.* of the order of 1000 Hz).[100] Therefore, newer systems have increased energy storage density,[97,98] making them ideally suited for energy storage from energy recovery systems in cars, trains and ships,[100–102] UPS backup[87,103] and fast frequency-response stabilisation of electricity grids.[89,104]

New flywheel systems are often buried in the ground in order to contain the flywheel in the case of intact rotor failure and are designed to self-destruct and disintegrate to powder if the tensile strength is exceeded.[105] These flywheels commonly use magnetic bearings and operate in a vacuum,[97,106] greatly reducing wear and tear and energy losses due to friction and drag. However, since these systems are operating in vacuum, they require pumps to maintain the vacuum[87,101] and, for high-speed devices, they often require cooling, either to compensate for heat due to friction or, in the case of some magnetic bearings, to maintain low temperatures for optimum operation.[97,101,106] Therefore, although these recent improvements have increased the durability, reliability and safety of flywheels, they can also add significant running costs. There are possible applications for such systems at wind farms and other locations where curtailed or excess electricity is cheap enough to power the continuous charging of the system to compensate for losses to inefficiency and system operation. Alternatively, they are ideal for frequently storing energy for short durations, returning that energy quickly. Examples of such applications include the traditional uses of smoothing of running of reciprocating engines and storing of energy in riveting machines to be delivered as a pulse, but also modern applications in kinetic energy storage and grid stability.

Response times for flywheel batteries are currently faster than those for electrochemical batteries. However, except in synchronous operation where the response is controlled by RoCoF, the response from flywheel batteries is not instantaneous. Patents on rotating stabilisers for electricity grids[107] have shown that, in standby mode, these flywheels can be run as synchronous machines, providing instantaneous support and stability to the grid but, when needed, their connection to the grid can be redirected *via* power electronics. This would allow the majority of the rotational energy stored in the flywheel to be discharged to the grid over a short period, giving such systems the advantages of both flywheel batteries and synchronous condensers. Such systems and modern flywheels facilitate low-cost systems[108] that exhibit high power and energy density and reliable, safe and long operational lives.[109,110]

References

1. G. Z. Chen, presented at 220th ECS Meeting, Perception of super-capacitor and supercapattery, Boston, October, 2011.
2. H. Shao, N. Padmanathan, D. McNulty, C. O'Dwyer and K. M. Razeeb, *ACS Appl. Mater. Interfaces*, 2016, **8**, 28592.
3. D. P. Dubal, O. Ayyad, V. Ruiz and P. Gomez-Romero, *Chem. Soc. Rev.*, 2015, **44**, 1777.
4. K. R. Bullock, *J. Power Sources*, 1994, **51**, 1.
5. I. H. Henderson and S. G. Ladan, *US Pat.* US3326721A, 1967.
6. B. R. Reddy and D. N. Priya, *J. Power Sources*, 2006, **161**, 1428.
7. V. Etacheri, R. Marom, R. Elazari, G. Salitra and D. Aurbach, *Energy Environ. Sci.*, 2011, **4**, 3243.
8. D. Aurbach, *J. Power Sources*, 2000, **89**, 206.
9. D. McNulty, D. N. Buckley and C. O'Dwyer, *J. Power Sources*, 2014, **267**, 831.
10. M. R. Palacin, *Chem. Soc. Rev.*, 2009, **38**, 2565.
11. S. Ardizzone, G. Fregonara and S. Trasatti, *Electrochim. Acta*, 1990, **35**, 263.
12. J. Zheng, P. Cygan and T. Jow, *J. Electrochem. Soc.*, 1995, **142**, 2699.
13. H. Y. Lee and J. B. Goodenough, *J. Solid State Chem.*, 1999, **144**, 220.
14. T. Brousse, M. Toupin, R. Dugas, L. Athouël, O. Crosnier and D. Bélanger, *J. Electrochem. Soc.*, 2006, **153**, A2171.
15. S. Mohapatra, A. Acharya and G. Roy, *Lat. Am. J. Phys. Educ.*, 2012, **6**, 380.
16. S.-M. Chen, R. Ramachandran, V. Mani and R. Saraswathi, *Int. J. Electrochem. Sci.*, 2014, **9**, 4072.
17. M. Beidaghi and C. Wang, *Adv. Funct. Mater.*, 2012, **22**, 4501.
18. T. Chen and L. Dai, *Mater. Today*, 2013, **16**, 272.
19. L. Fan, L. Tang, H. Gong, Z. Yao and R. Guo, *J. Mater. Chem.*, 2012, **22**, 16376.
20. Z. Lu, Z. Chang, J. Liu and X. Sun, *Nano Res.*, 2011, **4**, 658.
21. R. Rakhi, W. Chen, D. Cha and H. N. Alshareef, *Nano Lett.*, 2012, **12**, 2559.
22. G. Z. Chen, *Prog. Nat. Sci.: Mater. Int.*, 2013, **23**, 245.
23. N. Padmanathan, H. Shao, D. McNulty, C. O'Dwyer and K. M. Razeeb, *J. Mater. Chem. A*, 2016, **4**, 4820.

24. T. Brousse, D. Bélanger and J. W. Long, *J. Electrochem. Soc.*, 2015, **162**, A5185.
25. G. Z. Chen, *Int. Mater. Rev.*, 2017, **62**, 173.
26. M. D. Stoller, S. Park, Y. Zhu, J. An and R. S. Ruoff, *Nano Lett.*, 2008, **8**, 3498.
27. S. Vivekchand, C. S. Rout, K. Subrahmanyam, A. Govindaraj and C. Rao, *J. Chem. Sci.*, 2008, **120**, 9.
28. L. L. Zhang and X. Zhao, *Chem. Soc. Rev.*, 2009, **38**, 2520.
29. Y. Ma, P. Li, J. W. Sedloff, X. Zhang, H. Zhang and J. Liu, *ACS Nano*, 2015, **9**, 1352.
30. M. Härmas, T. Thomberg, H. Kurig, T. Romann, A. Jänes and E. Lust, presented at 232 nm ECS Meeting, D-Glucose Derived Carbon Materials Activated by Zinc Chloride, Potassium Hydroxide or Mixture of Them for Supercapacitor Electrodes, National Harbor, October, 2017.
31. A. Yamada, S.-C. Chung and K. Hinokuma, *J. Electrochem. Soc.*, 2001, **148**, A224.
32. K.-S. Park, A. Benayad, D.-J. Kang and S.-G. Doo, *J. Am. Chem. Soc.*, 2008, **130**, 14930.
33. H.-W. Lee, P. Muralidharan, R. Ruffo, C. M. Mari, Y. Cui and D. K. Kim, *Nano Lett.*, 2010, **10**, 3852.
34. N. Padmanathan, S. Selladurai, K. M. Rahulan, C. O'Dwyer and K. M. Razeeb, *Ionics*, 2015, **21**, 2623.
35. N. Padmanathan, S. Selladurai and K. M. Razeeb, *RSC Adv.*, 2015, **5**, 12700.
36. N. Padmanathan, H. Shao, S. Selladurai, C. Glynn, C. O'Dwyer and K. M. Razeeb, *Int. J. Hydrogen Energy*, 2015, **40**, 16297.
37. S. Dsoke, B. Fuchs, E. Gucciardi and M. Wohlfahrt-Mehrens, *J. Power Sources*, 2015, **282**, 385.
38. N. Arun, A. Jain, V. Aravindan, S. Jayaraman, W. C. Ling, M. P. Srinivasan and S. Madhavi, *Nano Energy*, 2015, **12**, 69.
39. Z. Jian, V. Raju, Z. Li, Z. Xing, Y. S. Hu and X. Ji, *Adv. Funct. Mater.*, 2015, **25**, 5778.
40. J. Ding, H. Wang, Z. Li, K. Cui, D. Karpuzov, X. Tan, A. Kohandehghan and D. Mitlin, *Energy Environ. Sci.*, 2015, **8**, 941.
41. S. Gao, J. Zhao, Y. Zhao, Y. Wu, X. Zhang, L. Wang, X. Liu, Y. Rui and J. Xu, *Mater. Lett.*, 2015, **158**, 300.
42. T. Lin, I.-W. Chen, F. Liu, C. Yang, H. Bi, F. Xu and F. Huang, *Science*, 2015, **350**, 1508.
43. M. Cakici, R. R. Kakarla and F. Alonso-Marroquin, *Chem. Eng. J.*, 2017, **309**, 151.
44. L. Yu and G. Z. Chen, *J. Power Sources*, 2016, **326**, 604.
45. S. Liu, Q. Zhao, M. Tong, X. Zhu, G. Wang, W. Cai, H. Zhang and H. Zhao, *J. Mater. Chem. A*, 2016, **4**, 17080.
46. E. Frackowiak, V. Khomenko, K. Jurewicz, K. Lota and F. Béguin, *J. Power Sources*, 2006, **153**, 413.
47. W. He, C. Wang, F. Zhuge, X. Deng, X. Xu and T. Zhai, *Nano Energy*, 2017, **35**, 242.

48. L. Wang, Y. Han, X. Feng, J. Zhou, P. Qi and B. Wang, *Coord. Chem. Rev.*, 2016, **307**, 361.
49. B. Liu, H. Shioyama, H. Jiang, X. Zhang and Q. Xu, *Carbon*, 2010, **48**, 456.
50. D. Sheberla, J. C. Bachman, J. S. Elias, C.-J. Sun, Y. Shao-Horn and M. Dincă, *Nat. Mater.*, 2017, **16**, 220.
51. R. Liu, J. Duay and S. B. Lee, *Chem. Commun.*, 2011, **47**, 1384.
52. D. Dubal, V. Fulari and C. Lokhande, *Microporous Mesoporous Mater.*, 2012, **151**, 511.
53. H. Pang, S. Wang, W. Shao, S. Zhao, B. Yan, X. Li, S. Li, J. Chen and W. Du, *Nanoscale*, 2013, **5**, 5752.
54. D.-S. Kong, J.-M. Wang, H.-B. Shao, J.-Q. Zhang and C.-n. Cao, *J. Alloys Compd.*, 2011, **509**, 5611.
55. M.-J. Deng, F.-L. Huang, I.-W. Sun, W.-T. Tsai and J.-K. Chang, *Nanotechnology*, 2009, **20**, 175602.
56. G. S. Gund, D. P. Dubal, S. B. Jambure, S. S. Shinde and C. D. Lokhande, *J. Mater. Chem. A*, 2013, **1**, 4793.
57. K. Liang, X. Tang and W. Hu, *J. Mater. Chem.*, 2012, **22**, 11062.
58. J. Xu, L. Gao, J. Cao, W. Wang and Z. Chen, *Electrochim. Acta*, 2010, **56**, 732.
59. Z.-Y. Yu, L.-F. Chen and S.-H. Yu, *J. Mater. Chem. A*, 2014, **2**, 10889.
60. S. G. Mohamed, C.-J. Chen, C. K. Chen, S.-F. Hu and R.-S. Liu, *ACS Appl. Mater. Interfaces*, 2014, **6**, 22701.
61. X. Lu, X. Huang, S. Xie, T. Zhai, C. Wang, P. Zhang, M. Yu, W. Li, C. Liang and Y. Tong, *J. Mater. Chem.*, 2012, **22**, 13357.
62. L. Li, Y. Cheah, Y. Ko, P. Teh, G. Wee, C. Wong, S. Peng and M. Srinivasan, *J. Mater. Chem. A*, 2013, **1**, 10935.
63. F. Yang, J. Yao, F. Liu, H. He, M. Zhou, P. Xiao and Y. Zhang, *J. Mater. Chem. A*, 2013, **1**, 594.
64. Y. Zhao, L. Hu, S. Zhao and L. Wu, *Adv. Funct. Mater.*, 2016, **26**, 4085.
65. W. Kong, C. Lu, W. Zhang, J. Pu and Z. Wang, *J. Mater. Chem. A*, 2015, **3**, 12452.
66. J. Li, Y. Cheng, M. Jia, Y. Tang, Y. Lin, Z. Zhang and Y. Liu, *J. Power Sources*, 2014, **255**, 130.
67. Y. Ye, Y. Shi and A. A. Tay, *J. Power Sources*, 2012, **217**, 509.
68. X. Li, A. M. Elshahawy, C. Guan and J. Wang, *Small*, 2017, **13**, 1701530.
69. F. S. Omar, A. Numan, N. Duraisamy, S. Bashir, K. Ramesh and S. Ramesh, *RSC Adv.*, 2016, **6**, 76298.
70. R. Bendi, V. Kumar, V. Bhavanasi, K. Parida and P. S. Lee, *Adv. Energy Mater.*, 2016, **6**, 1501833.
71. H. Pang, Z. Yan, Y. Ma, G. Li, J. Chen, J. Zhang, W. Du and S. Li, *J. Solid State Electrochem.*, 2013, **17**, 1383.
72. H. Pang, C. Wei, Y. Ma, S. Zhao, G. Li, J. Zhang, J. Chen and S. Li, *ChemPlusChem*, 2013, **78**, 546.
73. M.-C. Liu, J.-J. Li, Y.-X. Hu, Q.-Q. Yang and L. Kang, *Electrochim. Acta*, 2016, **201**, 142.

74. J. Yang, C. Fang, C. Bao, W. Yang, T. Yu, W. Zhu, F. Li, J. Liu and Z. Zou, *RSC Adv.*, 2016, **6**, 75186.
75. X. Wang, J. Hao, Y. Su, F. Liu, J. An and J. Lian, *J. Mater. Chem. A*, 2016, **4**, 12929.
76. X. Chen, M. Cheng, D. Chen and R. Wang, *ACS Appl. Mater. Interfaces*, 2016, **8**, 3892.
77. N. Nitta, F. Wu, J. T. Lee and G. Yushin, *Mater. Today*, 2015, **18**, 252.
78. M. R. Lukatskaya, B. Dunn and Y. Gogotsi, *Nat. Commun.*, 2016, **7**, 12647.
79. N. Padmanathan, H. Shao and K. Razeeb, *ACS Appl. Mater. Interfaces*, 2018, **10**, 8599.
80. Y. Wang, Y. Song and Y. Xia, *Chem. Soc. Rev.*, 2016, **45**, 5925.
81. L. Xia, L. Yu, D. Hu and G. Z. Chen, *Mater. Chem. Front.*, 2017, **1**, 584.
82. V. A. Biocea, *P. IEEE*, 2014, **102**, 1777.
83. H. S. Chen, T. N. Cong, W. Yang, C. Q. Tan, Y. L. Li and Y. L. Ding, *Prog. Nat. Sci.*, 2009, **19**, 291.
84. S. Nishijima1, S. Eckroad, A. Marian, K. Choi, W. S. Kim, M. Terai, Z. Deng, J. Zheng, J. Wang, K. Umemoto, J. Du, P. Febvre, S. Keenan, O. Mukhanov, L. D. Cooley, C. P Foley, W. V. Hassenzahl and M. Izumi, *Supercond. Sci. Technol.*, 2013, **26**, 113001.
85. P. F. Ribeiro, B. K. Johnson, M. L. Crow, A. Arsoy and Y. L. Liu, *P. IEEE*, 2001, **89**, 1744.
86. S. M. Mousavi, G. F. Faraji, A. Majazi and K. Al-Haddad, *Renewable Sustainable Energy Rev.*, 2017, **67**, 477.
87. M. Amiryar and K. Pullen, *Appl. Sci.*, 2017, **7**, 286.
88. R. Pena-Alzola, R. Sebastian, J. Quesada and A. Colmenar, in *2011 International Conference on Power Engineering, Energy and Electrical Drives, IEEE*, 2011, 1.
89. C. Szabo, *Power Eng. Int.*, 2017, **25**(11), 6.
90. S. Grimbly, *Encyclopedia of the Ancient World*, Fitzroy Dearborn, London, 2000.
91. *Machinery's Reference Series No. 70: Steam Engines*, The Industrial Press, New York, 1911.
92. P. Daly, D. Flynn and N. Cunniffe, in *2015 IEEE Eindhoven PowerTech*, IEEE, 2015, 1.
93. V. Knap, S. K. Chaudhary, D.-I. Stroe, M. Swierczynski, B.-I. Craciun and R. Teodorescu, *IEEE Trans. Power Syst.*, 2016, **31**, 3447.
94. S. H. Kim, *Electric Motor Control: DC, AC, and BLDC Motors*, Elsevier Science, 2017.
95. *Electric Power Systems*, ed. B. M. Weedy, John Wiley & Sons, Ltd, Chichester, West Sussex, UK, 5th edn, 2012.
96. Y. Liu, S. Yang, S. Zhang and F. Z. Peng, in *2014 IEEE Energy Conversion Congress and Exposition (ECCE)*, IEEE, 2014, 2684.
97. S. Mukoyama, T. Matsuoka, M. Furukawa, K. Nakao, K. Nagashima, M. Ogata, T. Yamashita, H. Hasegawa, K. Yoshizawa, Y. Arai,

K. Miyazaki, S. Horiuchi, T. Maeda and H. Shimizu, *Phys. Procedia*, 2015, **65**, 253.

98. A. Rupp, H. Baier, P. Mertiny and M. Secanell, *Energy*, 2016, **107**, 625.
99. T. A. Baby, T. Kurian and M. Eldho Shibu, *Int. J. Adv. Res. Educ. Technol.*, 2015, **2**, 108.
100. M. Krack, M. Secanell and P. Mertiny, in *Energy Storage in the Emerging Era of Smart Grids*, ed. R. Carbone, InTech, 2011.
101. H. Lee, S. Jung, Y. Cho, D. Yoon and G. Jang, *Phys. C Supercond.*, 2013, **494**, 246.
102. J. Hou, J. Sun and H. Hofmann, *Appl. Energy*, 2018, **212**, 919.
103. D. R. Brown and W. D. Chvala, *Energy Eng.*, 2005, **102**, 7.
104. N. Hamsic, A. Schmelter, D. A. Mohd, E. Ortjohann, E. Schultze, A. Tuckey and J. Zimmermann, in *The Great Wall World Renewable Energy Forum*, Beijing, China, 2006.
105. D. Bender, *Recommended Practices for the Safe Design and Operation of Flywheels*, Sandia National Laboratories, Albuquerque, New Mexico, 2015.
106. J. E. Martin, L. E. S. Rohwer and J. Stupak, *Composites. Part B*, 2016, **97**, 141.
107. L. Gertmar, A. Nysveen, Per-Anders Löf and T. F. Nestli, *World Pat. Appl.*, WO2000067358 A1, 2000.
108. A. Buchroithner, A. Haan, R. Pressmair, M. Bader, B. Schweighofer, H. Wegleiter and H. Edtmayer, in *2016 International Conference on Sustainable Energy Engineering and Application (ICSEEA)*, IEEE, 2016, 41.
109. P. Yulong, A. Cavagnino, S. Vaschetto, C. Feng and A. Tenconi, in *2017 6th International Conference on Clean Electrical Power (ICCEP)*, IEEE, 2017, 492.
110. R. H. Byrne, T. A. Nguyen, D. A. Copp, B. R. Chalamala and I. Gyuk, *IEEE Access*, 2018, **6**, 13231.

Photochemical Energy Storage

GAIA NERI, MARK FORSTER AND ALEXANDER J. COWAN*

ABSTRACT

This chapter highlights energy storage strategies that utilise solar energy to drive the formation of chemicals, fuels and feedstocks. The production of solar fuels that can be stored and transported is an attractive way to address the intermittency of terrestrial solar energy and provide sustainable access to the fundamental feedstocks upon which society has come to rely. The solar energy-driven reactions considered here are the splitting of water to produce hydrogen and oxygen, and the coupled oxidation of water and reduction of CO_2 to produce a variety of higher value carbon products and oxygen. The chapter aims to provide an introductory overview of both direct (photochemical) and indirect solar (photovoltaic-enabled electrolysis) routes to these fuels.

1 Introduction

Enormous amounts of solar energy reach the surface of the Earth each year, *ca.* 3.8×10^6 EJ (1 EJ $= 5 \times 10^{18}$ J), and of this, *ca.* 5×10^4 EJ is estimated to be easily harvestable,[1] compared with an annual global energy consumption of *ca.* 5×10^2 EJ.[2] However, the intermittent nature of solar energy at the terrestrial surface necessitates a means of large-scale and potentially long-term energy storage if solar energy is to become the dominant renewable resource. Photochemical energy storage provides a potential solution to the current

---

*Corresponding author.

Issues in Environmental Science and Technology No. 46
Energy Storage Options and Their Environmental Impact
Edited by R.E. Hester and R.M. Harrison
Published by the Royal Society of Chemistry, www.rsc.org

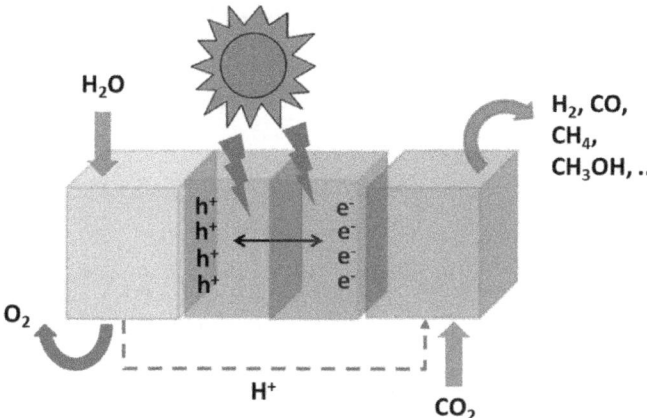

Figure 1 Schematic of a solar-to-fuels device where a light-absorbing centre, here shown in the middle of the device, absorbs photons and generates high-energy electrons and holes that are transferred to the catalytic centres for oxidation of water (left) and reduction of either protons to H_2 or CO_2 to carbon-based fuels (right). Based on ref. 3.

challenges associated with the gathering and storage of energy on the global scale. Inspired by photosynthesis, solar energy is harnessed and used to drive chemical reactions, producing useful fuels and feedstocks such as H_2 and CO and CH_3OH, effectively storing the solar energy in the form of chemical bonds (see Figure 1). These stored chemicals can then be transported and used as and when required, thereby overcoming the problem of intermittency typically associated with solar technologies.

There are numerous approaches under consideration to utilising solar energy to produce fuels, including thermochemical transformations (see the chapter Thermal and Thermochemical Storage),[4] photovoltaic (PV) systems coupled to electrolysers,[5] integrated photoelectrochemical (PEC) cells,[6] semiconductor photocatalysts,[7] molecular photocatalysts[8] and photo-biological organisms.[9] The examination of all possible routes is beyond the scope of this chapter and readers are directed to several other excellent reviews.[4,5,7–10] Here we focus on systems in which the light-absorbing unit in Figure 1 is one or more semiconductor materials. In these cases, absorption of photons that have an energy greater than the electronic bandgap of the semiconductor leads to the generation of photoelectrons in the conduction band and photoholes (the positive counterpart) in the valence band.[11] These high-energy reducing and oxidising equivalents can be directly utilised in a chemical (reduction or oxidation) reaction for fuel generation and semiconductor photocatalysts and photoelectrodes are described in Section 4.3 and Figure 2b and c. Alternatively, when light absorption occurs in a semiconductor photovoltaic device, the generated photovoltage can be used indirectly to drive a remote electrolyser and such systems will be considered in Sections 4.1 and 4.2 and Figure 2a.

2   Classes of Solar Fuels and Feedstocks

Desirable products take the form of combustible, energy-dense fuels. Figure 3 shows a Ragone plot, where the specific energy (density) of a

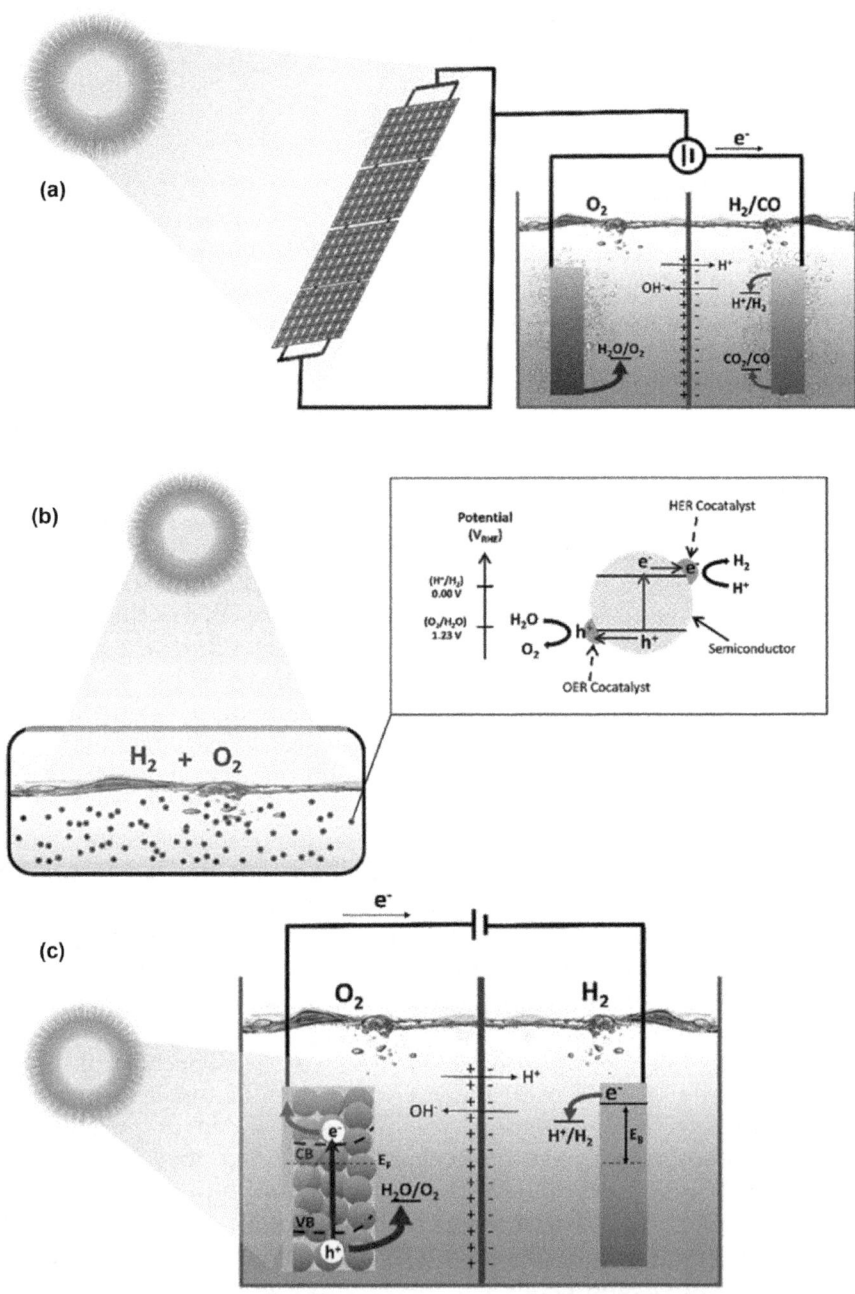

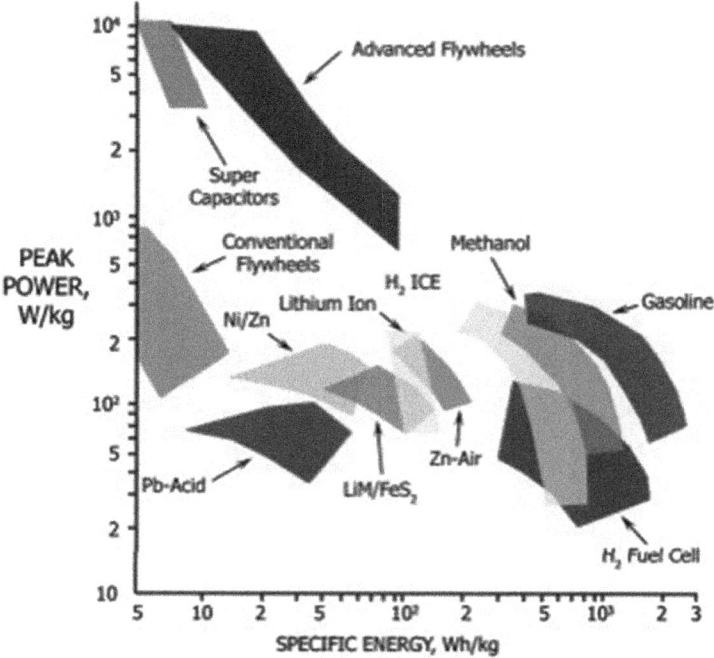

Figure 3 Ragone plot of energy density *versus* peak power for some energy storage technologies, with focus on technologies suitable for transportation. Reproduced from ref. 14 with permission from Elsevier, Copyright 2011.

material is plotted against the achievable peak power. It is clear that although electrochemical and mechanical energy storage devices are ideally placed for fast-response, high peak power applications, the relatively low energy density means that they may not be suitable for all scenarios. In particular, if renewables such as solar energy are to become our dominant resource, then there will be a need to achieve seasonal time-scale (>3 months) storage. Such a case would require storage on a massive scale, making high energy density fuels an attractive option.

High energy density fuels such as gasoline have a second critical advantage as the infrastructure for their distribution and utilisation, a vast investment across the globe, already exists. However, if chemical fuels are to

Figure 2 Multiple routes to generating solar fuels exist, including the following. (a) A PV system coupled to an electrolyser. One or multiple PV cells convert sunlight into electrical energy, which is used to power an electrolyser cell. (b) Semiconductor photocatalysts where a light-absorbing semiconductor can be coupled to selective catalysts, such as those for the oxygen evolution reaction (OER) and hydrogen evolution reaction (HER), are often bound to the surface to enhance performance.[12,13] (c) Photoelectrochemical (PEC) systems such as those for water splitting shown here, where a light-absorbing photoanode is being used to oxidise water, with H_2 evolution occurring at a counter electrode.

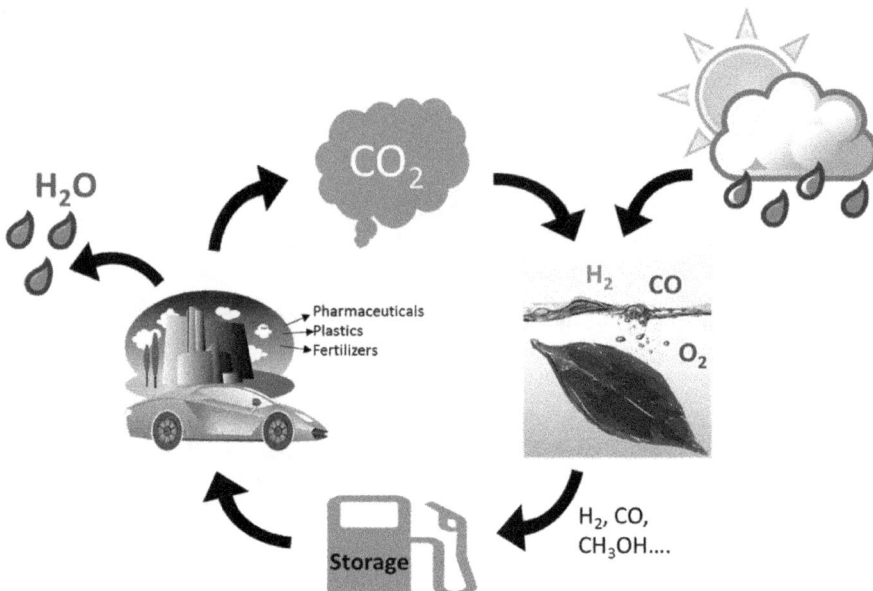

Figure 4 Simplified version of a photochemical energy storage cycle. Solar energy is captured and used to drive chemical reactions to produce energy-dense fuels and feedstocks such as H_2 and CO. The products can then be used immediately or stored and transported for later use. Utilisation of the fuels and feedstocks completes the cycle.

remain a key part of our energy landscape, it is important that their generation and utilisation are a circular process, with only solar energy being the net input (see Figure 4). For example, in the systems considered here, H_2 production can be achieved with only water and sunlight as inputs and the consumption of H_2 will yield only water and stored energy, thus closing the circuit. Although more challenging, similar clean cycles could also be envisaged for carbon-based solar fuels involving CO_2 emission and recapture, with it even being viable to produce a low carbon intensity gasoline using only CO_2, water and sunlight as the input materials (see Figure 4).

2.1 Sustainable H_2 Production

Before going on to describe the state-of-the-art for each solar fuel, we initially explore the two highest profile classes of solar fuels. The simplest concept for photogenerated energy storage is the electrolysis of water to split it into its constituent parts, generating O_2 in one half-reaction and H_2, an energy-rich fuel/feedstock, in the other half. Water splitting is an energetically uphill process (237 kJ mol^{-1} of liquid water at 298 K) corresponding to 1.23 eV per electron transferred; see eqn (1)–(3).[15]

$$2H_2O \rightarrow O_2 + 4H^+ + 4e^- \quad (E^\circ = 1.23\ V_{RHE}) \text{ (oxygen evolution reaction)} \quad (1)$$

$$4H^+ + 4e^- \rightarrow 2H_2 \quad (E^\circ = 0.00 \ V_{RHE}) \ \text{(hydrogen evolution reaction)} \qquad (2)$$

$$2H_2O \rightarrow O_2 + 2H_2 \quad (E^\circ_{cell} = 1.23 \ V_{RHE}) \qquad (3)$$

H_2 is a feedstock in a wide range of industrial processes, including the production of fertilisers, plastics and pharmaceuticals and in the hydrogenation of oils to form fats.[16] As a fuel it can be utilised in multiple ways, including through combustion[17] or through an electrochemical fuel cell that takes in oxygen and compressed hydrogen to generate electricity and heat.[18] H_2 fuel-cell technology is usually modular, giving it the flexibility to cover a range of applications. H_2 fuel cells can be stationary or portable (hand-held devices, portable generators and vehicle engines) and can be used as an alternative to, or in combination with, energy supplies such as batteries or supercapacitors. Fuel cells offer low emissions (only water), low noise outputs, high electrical efficiency and flexible fuel use. The heat generated from fuel cells can also be harnessed in a combined heat and power system.[19] Based on currently available technology for sustainable electric transportation, the use of a fuel cell in addition to or in place of battery technology currently allows for faster refill rates, longer operating times (increased range) and reduced weight.[20] There is a wide range of fuel-cell technologies available and their energy efficiencies typically lie between 40 and 60% (compared with *ca.* 25% for an internal combustion engine).[19,21] Combined heat and power systems can increase this efficiency up to *ca.* 85–90%.[19] Although fuel-cell efficiencies are enticing, they are typically associated with high costs. It is also possible to introduce H_2 into an existing combustion engine to form a co-combustion system with reduced emissions.[22]

A significant drawback with H_2 systems relates to the challenges of storage and transport, particularly at ambient temperatures. Current methods of hydrogen storage include reinforced high-pressure gas tanks, liquid storage by cryogenic cooling ($-253 \ ^\circ$C) and solid storage using high surface area materials that act as chemical or physical sorbents.[23] If H_2 is to become successful as a fuel for transportation, a distribution infrastructure also needs to be put in place or existing gas networks modified. The anticipated environmental effects related to unintended emissions due to H_2 leakage into the atmosphere are estimated to be relatively low; however, further detailed studies may be required prior to H_2 achieving widespread use.[24]

2.2 Sustainable Carbon Fuels Through CO_2 Reduction

Although H_2 from water splitting is an attractive fuel option, issues with storage and the costs of new infrastructure may present a significant barrier to widespread implementation. In contrast, society has already invested vast wealth into systems for the storage, transportation and utilisation of carbon fuels. Therefore, the ability to generate such products in a sustainable manner from only water, CO_2 and sunlight is an extremely attractive proposition. The ability to convert carbon dioxide also has wider implications for the energy sector. Much research has been focused on CO_2 capture in the past years as a means of preventing CO_2 emissions and in this context the reduction of CO_2, a waste molecule, to produce energy-rich fuels and high-value feedstocks is an attractive route to offsetting the cost of carbon sequestration. Potentially, if air capture[25] of CO_2 becomes economically viable, the light-driven reduction of CO_2 to stable carbon products even offers a way to achieve negative greenhouse gas emissions.

However, the reduction of CO_2 to fuels or feedstocks is challenging. Existing thermal processes exist, such as the reverse water gas shift reaction ($CO_2 + H_2 \rightarrow CO + H_2O$), and multiple examples of CO_2 hydrogenation catalysts to alternative carbon products (e.g. methanol) are known in the literature and have been reviewed elsewhere;[26] however, to achieve significant reaction rates, relatively high operating temperatures are often required. Alternatively, CO_2 can be reduced using solar energy either using a highly reducing photoelectron or at a cathode in an electrolyser coupled to a photovoltaic (PV) cell. The direct one-electron reduction of CO_2 to the radical anion $CO_2^{\bullet-}$ is energetically demanding [see eqn (4); all standard potentials were converted from NHE at pH 7][27] as it requires an unfavourable geometric rearrangement from a stable linear geometry to a bent one; however, the proton-assisted, multi-electron reductions [see eqn (5)–(8)], have more accessible thermodynamic reduction potentials and this review focuses on carbon fuels and feedstocks formed via these reactions.

$$CO_2 + e^- \rightarrow CO_2^{\bullet-} \quad (E^\circ = -1.90\ V_{RHE}) \tag{4}$$

$$CO_2 + 2H^+ + 2e^- \rightarrow CO + H_2O \quad (E^\circ = -0.12\ V_{RHE}) \tag{5}$$

$$CO_2 + 2H^+ + 2e^- \rightarrow HCO_2H \quad (E^\circ = -0.20\ V_{RHE}) \tag{6}$$

$$CO_2 + 6H^+ + 6e^- \rightarrow CH_3OH + H_2O \quad (E^\circ = +0.03\ V_{RHE}) \tag{7}$$

$$CO_2 + 8H^+ + 8e^- \rightarrow CH_4 + 2H_2O \quad (E^\circ = +0.17\ V_{RHE}) \tag{8}$$

Eqn (4)–(8) represent the simplest C_1 products achievable via the electro- or photoelectrochemical reduction of CO_2 and it is apparent that a wide variety of products can be achieved. Figure 5 shows a Latimer–Frost diagram for the reduction of CO_2 to various products. This diagram highlights the difficulty in obtaining the more complex multi-electron, multi-proton products (e.g. methanol, methane). Although these products both have lower standard reduction potentials than the simpler two-electron products (carbon monoxide, formate), there is a need to deliver concertedly six (methanol) or eight (methane) electrons and protons if the formation of higher energy intermediates is to be avoided. Although catalytic centres that can achieve such a feat are being studied, work remains at an early stage and by far the most widely studied CO_2 electro/photocatalytic reduction reactions are the simplest $2e^-/2H^+$ processes to give carbon monoxide or formate [see eqn (5) and (6)].

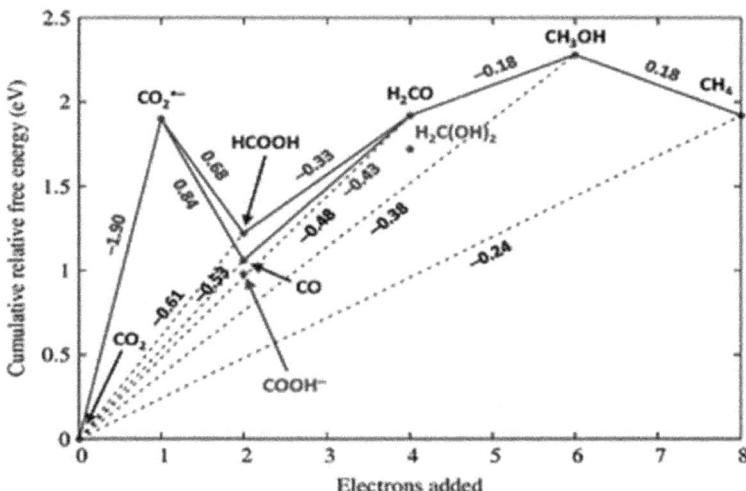

Figure 5 Latimer–Frost diagram for the various possible reduction reactions of CO_2 in water at pH 7. The dotted lines represent the standard potentials for the lowest energy pathways to the products indicated.
Reproduced from ref. 28 with permission from The Royal Society of Chemistry.

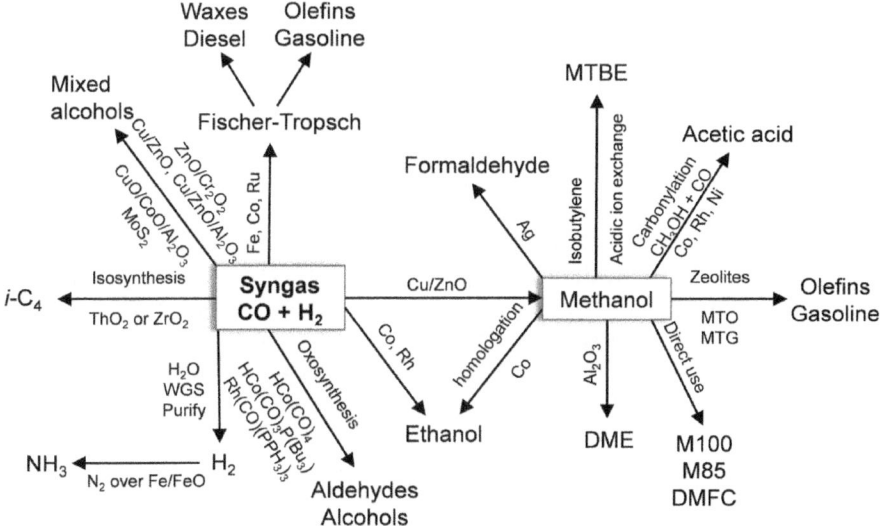

Figure 6 Various industrial processes using syngas. Based on ref. 32.

When coupled to water oxidation at the anode, the two-electron reduction of CO_2 to CO requires 259 kJ mol^{-1}, corresponding to 1.35 eV per electron.[29] CO is a high-value feedstock as it is currently used in various industrial processes to access a variety of high carbon content materials, and it can therefore be used with existing technologies (see Figure 6). One such process is the Fischer–Tropsch (F–T) synthesis, which takes a mixture of CO and H_2 (synthesis gas or syngas) and transforms it into hydrocarbons. The nature of the hydrocarbon, such as the number of C atoms or the saturation, are determined by the F–T catalyst, the reactor conditions and the $CO : H_2$ ratios used.[30,31] This process is particularly attractive as the synthetic fuels for high-value applications, such as in the aviation and marine sectors that will be challenging to decarbonise, can be produced with existing chemical industry infrastructure.

The main issue with syngas transformations currently is the need to derive high-purity CO and H_2 from fossil resources, primarily coal and natural gas, which has environmental and cost implications; for example, purification of the input syngas represents 60–70% of the total cost of the hydrocarbons produced in the F–T process.[33] The light-driven reduction of CO_2 to CO and reduction of H^+ to H_2 coupled to water oxidation is therefore a particular focus of many research groups as it would permit the direct generation of high-purity syngas.[34–38]

$$2CO_2 + 10H^+ + 10e^- \rightarrow CH_3CHO + 3H_2O \quad (E° = 0.05 \ V_{RHE}) \qquad (9)$$

$$2CO_2 + 12H^+ + 12e^- \rightarrow C_2H_5OH + 3H_2O \quad (E° = 0.09 \ V_{RHE}) \qquad (10)$$

$$2CO_2 + 12H^+ + 12e^- \rightarrow C_2H_4 + 4H_2O \quad (E° = 0.06 \ V_{RHE}) \qquad (11)$$

Beyond the C_1 products of eqn (4)–(8), it is also possible to identify multi-electron, multi-proton, pathways to a range of C_{2+} products [*i.e.* ethylene, ethanol, acetaldehyde; see eqn (9)–(11); all standard potentials were converted from NHE at pH 7].[39,40] The successful reduction of CO_2 to any of these materials would be an important milestone, as some of them (*e.g.* ethanol) can be used directly as fuels in place of fossil resources without any other manipulation. However, the higher number of electrons required per molecule, together with the increased number of steps in the catalytic cycle, represents a significant challenge for catalyst design. Additionally, the often similar mechanistic pathways between more complex products usually leads to a selectivity issue, where a distribution of products is obtained, and this causes problems with product purification.

3   Reaction Enhancement and Selectivity by Catalysis

Although there are some materials that are capable of both absorbing light and catalysing water splitting or CO_2 reduction unassisted (*e.g.* haematite for water oxidation[41] and C_3N_4 for proton reduction;[42] see Section 4.3), most of the systems under consideration have distinct materials for catalysis and light absorption. Such a modular approach allows for uncoupling of the two roles and the development of the two components independently. The primary role of the catalytic centre is to lower the activation barrier for reaction intermediates, thereby lowering the overpotential for an electrochemical reaction and potentially increasing selectivity towards a single product. Both homogeneous and heterogeneous catalysts are under development for solar fuel production. Homogeneous catalysts are generally transition metal complexes dissolved in the electrolyte solution (see Figure 7b) and have the ability to bind either a proton, to form a metal hydride, or CO_2, generally following an electrochemical step to generate the active form of the catalyst. Heterogeneous systems include metal and metal oxide materials, either in bulk or nanostructured (see Figure 7a), and also molecular catalysts heterogenised by immobilising the catalyst on the electrode surface or the light absorber *via* various methods (*e.g.* covalent binding, incorporation on a membrane); see Figure 7c.

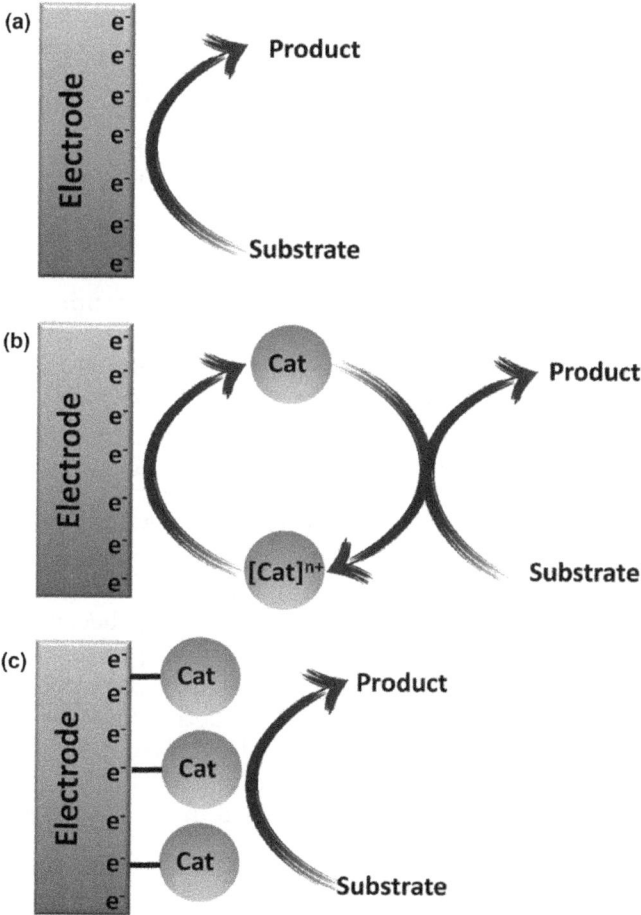

Figure 7   Various types of catalysis: (a) heterogeneous catalysis (typically either a metal or a metal oxide); (b) homogeneous catalysis, where the catalyst is dissolved in the electrolyte solution; (c) the catalyst is bound to the electrode/light absorber.

4 Current Status of Light-driven Fuel Production

As outlined in Section 1, multiple routes to generating solar fuels exist; here we focus on systems where the light-absorbing unit in Figure 1 is one or more semiconductor materials. Initially the general technology is described before selected examples of systems and catalysts for H_2 or carbon-based fuel generation are examined.

4.1 PV-driven Electrolysis of Water to Generate H_2

A summary of the field of photovoltaics is not included here; instead, readers are directed to recent reviews[43–45] and the focus here is on the chemistry of the electrolysis cell itself, which will use electrical energy provided from one or more PV devices. In its simplest configuration, an electrolyser can be considered to consist of two electrodes immersed in an electrolyte solution, separated by an ion-permeable membrane of some form and a route to removing the gases generated during electrolysis. The cathode will carry out proton reduction to generate H_2 [see eqn (2)] whereas water oxidation to produce O_2 [eqn (1)] occurs at the anode. Theoretically, an electrolyser can reach efficiencies of up to 90–95% of the energy from PV, which would translate to an overall PV electrolyser efficiency (solar-to-hydrogen or STH efficiency) of 57% for a 3J (triple junction) cell and 62% for a 4J or 5J cell.[46] For PV electrolyser-driven water splitting, the maximum STH efficiency reached by a test system is 30%,[47] averaged over 48 h, using a commercial 3J solar cell coupled to two polymer electrolyte membrane (PEM) electrolysers.

Water-splitting electrolysers are often grouped into one of three classes: alkaline water electrolysers and PEM electrolysers, which operate at low temperatures, and solid oxide cells, which require higher temperatures,[48,49] and systems are being explored for commercial solar-driven water splitting. However, the current high cost of H_2 produced by PV electrolysis is a major barrier to industrial scale-up, as steam reforming to produce hydrogen is comparatively cheap. The US Department of Energy estimated the threshold cost of solar H_2 (the cost below which the technology becomes competitive with currently employed fuels) to be \$2.00–4.00 per gallon of gasoline equivalent (gge),[50] while the current cost of hydrogen produced *via* electrolysis (not necessarily PV) is \$3.26–6.62 gge (at the time of writing, 2018).[51] Although it is predicted that the cost could decrease dramatically with PV electrolysers,[52–54] in large part due to rapidly decreasing PV prices, another major factor is the cost of the electrodes used in the electrolysis cell.

PEM electrolysers have attracted particular attention owing to their compatibility with the variable current input that will be achieved when coupled to solar PV. PEM electrolysers make use of platinum as a cathode material,[55] and the most commonly employed anode materials are IrO_2 and RuO_2.[56] For very large scale-up, the use of such precious metals is undesirable, owing to their high cost and low abundance. For this reason, much research has been dedicated to finding alternative electrode materials made of Earth-abundant elements that can perform water reduction or oxidation with a similar level of electrocatalytic activitiy.[55,56] In recent years, first-row transition metal materials based on Ni,[57] Mo[58] and Co[59] have been found to be efficient electrode materials for water splitting;[55] however, performances remain weak compared with noble metal electrodes. The reason for the dominance of Pt can be readily illustrated through a volcano plot (Figure 8). Volcano plots for the hydrogen evolution reaction (HER) were introduced by Trasatti in the 1970s[60] and relate the exchange current densities of a catalyst (that is, how efficient a catalyst is) with the adsorption energy of hydrogen on the material. The volcano shape of these plots indicates that for monometallic systems, the catalytic activity of Pt will be challenging to surpass. Instead, efforts to develop water oxidation catalysts on oxides of abundant elements have led to some promising results. One of the benchmark Earth-abundant electrocatalysts for the oxygen evolution reaction (OER) is nickel–iron oxyhydride ($Ni_{1-x}Fe_xOOH$),[61] reported to be highly active, particularly in alkaline media. Several studies have explored the activity of NiFe oxyhydrides, in particular with regard to the iron content in the catalyst, which, although present at low concentrations, is believed to be the active site.[62–64] However, a limiting factor with these materials, and many other mixed metal oxides/oxyhydrides, is the need for basic conditions for them to operate efficiently. Interestingly, Nocera and co-workers[65] reported that a Co oxy-hydroxide catalyst, formed *in situ* in the presence of phosphate, is able to oxidise water at a relatively low overpotential at a range of pHs, including in neutral buffered solutions. Furthermore, they also showed[66] that this catalyst is able to maintain catalytic activity even in impure waters such as that collected from a river, in contrast to most electrocatalysts, which require very pure water to operate with a high Faradaic efficiency.

Although PV-driven electrolysis of water is a relatively mature field, there remains intense interest in developing new approaches to making lower cost systems. One interesting approach is to develop integrated PV absorbers where the catalyst for either water oxidation or hydrogen evolution is deposited on the terminals of the PV device. Such 'artificial leaves' have been developed extensively by Nocera's group,[68] where both the anode and the cathode were deposited wirelessly on a triple-junction silicon-based solar cell. Degradation of the cell is prevented by depositing a layer of indium tin oxide (ITO) between the solar cell and the electroactive materials. The anode was made of self-assembling, self-repairing Co phosphate and the cathode was made of a ternary alloy of Ni, Mo and Zn. Both electrocatalysts are able to work at mild pH values and at room temperature, are constituted only of Earth-abundant catalysts and achieve a solar-to-hydrogen efficiency of 2.5%, when using a solar cell with a light-to-electricity efficiency of 6.2%.

Another approach to obtaining low-cost proton reduction catalysts is the use of homogeneous molecular catalysts. There are a few examples of molecular catalysts for the reduction of protons, of which the most notable ones are a class of Co complexes known as cobaloximes.[69] These complexes have been used in both electrochemical and photochemical reactions (in the

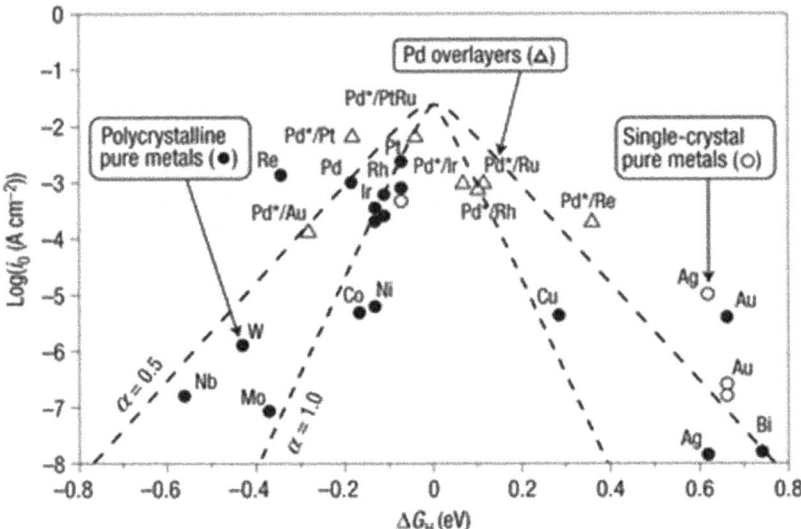

Figure 8 Volcano plot for the HER. The highest exchange current densities for H_2 evolution are achieved with metals, with a metal indicated with a free energy for hydrogen adsorption of *ca.* 0.
Reproduced from ref. 67 with permission from Springer Nature, Copyright 2006.

latter case, when covalently bound to either Ru^{70} or Ir^{71} dyes); however, they have been studied in organic solvents with added proton sources. Ni phosphine catalysts have been extensively studied for electro- and photochemical proton reduction[72] since the first report by DuBois and co-workers,[73] and water solubility has been imparted by modification of the ligand framework.[74] A significant effort has been made in recent years towards the modification of the ligand framework to impart specific qualities, based on the knowledge of proton reduction mechanisms, both in enzymes and in molecular catalysis. In particular, the addition of proton relays in the second coordination sphere has been shown to be a promising strategy to increase the activity of proton reduction catalysts.[75,76]

4.2 PV-driven Electrolysis for CO_2 Reduction

The electrocatalytic reduction of CO_2 is at an early stage of development. For CO_2 reduction to be economically and environmentally viable, it will need to be coupled to a sustainable oxidation reaction, probably water oxidation, making the development of catalysts and electrodes that operate in water necessary. This causes problems, as proton reduction to produce hydrogen competes owing to its comparable thermodynamic potential and the low solubility of CO_2 in water (*ca.* 0.03 M at 298 K, 1 atm). Historically, metal electrodes for CO_2 reduction tend to show high current densities and Faradaic efficiencies (FE) (percentage of conversion of the charge passed into products); however, selectivity to a particular product can be difficult to achieve and the current focus on high-value metals (*e.g.* Au, Pd, Ag) may be limiting. Currently, Pd is receiving particular attention for the reduction of CO_2 to formate[77] and gold and silver show high levels of activity for CO_2 reduction to CO. Seminal work by Hori surveyed many of the metal electrodes[78] and much current research is focused on nanostructuring of these materials and on the inclusion of grain boundaries in a controlled manner in order to increase the number of active sites and improve the performance of the electrodes.[79] For example, nanoporous Ag electrodes, made by dealloying a Ag–Al precursor, for CO production at low overpotentials (<0.5 V) and with an FE of 92%,[80] show a significantly greater current density at the same potential as a Ag polycrystalline electrode. Remarkably, Au nanoneedle electrodes have shown CO production at overpotentials as low as 0.07 V, albeit with low current density (0.6 mA cm^{-2}) and 40% FE.[81] When the overpotential was increased to 0.19 V, the current was increased to 7 mA cm^{-2} and the FE reached 90%, highlighting the ability of these structured metal electrodes to achieve excellent selectivities and activities at moderate overpotentials. Accessing C_2 and higher products is challenging, but Cu electrodes stand out as they are able to reduce CO_2 not only to CO but also to a variety of C_1 and C_2 products;[78] however, historically low selectivities have been achieved, which would make purification extremely challenging and costly.

In an attempt to deliver higher catalytic current densities and improved selectivities, significant research effort has been directed towards the development of well-defined single-site molecular electrocatalysts.

Complexes of Mn,[82] Fe,[83] Co,[84] Ni[85] and Mo[86] have been found to be excellent catalysts for the reduction of CO_2;[87] however, only a few examples have been reported to be selective catalysts in water. This is partly due to the insolubility of many of the known catalysts that utilise ligands that are more suited to organic solvents. This issue has been addressed in two main ways: (i) by adding hydrophilic functional groups to the ligand framework of known catalysts for CO_2 reduction to impart water solubility,[88,89] or (ii) by immobilising insoluble catalysts on the electrode surface and using the electrode in aqueous solution.[90,91] In the latter approach, low solubility of the catalyst in water is actually desirable as it reduces the chances of the catalyst leaching out of the electrode and into the electrolyte solution. Amongst the homogeneous catalysts, Fe porphyrins stand out as the most efficient catalysts in terms of low overpotentials and current densities,[92] and recently a few examples of water-soluble Fe porphyrins have been published that show retention of selectivity in aqueous electrolytes.[83,89] In a similar approach to the studies described for molecular catalysts for proton reduction, the studies by Savéant and co-workers[89] showed how the addition of proton relays, or functional groups imparting 'through-space' properties, can also be beneficial to CO_2 reduction, and similar efforts have shown good results for catalysts of Ni[93] and Mn.[94] The immobilisation of molecular catalysts on an electrode surface is a relatively new strategy for CO_2 reduction electrocatalysis; however, excellent results have been achieved already using $Mn(bpy)(CO)_3X$-type catalysts, either immobilised using Nafion as a polymer support[91,95] or modified with a pyrene group[96] or a Co phthalocyanine/carbon nanotube assembly.[97] However, much work still needs to be done in terms of stability, as transition metal complexes are generally more susceptible to degradation.

Although CO_2 electrolysers are becoming more widely studied in the research arena, we are aware of very few examples where they have been coupled to a PV array.[29,98] Grätzel and co-workers built an electrolyser composed by a porous gold cathode and an IrO_2 anode in an aqueous solution of $NaHCO_3$ as the electrolyte.[29] CO production was observed on the gold electrode with a 290 mV overpotential and an FE of 90% (the remaining 10% being residual proton reduction). The electrochemical cell was then powered by connecting it to three perovskite PV cells in series, and the cell ran at a constant current density of 5.8 mA cm^{-2} for more than 18 h. In a very recent report, Schmid and co-workers[98] demonstrated a scalable electrolyser for syngas production consisting of a commercially available gas diffusion cathode (silver based) and an IrO_2-coated titanium anode. At the cathode, exclusively CO and hydrogen were produced with near 100% FE and sustained current densities of 300 mA cm^{-2} for over 1200 h. The electrolyser was powered by electricity supplied from a commercial PV module and was further coupled with a fermentation module transforming the syngas produced by the electrolyser into higher molecular weight products, mainly hexanol and butanol. The authors calculated the price of the resulting butanol to be 2.2 times higher than the price of the PV electricity supplied. These results are encouraging for a large-scale PV electrolyser for CO_2 reduction and plans are in place to realise a test plant, with recent announcements including the scaling up of the electrolyser described above by the end of 2021.[99]

4.3   *Photochemical and Photoelectrochemical Cells*

4.3.1   Photocatalytic Suspension Reactors for H_2 Production and CO_2 Reduction.   In contrast to PV electrolysers where the light absorber is separated from the reaction site, photochemical and photoelectrochemical approaches combine light-absorbing materials and catalysts in the device. In a photocatalytic system, light absorption by a semiconductor leads to electron and hole generation in the conduction and valence bands, respectively. The excited-state charges generated within the semiconductor are then transported to the semiconductor surface where catalysis can take place (either water oxidation and proton reduction or water oxidation and carbon dioxide reduction); see Figure 2b. As only photons of energy greater than the bandgap of the semiconductor can be absorbed, there is a desire to move towards narrower bandgap materials that are able to harvest a greater portion of the solar spectrum. However, in addition to maximising light absorption, the bandgap and band edges of the semiconducting light absorber should also be at appropriate positions to drive surface reactions; for example, photoelectrons within the conduction band need to be sufficiently reducing to drive one of the CO_2 reductions [see eqn (5)–(8)] or proton reduction [eqn (2)], while the valence band edge must be at a potential positive of the water oxidation potential [eqn (1)], leading to a theoretical minimum required bandgap for water splitting and hydrogen evolution of 1.23 eV and for CO_2 reduction to CO coupled to water oxidation of 1.35 eV.[100]

A key advantage of photocatalytic systems is that the reactor design is very inexpensive, with suspensions of light-absorbing semiconductor particles[13,101] placed in a simple transparent container from which product gases can be removed; see Figure 2b. Semiconductor particles are typically on the nanoscale or microscale to maximise the catalytic surface area and minimise the distance between the site of light absorption and catalysis, maximising the probability of charges reaching the surface prior to their recombination (Figure 9).

Semiconductor photocatalysts have been studied for over 40 years for both hydrogen evolution from water[102] and for CO_2 reduction in water.[103] However, to date, relatively few materials have managed to meet the requirements for the energetics of the band edge, the need for catalytic activity, high internal efficiency for charge separation and an ability to operate stably under visible light. Titanium dioxide remains the most extensively studied photocatalyst for both reactions; however, its wide bandgap (*ca.* 3.2 eV, anatase) limits activity to the UV region and typical STH efficiencies for single absorber photocatalysts are <2%.[13] Nonetheless progress is being made for both CO_2 reduction[104,105] and hydrogen evolution, with particularly interesting reports on a range of nitrides and oxides such as C_3N_4[42] and CoO.[106] The low STH efficiencies are in part due to the need to have bandgaps that are significantly greater than the minimum values outlined above. In reality, the presence of relatively large overpotentials for catalysis and the need to achieve a sufficient turnover frequency for the surface reactions mean that the practical minimum bandgap for water splitting and hydrogen production is often quoted as *ca.* 2 eV, significantly lowering the maximum achievable STH for a single absorber system.

In addition to the strict bandgap alignment criteria, single semiconductor photocatalysts often suffer from losses associated with electron–hole recombination and back-reactions at the surface due to both the oxidation and reduction reactions taking place in close proximity and no immediate way of isolating the oxidation and reduction products, which may then go on to react further. To improve the efficiency of desirable surface processes, many light-absorbing materials are modified with catalytic sites. Often the same catalysts that have been shown to be effective in electrolysers (either as heterogeneous electrodes or as a homogeneous electrocatalyst) are found to be suitable when coupled to a light absorber, provided that the band alignment of the semiconductor is such that electron transfer to (reduction) or from (oxidation) the catalyst is thermodynamically feasible. In addition to

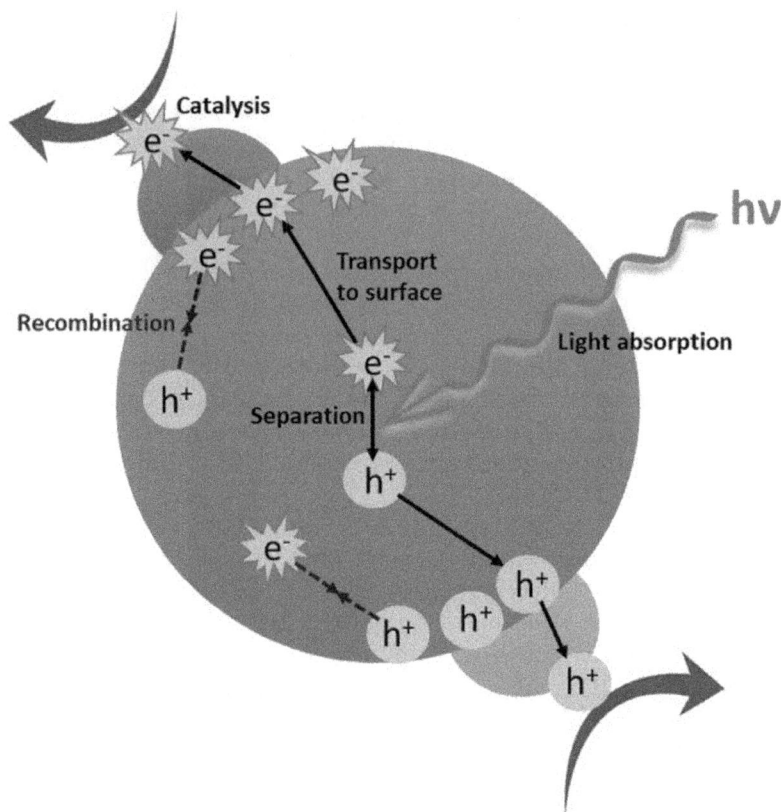

Figure 9   Key kinetic processes occurring in a semiconductor after illumination. Photons that are of sufficient energy are absorbed to promote electrons from the valence band into the conduction band. Photogenerated charges can either recombine or move towards the surface of the semiconductor and further react.

providing sites with lower onset potentials, catalytic centres also act as localised stores for multiple electron/hole equivalents, allowing access to more complex fuel products [*e.g.* eqn (7)–(11)] and they can prevent electron–hole recombination by rapidly removing one of the charges from the light-absorbing material. Examples of co-catalysts used in photocatalytic suspension cells include Pt as a hydrogen evolution[107] or CO_2 reduction[108] catalyst and IrO_x as an oxygen evolution catalyst,[109] and Ag, Cu and a range of molecular co-catalysts have been explored for CO_2 reduction.[110]

    One approach to relaxing the bandgap criteria described above is to include two different light-absorbing semiconductors within the suspension reactor to produce a Z-scheme system; see Figure 10.[12,13,101] The two half-reactions (*e.g.* water oxidation and water reduction) are carried out on different semiconductors, thus relaxing the requirements on each individual material and allowing for a wider range of materials from which to choose.

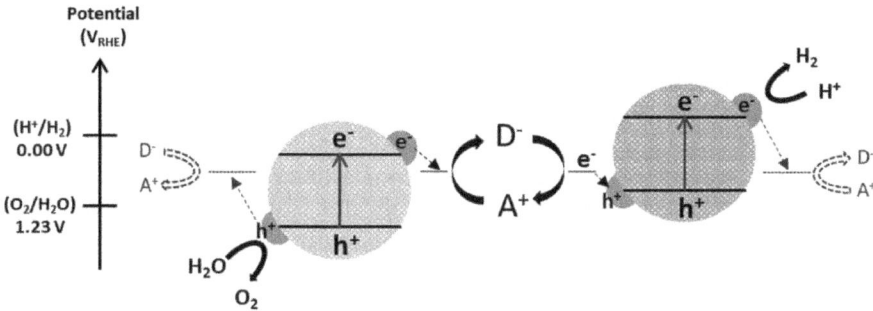

Figure 10 A Z-scheme photocatalytic system for water splitting consisting of two complementary semiconductor photocatalysts with overlapping bandgaps. A shuttle electron mediator is employed to transfer charges between catalysts, effectively expanding the bandgap of the system to be a product of the two individual photocatalysts. Usually, one material is dedicated to the reduction half-reaction, while the other is dedicated to the oxidation half-reaction. The presence of a redox shuttle can lead to unwanted side reactions, which reduce efficiency (dotted lines).[12,13]

In such a device, charges are transferred between the two materials by a redox shuttle. This shuttle can be an acceptor/donor pair in solution such as Fe_3^+/Fe_2^{+111} and IO_3^-/I^{-112} or a solid-state interface such as graphene oxide.[113] Two of the most promising examples of Z-scheme systems are ZrO_2–TaON/WO_3[114] and $SrTiO_3$: Rh/$BiVO_4$.[115] In single-absorber semiconductor photocatalysts, production of both oxygen and the product fuel/feedstock (*e.g.* H_2, CO) occurs in a single reactor, generating a mixed environment that can lead to safety and cost implications. Although a similar single-compartment reactor is also often used with Z-scheme photocatalysts, it is also possible to implement configurations in which the redox shuttle is flowed between two different chambers where each photocatalyst is situated so as to prevent mixing of products, albeit with additional costs associated with the more complex reactor design and the need for pumping.[116] Alternatively, Z-scheme photocatalytic sheets have been demonstrated that consist of water oxidation and H_2 production photocatalyst powders in close proximity in a conducting medium such as gold or carbon for efficient electron transfer.[117,118] This configuration eliminates the need for a conducting electrolyte and redox mediator and can suppress the generation of concentration overpotentials of H^+/OH^- and ohmic potential drops between the hydrogen and oxygen evolution sites that are commonly observed in (photo)electrochemical water splitting; efficiencies are now starting to match those required for niche commercialisation.[117,118]

4.3.2 Photoelectrochemical Cells. Photoelectrochemical (PEC) cells[119–122] represent a hybridisation of semiconductor photocatalysis and conventional electrolysis devices. A PEC cell consists of two electrodes that are electrically connected and immersed in an electrolyte with the electrodes separated by a conducting membrane to isolate the oxidation and reduction products.[116] Typically one (although in some cases both) of the electrodes is a photoactive semiconductor (photoelectrode) which can absorb light and has a suitable valence band edge for the oxidation of water (a photoanode) or a suitable conduction band edge to permit either the reduction of CO_2 or evolution of H_2 (a photocathode); see Figure 2c. A wide range of n-type semiconductors suitable for use as photoanodes are known and the discussion of the physical model of activity will focus on the behaviour of this class of materials. When an n-type semiconductor is immersed in an aqueous electrolyte, electrons flow from the semiconductor (which, in the dark, will typically have a Fermi level above that of the aqueous electrolyte) into the solution in order to achieve electrostatic equilibrium. This leads to a depleted region within the semiconductor close to the electrolyte junction, generating an electric field; the band energies close to the surface are different from those in the bulk. The presence of upwards band bending close to the semiconductor/electrolyte interface allows efficient separation of electrons and holes following light absorption, with holes being swept towards the surface of the electrode to carry out water oxidation while the electrons are transported into the bulk of the material and to a collecting conductor for transport to the cathode for hydrogen production or CO_2 reduction. Under strong illumination, the increase in the quasi-electron Fermi level leads to a flattening of the bands in the absence of an externally applied potential, usually making it necessary to apply an external bias between the electrodes to maintain efficient charge separation; see Figure 11.[121] The use of an additional electrical energy input also allows the use of a wider range of materials with narrower bandgaps and hence better light-harvesting properties.[123] For example, α-Fe_2O_3,[41] a commonly used photoelectrode for water splitting, has a valence band position of sufficient energy to facilitate water oxidation but does not have a conduction band position capable of facilitating H_2 production. However, the addition of a modest voltage between the α-Fe_2O_3 photoelectrode and a suitable cathode can lead to the electrons being capable of hydrogen production.[123] A similar model can also be derived for p-type photocathodes, where the accumulation of charge within the semiconductor close to the electrolyte interface leads to a downwards band bending and the movement of photoelectrons towards the semiconductor surface for catalysis and holes towards the bulk of the material.

Numerous light-absorbing photoanodes have been examined with metal oxides such as α-Fe_2O_3,[41] TiO_2,[124] WO_3 and $BiVO_4$[125] attracting interest for water oxidation owing to the abundance of the elements, their low cost, ease of synthesis and excellent stability. In principle, very high STH efficiencies (ca. 15% for Fe_2O_3) can be achieved; however, to date, for α-Fe_2O_3 efficiencies have rarely exceeded 1%, often because of the need for an undesirably large electrical energy input.[126] An attractive solution, therefore, is also to generate the electrical bias using solar energy and a range of tandem PV–PEC devices

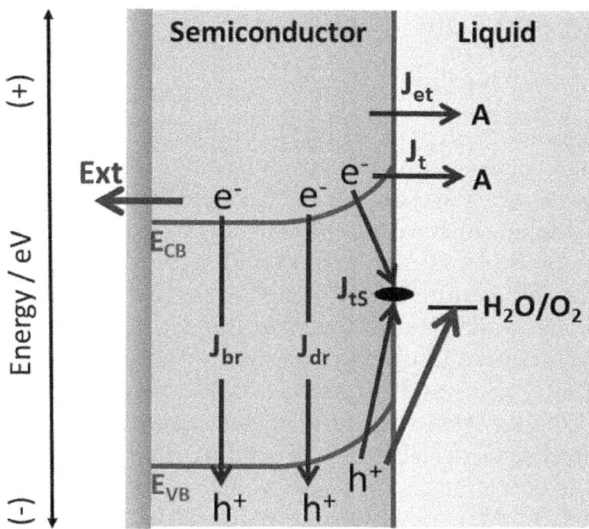

Figure 11 Recombination pathways for photoexcited carriers in an n-type semi-conductor photoelectrochemical cell can be broken down into different categories, represented by the arrows in the diagram. The electron–hole pairs can recombine *via* radiative or non-radiative recombination in the bulk of the semiconductor (J_{br}), depletion-region recombination (J_{dr}), trap-state recombination due to defects (J_{ts}), indirect surface recombination following tunnelling (J_t), and indirect recombination following majority carriers traversing the interfacial barrier (J_{et}). Electron collection by the back-contact and hole reaction with the redox couple (*e.g.* oxidation of water to O_2) are processes that contribute positively to device efficiency (thick arrows). Based on ref. 123.

have now been studied where a PV device with a complementary absorption profile to the semiconductor photoanode is used.[127,128] Various approaches to addressing the photoelectrode's internal efficiency have also been explored, including doping to improve electrical properties and increase the charge separation efficiency,[129] nanostructuring to facilitate transport of holes to the electrode surface while maintaining light absorption[130] and the addition of catalysts such as those described in Section 4.1 (*e.g.* NiFeOOH) to aid the utilisation of photoholes.[41,131] A wide range of photocathode materials (*e.g.* Si, GaP, ZnTe, GaAs, InP, CdTe) are also known.[132] In particular, several Cu(i)-containing oxides have shown promising results, exemplified by p-Cu_2O, which has been used both for the HER[133] and for CO_2 reduction,[134] again with both molecular and metal co-catalysts being used to aid activity and selectivity. Although significant issues exist with the stability of many leading p-type photocathodes in aqueous electrolytes under illumination, great advances have recently been made, particularly through the use of very thin (*ca.* 1–10 nm) protective oxide overlayers, which increase the stability of the photocathodes,[135,136] suggesting that photocathodes for H_2 evolution and CO_2 reduction are a long-term viable option.

5 Summary and Conclusions

Converting solar energy into chemical bonds is an attractive complementary technology to existing electrochemical and mechanical energy storage systems. The high energy density of fuels coupled to the existing storage and transport infrastructure makes them a feasible way to achieve long-term (months) storage. We have examined the case for, and routes to, the production of hydrogen- and carbon-based fuels in a sustainable manner, *i.e.* using only the products of combustion and solar energy as inputs. It is apparent that the key components for light-driven hydrogen evolution from water are already available and that costs and the competitiveness of PV–electrolyser systems are likely to continue to improve. Similarly, PV-driven electrolysis of CO_2 and water for carbon fuel formation (probably *via* initial formation of CO in a syngas mixture) is close to realisation. Indeed, the key technology for electrochemical CO_2 reduction is beginning to be translated into industry-led demonstrator projects. However, improvements in electrodes/catalysts for CO_2 reduction are required if competitiveness with existing fossil carbon fuels is to be achieved. In the longer term, further improvements in efficiency/cost for both solar-driven H_2 production and CO_2 reduction will be needed and an exciting option is to integrate light absorption and catalysis in a single device. Such photocatalytic and photo-electrochemical systems are at an earlier stage of development; however, the pace of development provides optimism that the research community will deliver a sustainable route to the fuels and feedstocks upon which society has come to rely.

References

1. E. Kabir, P. Kumar, S. Kumar, A. A. Adelodun and K.-H. Kim, *Renewable Sustainable Energy Rev.*, 2018, **82**, 894–900.
2. Int. energy agency, www.iea.org, accessed 27/03/2018.
3. N. S. Lewis and D. G. Nocera, *Proc. Natl. Acad. Sci. U. S. A.*, 2006, **103**, 15729–15735.
4. A. Meier and A. Steinfeld, *Adv. Sci. Technol.*, 2010, **74**, 303–312.
5. T. J. Jacobsson, V. Fjällström, M. Edoff and T. Edvinsson, *Energy Environ. Sci.*, 2014, 7, 2056–2070.
6. C. Jiang, S. J. A. Moniz, A. Wang, T. Zhang and J. Tang, *Chem. Soc. Rev.*, 2017, **46**, 4645–4660.
7. K. Takanabe, *Top. Curr. Chem.*, 2015, **371**, 73–104.
8. Y. Tamaki and O. Ishitani, *ACS Catal.*, 2017, 7, 3394–3409.
9. H. Wang, F. Qian and Y. Li, *Nano Energy*, 2014, **8**, 264–273.
10. C. Jiang, S. J. A. Moniz, A. Wang, T. Zhang and J. Tang, *Chem. Soc. Rev.*, 2017, **46**, 4645–4660.
11. P. Y. Yu and M. Cardona, *Fundamentals of Semiconductors*, Springer, Berlin, Heidelberg, 2010.
12. K. Maeda and K. Domen, *J. Phys. Chem. Lett.*, 2010, **1**, 2655–2661.

13. D. M. Fabian, S. Hu, N. Singh, F. A. Houle, T. Hisatomi, K. Domen, F. E. Osterloh and S. Ardo, *Energy Environ. Sci.*, 2015, **8**, 2825–2850.
14. A. F. Ghoniem, *Prog. Energy Combust. Sci.*, 2011, **37**, 15–51.
15. R. M. Navarro Yerga, M. C. Álvarez Galván, F. del Valle, J. A. Villoria de la Mano and J. L. G. Fierro, *ChemSusChem*, 2009, **2**, 471–485.
16. R. Ramachandran, *Int. J. Hydrogen Energy*, 1998, **23**, 593–598.
17. J. Li, Z. Zhao, A. Kazakov and F. L. Dryer, *Int. J. Chem. Kinet.*, 2004, **36**, 566–575.
18. M. Z. Jacobson, W. G. Colella and D. M. Golden, *Science*, 2005, **308**, 1901–1905.
19. T. Elmer, M. Worall, S. Wu and S. B. Riffat, *Renewable Sustainable Energy Rev.*, 2015, **42**, 913–931.
20. C. E. Sandy Thomas, *Int. J. Hydrogen Energy*, 2009, **34**, 9279–9296.
21. U.S. Dep. Energy, https//www.fueleconomy.gov/feg/atv.shtml, accessed 27/03/2018.
22. G. K. Lilik, H. Zhang, J. M. Herreros, D. C. Haworth and A. L. Boehman, *Int. J. Hydrogen Energy*, 2010, **35**, 4382–4398.
23. B. Sakintuna, F. Lamari-Darkrim and M. Hirscher, *Int. J. Hydrogen Energy*, 2007, **32**, 1121–1140.
24. T. K. Tromp, *Science*, 2003, **300**, 1740–1742.
25. E. S. Sanz-Pérez, C. R. Murdock, S. A. Didas and C. W. Jones, *Chem. Rev.*, 2016, **116**, 11840–11876.
26. H. Yang, C. Zhang, P. Gao, H. Wang, X. Li, L. Zhong, W. Wei and Y. Sun, *Catal. Sci. Technol.*, 2017, 7, 4580–4598.
27. B. Kumar, M. Llorente, J. Froehlich, T. Dang, A. Sathrum and C. P. Kubiak, *Annu. Rev. Phys. Chem.*, 2012, **63**, 541–569.
28. J. Schneider, H. Jia, J. T. Muckerman and E. Fujita, *Chem. Soc. Rev.*, 2012, **41**, 2036–2051.
29. M. Schreier, L. Curvat, F. Giordano, L. Steier, A. Abate, S. M. Zakeeruddin, J. Luo, M. T. Mayer and M. Grätzel, *Nat. Commun.*, 2015, **6**, 7326.
30. Krylova, *Solid Fuel Chem.*, 2014, **48**, 22–35.
31. Q. Zhang, J. Kang and Y. Wang, *ChemCatChem*, 2010, **2**, 1030–1058.
32. P. L. Spath, *Preliminary screening–technical and economic assessment of synthesis gas to fuels and chemicals with emphasis on the potential for biomass-derived syngas [electronic resource]/P.L. Spath and D.C. Dayton.*, Golden, Colo: National Renewable Energy Laboratory, 2003.
33. M. E. Dry, *Catal. Today*, 2002, **71**, 227–241.
34. Y. Hori, H. Wakebe, T. Tsukamoto and O. Koga, *Electrochim. Acta*, 1994, **39**, 1833–1839.
35. F. Li, S.-F. Zhao, L. Chen, A. Khan, D. R. MacFarlane and J. Zhang, *Energy Environ. Sci.*, 2016, **9**, 216–223.
36. M. Ma, H. A. Hansen, M. Valenti, Z. Wang, A. Cao, M. Dong and W. A. Smith, *Nano Energy*, 2017, **42**, 51–57.
37. F. Sastre, M. J. Muñoz-Batista, A. Kubacka, M. Fernández-García, W. A. Smith, F. Kapteijn, M. Makkee and J. Gascon, *ChemElectroChem*, 2016, **3**, 1497–1502.

38. P. Kang, Z. Chen, A. Nayak, S. Zhang and T. J. Meyer, *Energy Environ. Sci.*, 2014, **7**, 4007–4012.
39. J. Hong, W. Zhang, J. Ren and R. Xu, *Anal. Methods*, 2013, **5**, 1086.
40. Z. Sun, T. Ma, H. Tao, Q. Fan and B. Han, *Chem*, 2017, **3**, 560–587.
41. K. Sivula, F. Le Formal and M. Grätzel, *ChemSusChem*, 2011, **4**, 432–449.
42. J. Liu, Y. Liu, N. Liu, Y. Han, X. Zhang, H. Huang, Y. Lifshitz, S.-T. Lee, J. Zhong and Z. Kang, *Science*, 2015, **347**, 970–974.
43. R. W. Miles, G. Zoppi and I. Forbes, *Mater. Today*, 2007, **10**, 20–27.
44. T. D. Lee and A. U. Ebong, *Renewable Sustainable Energy Rev.*, 2017, **70**, 1286–1297.
45. B. Parida, S. Iniyan and R. Goic, *Renewable Sustainable Energy Rev.*, 2011, **15**, 1625–1636.
46. A. De Vos, *J. Phys. D: Appl. Phys.*, 1980, **13**, 839–846.
47. J. Jia, L. C. Seitz, J. D. Benck, Y. Huo, Y. Chen, J. W. D. Ng, T. Bilir, J. S. Harris and T. F. Jaramillo, *Nat. Commun.*, 2016, **7**, 13237.
48. M. Carmo, D. L. Fritz, J. Mergel and D. Stolten, *Int. J. Hydrogen Energy*, 2013, **38**, 4901–4934.
49. C. Xiang, K. M. Papadantonakis and N. S. Lewis, *Mater. Horizons*, 2016, **3**, 169–173.
50. M. Ruth and F. Joseck, *Prog. Rec. Offices Fuel Cell Technol.*, 2011, 1–8.
51. O. E. Miller, C. Ainscough and A. Talapatra, *United States Dep. Energy*, 2011, **10**, 1–9, 2014.
52. M. Dumortier, S. Tembhurne and S. Haussener, *Energy Environ. Sci.*, 2015, **8**, 3614–3628.
53. B. A. Pinaud, J. D. Benck, L. C. Seitz, A. J. Forman, Z. Chen, T. G. Deutsch, B. D. James, K. N. Baum, G. N. Baum, S. Ardo, H. Wang, E. Miller and T. F. Jaramillo, *Energy Environ. Sci.*, 2013, **6**, 1983.
54. S. A. Bonke, M. Wiechen, D. R. MacFarlane and L. Spiccia, *Energy Environ. Sci.*, 2015, **8**, 2791–2796.
55. I. Roger, M. A. Shipman and M. D. Symes, *Nat. Rev. Chem.*, 2017, 1.
56. B. M. Hunter, H. B. Gray and A. M. Müller, *Chem. Rev.*, 2016, **116**, 14120–14136.
57. L. Fan, P. F. Liu, X. Yan, L. Gu, Z. Z. Yang, H. G. Yang, S. Qiu and X. Yao, *Nat. Commun.*, 2016, **7**, 10667–10674.
58. D. Merki, S. Fierro, H. Vrubel and X. Hu, *Chem. Sci.*, 2011, **2**, 1262.
59. E. J. Popczun, C. G. Read, C. W. Roske, N. S. Lewis and R. E. Schaak, *Angew. Chem., Int. Ed.*, 2014, **53**, 5427–5430.
60. S. Trasatti, *J. Electroanal. Chem.*, 1972, **39**, 163–184.
61. M. Merrill and R. Dougherty, *J. Phys. Chem. C*, 2008, **112**, 3655–3666.
62. X. Lu and C. Zhao, *Nat. Commun.*, 2015, **6**.
63. M. Gong, Y. Li, H. Wang, Y. Liang, J. Z. Wu, J. Zhou, J. Wang, T. Regier, F. Wei and H. Dai, *J. Am. Chem. Soc.*, 2013, **135**, 8452–8455.
64. X. Li, F. C. Walsh and D. Pletcher, *Phys. Chem. Chem. Phys.*, 2011, **13**, 1162–1167.
65. M. W. Kanan and D. G. Nocera, *Science*, 2008, **321**, 1072–1075.

66. A. J. Esswein, Y. Surendranath, S. Y. Reece and D. G. Nocera, *Energy Environ. Sci.*, 2011, **4**, 499–504.
67. J. Greeley, T. F. Jaramillo, J. Bonde, I. Chorkendorff and J. K. Nørskov, *Nat. Mater.*, 2006, **5**, 909–913.
68. S. Y. Reece, J. A. Hamel, K. Sung, T. D. Jarvi, A. J. Esswein, J. J. H. Pijpers and D. G. Nocera, *Science*, 2011, **334**, 645–648.
69. J. L. Dempsey, B. S. Brunschwig, J. R. Winkler and H. B. Gray, *Acc. Chem. Res.*, 2009, **42**, 1995–2004.
70. C. Li, M. Wang, J. Pan, P. Zhang, R. Zhang and L. Sun, *J. Organomet. Chem.*, 2009, **694**, 2814–2819.
71. A. Fihri, V. Artero, A. Pereira and M. Fontecave, *Dalton Trans.*, 2008, 5567.
72. M. A. Gross, A. Reynal, J. R. Durrant and E. Reisner, *J. Am. Chem. Soc.*, 2014, **136**, 356–366.
73. C. J. Curtis, A. Miedaner, R. Ciancanelli, W. W. Ellis, B. C. Noll, M. Rakowski DuBois and D. L. DuBois, *Inorg. Chem.*, 2003, **42**, 216–227.
74. A. Dutta, S. Lense, J. Hou, M. H. Engelhard, J. A. S. Roberts and W. J. Shaw, *J. Am. Chem. Soc.*, 2013, **135**, 18490–18496.
75. M. L. Helm, M. P. Stewart, R. M. Bullock, M. R. DuBois and D. L. DuBois, *Science*, 2011, **333**, 863–866.
76. A. D. Wilson, R. H. Newell, M. J. McNevin, J. T. Muckerman, M. R. DuBois and D. L. DuBois, *J. Am. Chem. Soc.*, 2006, **128**, 358–366.
77. B. Jiang, X.-G. Zhang, K. Jiang, D.-Y. Wu and W.-B. Cai, *J. Am. Chem. Soc.*, 2018, **140**, 2880–2889.
78. Y. Hori, *Mod. Aspects Electrochem.*, 2008, 89–189.
79. L. Zhang, Z. J. Zhao and J. Gong, *Angew. Chem., Int. Ed.*, 2017, **56**, 11326–11353.
80. Q. Lu, J. Rosen, Y. Zhou, G. S. Hutchings, Y. C. Kimmel, J. G. Chen and F. Jiao, *Nat. Commun.*, 2014, **5**.
81. M. Liu, Y. Pang, B. Zhang, P. De Luna, O. Voznyy, J. Xu, X. Zheng, C. T. Dinh, F. Fan, C. Cao, F. P. G. de Arquer, T. S. Safaei, A. Mepham, A. Klinkova, E. Kumacheva, T. Filleter, D. Sinton, S. O. Kelley and E. H. Sargent, *Nature*, 2016, **537**, 382–386.
82. M. Bourrez, F. Molton, S. Chardon-Noblat and A. Deronzier, *Angew. Chem., Int. Ed.*, 2011, **50**, 9903–9906.
83. C. Costentin, G. Passard, M. Robert and J.-M. Savéant, *Proc. Natl. Acad. Sci.*, 2014, **111**, 14990–14994.
84. J. Shen, R. Kortlever, R. Kas, Y. Y. Birdja, O. Diaz-Morales, Y. Kwon, I. Ledezma-Yanez, K. J. P. Schouten, G. Mul and M. T. M. Koper, *Nat. Commun.*, 2015, **6**, 8177.
85. J. Schneider, H. Jia, K. Kobiro, D. E. Cabelli, J. T. Muckerman and E. Fujita, *Energy Environ. Sci.*, 2012, **5**, 9502.
86. J. Tory, B. Setterfield-Price, R. A. W. Dryfe and F. Hartl, *ChemElectroChem*, 2015, **2**, 213–217.
87. R. Francke, B. Schille and M. Roemelt, *Chem. Rev.*, 2018. DOI: 10.1021/acs.chemrev.7b00459.

88. I. Bhugun, D. Lexa and J.-M. Saveant, *J. Am. Chem. Soc.*, 1994, **116**, 5015–5016.
89. I. Azcarate, C. Costentin, M. Robert and J. M. Savéant, *J. Am. Chem. Soc.*, 2016, **138**, 16639–16644.
90. J. J. Walsh, C. L. Smith, G. Neri, G. F. S. Whitehead, C. M. Robertson and A. J. Cowan, *Faraday Discuss.*, 2015, **183**, 147–160.
91. J. J. Walsh, G. Neri, C. L. Smith and A. J. Cowan, *Chem. Commun.*, 2014, **50**, 12698–12701.
92. I. Bhugun, D. Lexa and J. Savéant, *J. Am. Chem. Soc.*, 1996, **118**, 1769–1776.
93. G. Neri, I. M. Aldous, J. J. Walsh, L. J. Hardwick and A. J. Cowan, *Chem. Sci.*, 2016, **7**, 1521–1526.
94. F. Franco, C. Cometto, F. Ferrero Vallana, F. Sordello, E. Priola, C. Minero, C. Nervi and R. Gobetto, *Chem. Commun.*, 2014, **50**, 14670–14673.
95. J. J. Walsh, C. L. Smith, G. Neri, G. F. S. Whitehead, C. M. Robertson and A. J. Cowan, *Faraday Discuss.*, 2015, **183**, 147–160.
96. B. Reuillard, K. H. Ly, T. E. Rosser, M. F. Kuehnel, I. Zebger and E. Reisner, *J. Am. Chem. Soc.*, 2017, **139**, 14425–14435.
97. X. Zhang, Z. Wu, X. Zhang, L. Li, Y. Li, H. Xu, X. Li, X. Yu, Z. Zhang, Y. Liang and H. Wang, *Nat. Commun.*, 2017, **8**, 14675.
98. T. Haas, R. Krause, R. Weber, M. Demler and G. Schmid, *Nat. Catal.*, 2018, **1**, 32–39.
99. Evonik, http://corporate.evonik.com/en/media/press_releases/Pages/news-details.aspx?newsid=72457, accessed 29/03/2018.
100. S. Chen and L.-W. Wang, *Chem. Mater.*, 2012, **24**, 3659–3666.
101. A. J. Bard, *J. Photochem.*, 1979, **10**, 59–75.
102. W. Fan, Q. Zhang and Y. Wang, *Phys. Chem. Chem. Phys.*, 2013, **15**, 2632.
103. T. Inoue, A. Fujishima, S. Konishi and K. Honda, *Nature*, 1979, **277**, 637–638.
104. M. Marszewski, S. W. Cao, J. G. Yu and M. Jaroniec, *Mater. Horizons*, 2015, **2**, 261–278.
105. J. L. White, M. F. Baruch, J. E. Pander III, Y. Hu, I. C. Fortmeyer, J. E. Park, T. Zhang, K. Liao, J. Gu, Y. Yan, T. W. Shaw, E. Abelev and A. B. Bocarsly, *Chem. Rev.*, 2015, **115**, 12888–12935.
106. L. Liao, Q. Zhang, Z. Su, Z. Zhao, Y. Wang, Y. Li, X. Lu, D. Wei, G. Feng, Q. Yu, X. Cai, J. Zhao, Z. Ren, H. Fang, F. Robles-Hernandez, S. Baldelli and J. Bao, *Nat. Nanotechnol.*, 2014, **9**, 69–73.
107. J. O. M. Bockris, I. A. Ammar and A. K. M. S. Huq, *J. Phys. Chem.*, 1957, **61**, 879–886.
108. Y.-X. Pan, Y. You, S. Xin, Y. Li, G. Fu, Z. Cui, Y.-L. Men, F.-F. Cao, S.-H. Yu and J. B. Goodenough, *J. Am. Chem. Soc.*, 2017, **139**, 4123–4129.
109. S. Trasatti, *J. Electroanal. Chem. Interfacial Electrochem.*, 1980, **111**, 125–131.
110. A. J. Morris, G. J. Meyer and E. Fujita, *Acc. Chem. Res.*, 2009, **42**, 1983–1994.
111. A. K. Padhi, *J. Electrochem. Soc.*, 1997, **144**, 1609.
112. K. Maeda, *ACS Catal.*, 2013, **3**, 1486–1503.

113. A. Iwase, Y. H. Ng, Y. Ishiguro, A. Kudo and R. Amal, *J. Am. Chem. Soc.*, 2011, **133**, 11054–11057.
114. K. Maeda, M. Higashi, D. Lu, R. Abe, K. Domen, M. Higash, D. Lu, R. Abe and K. Domen, *J. Am. Chem. Soc.*, 2010, **132**, 5858–5868.
115. H. Kato, Y. Sasaki, N. Shirakura and A. Kudo, *J. Mater. Chem. A*, 2013, **1**, 12327.
116. K. A. Mauritz and R. B. Moore, *Chem. Rev.*, 2004, **104**, 4535–4586.
117. Q. Wang, T. Hisatomi, Q. Jia, H. Tokudome, M. Zhong, C. Wang, Z. Pan, T. Takata, M. Nakabayashi, N. Shibata, Y. Li, I. D. Sharp, A. Kudo, T. Yamada and K. Domen, *Nat. Mater.*, 2016, **15**, 611–615.
118. Q. Wang, T. Hisatomi, Y. Suzuki, Z. Pan, J. Seo, M. Katayama, T. Minegishi, H. Nishiyama, T. Takata, K. Seki, A. Kudo, T. Yamada and K. Domen, *J. Am. Chem. Soc.*, 2017, **139**, 1675–1683.
119. L. M. Peter and K. G. Upul Wijayantha, *ChemPhysChem*, 2014, **15**, 1983–1995.
120. E. L. Miller, *Energy Environ. Sci.*, 2015, **8**, 2809–2810.
121. M. Grätzel, *Nature*, 2001, **414**, 338–344.
122. M. S. Prévot and K. Sivula, *J. Phys. Chem. C*, 2013, **117**, 17879–17893.
123. N. S. Lewis, *Inorg. Chem.*, 2005, **44**, 6900–6911.
124. J. Nowotny, T. Bak, M. Nowotny and L. Sheppard, *Int. J. Hydrogen Energy*, 2007, **32**, 2609–2629.
125. T. W. Kim and K.-S. Choi, *Science*, 2014, **343**, 990–994.
126. J. Y. Kim, G. Magesh, D. H. Youn, J.-W. Jang, J. Kubota, K. Domen and J. S. Lee, *Sci. Rep.*, 2013, **3**, 2681.
127. J. Brillet, J.-H. Yum, M. Cornuz, T. Hisatomi, R. Solarska, J. Augustynski, M. Graetzel and K. Sivula, *Nat. Photonics*, 2012, **6**, 824–828.
128. J.-W. Jang, C. Du, Y. Ye, Y. Lin, X. Yao, J. Thorne, E. Liu, G. McMahon, J. Zhu, A. Javey, J. Guo and D. Wang, *Nat. Commun.*, 2015, **6**, 7447.
129. M. Forster, R. J. Potter, Y. Ling, Y. Yang, D. R. Klug, Y. Li and A. J. Cowan, *Chem. Sci.*, 2015, **6**, 4009–4016.
130. R. van de Krol, Y. Liang and J. Schoonman, *J. Mater. Chem.*, 2008, **18**, 2311.
131. M. G. Walter, E. L. Warren, J. R. McKone, S. W. Boettcher, Q. Mi, E. A. Santori and N. S. Lewis, *Chem. Rev.*, 2010, **110**, 6446–6673.
132. B. Kumar, M. Llorente, J. Froehlich, T. Dang, A. Sathrum and C. P. Kubiak, *Annu. Rev. Phys. Chem.*, 2012, **63**, 541–569.
133. Y. Yang, D. Xu, Q. Wu and P. Diao, *Sci. Rep.*, 2016, **6**, 35158–35170.
134. M. Schreier, J. Luo, P. Gao, T. Moehl, M. T. Mayer and M. Grätzel, *J. Am. Chem. Soc.*, 2016, **138**, 1938–1946.
135. A. Paracchino, V. Laporte, K. Sivula, M. Grätzel and E. Thimsen, *Nat. Mater.*, 2011, **10**, 456–461.
136. C. G. Morales-Guio, S. D. Tilley, H. Vrubel, M. Grätzel and X. Hu, *Nat. Commun.*, 2014, **5**, 3059–3065.

Thermal and Thermochemical Storage

YUKITAKA KATO* AND TAKAHIRO NOMURA

ABSTRACT

The principles and potential of latent heat storage (LHS) and thermo-chemical energy storage (TCES) are introduced. LHS is a reliable technology for heat storage over a wide range of temperatures from low to high, and for cold storage such as ice storage. Phase-change material (PCM) development is a key technology for LHS. Technologies for overcoming the major PCM problems of low thermal conductivity and high corrosivity are reviewed. TCES can be operated as a thermally driven chemical heat pump and for heat transformation (upgrading) and cold heat production, and also to store heat with low loss, although TCES has not been commercialized. The principle of heat pump operation is outlined. TCES material generally has low thermal conductivity, and thermal conductivity enhancement methodology and the benefits of the methodology are reviewed.

1   Introduction

Surplus thermal energy storage is a useful way of saving energy. There is a variety of heat storage media that are able to recover and utilize surplus heat. Sensible heat storage (SHS) using water and some other fluids is the most popular way owing to its reliability and low cost. A higher heat storage density is accomplished by latent heat storage (LHS) using a phase-change material (PCM). For heat storage at temperatures higher than room temperature, organic materials such as waxes and metals such as copper are

---

*Corresponding author.

Issues in Environmental Science and Technology No. 46
Energy Storage Options and Their Environmental Impact
Edited by R.E. Hester and R.M. Harrison
© The Royal Society of Chemistry 2019
Published by the Royal Society of Chemistry, www.rsc.org

practical and appropriate media for PCM. For cold heat storage, water and some inorganic materials, including sodium sulfate, have commercial markets. Generally, the heat storage densities of SHS and LHS materials are <0.2 and <0.5 MJ L^{-1}, respectively. A higher energy storage density is required for automobile and massive heat storage sectors; thermochemical heat storage (TCES) has the potential to meet such a demand because of its higher density of >1 MJ L^{-1} and it is expected to widen LHS and TCES markets for enhancement of surplus thermal energy utilization. LHS and TCES materials are required to improve not only the heat storage density, but also the thermal conductivity for heat output power enhancement, material transportability and material adaptability for efficient connection between heat source and demand. This chapter introduces the key technologies of LHS and TCES for matching the requirements.

2   Latent Heat Storage

In the low-temperature range below 100 °C, various applications of LHS, such as ice thermal storage and building materials, have been commercialized. On the other hand, recently mid–high-temperature LHS above 100 °C has attracted considerable attention for reducing fossil fuel consumption and utilizing renewable energy. This section reviews the principle and applications of LHS, focusing especially on mid–high-temperature LHS.

2.1   Principle of LHS

The relationship between the cumulative heat storage capacity of a PCM and operating temperature (Figure 1) explains the principle of LHS. When a heat source for LHS has been targeted, an appropriate phase-change material (PCM) has to be selected. The temperature of the phase transformation should match the temperature of the heat source.

The cumulative heat storage capacity of PCM is given by

$$Q = m[C_{P,l}(T - T_{\text{transform}}) + L + C_{P,s}(T_{\text{transform}} - T_i)] \tag{1}$$

where $C_{P,l}$ is the sensible heat of liquid PCM, $C_{P,s}$ is the sensible heat of liquid PCM, $T_{\text{transform}}$ is the transformation temperature, T_i is the initial temperature and L is latent heat. An LHS material stores/releases heat when it changes phase (mainly solid $\leftrightarrow$ liquid); therefore, a stepwise change of Q occurs at $T_{\text{transform}}$. This simple principle provides three advantages for LHS: (1) its high heat capacity due to the latent heat of phase transformation; (2) it can be used as a constant-temperature heat absorber and source by utilizing the phase transformation; and (3) the repeatable use of a PCM since it does not involve a chemical reaction.

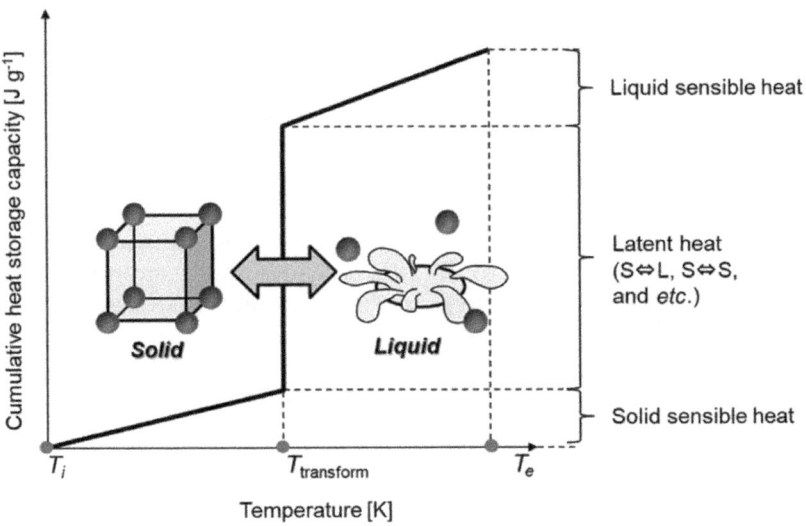

Figure 1   Principle of LHS: relationship between cumulative heat storage capacity of a PCM and temperature.

2.2   Materials for LHS

Simply described, every material that has a high latent heat, good chemical stability and sufficient thermal stability can be considered as a PCM candidate, although there are, of course, many other desirable properties for PCM candidates.[1] Various materials, such as paraffins, fatty acids, hydrated salts, sugar alcohols, molten salts, metals, alloys and ceramics, have been proposed. Table 1 lists the status of development of typical mid–high-temperature PCMs.[2] Each material group has a suitable temperature range. Sugar alcohols, molten salts and metals or alloys are suitable PCMs for mid–high-temperature applications at >100 °C. In this temperature range, solar cooling using a thermally driven heat pump, industrial waste heat recovery and concentrated solar power (CSP) are potential applications. Although there are a few commercialized LHS systems, mid–high-temperature PCMs are an underdeveloped technology. A major technical problem with mid–high-temperature PCMs is the high corrosivity of structural materials of liquid PCMs such as molten salts and metals.

2.3   Encapsulation and Composite Technology for LHS

2.3.1   Encapsulation Technology.   Encapsulation of PCMs is an important technology for LHS in all temperature ranges. Figure 2 shows the concept of encapsulating a PCM. Generally, a PCM with a high latent heat for its solid–liquid phase transition is typical for LHS and, therefore, encapsulation of the PCM by a solid shell is needed to maintain the shape and prevent leakage of liquid PCM. Encapsulation also provides other

Table 1 Status of mid–high-temperature LHS development.[2] Reprinted from ref. 2 with permission from John Wiley and Sons, Copyright © 2016 John Wiley & Sons, Ltd.

Property	Sugar alcohol		Molten salt		Metal or alloy
	Conventional	Future	Conventional	Future	Future
Melting point/°C	<120	150 +	200–400	500 +	500 +
PCM candidate	Erythritol (T_m: 118 °C)	Mannitol (T_m: 167 °C)	Nitrate (NaNO$_3$, T_m: 307 °C)	Chloride (NaCl, T_m: 800 °C), Carbonate (Na$_2$CO$_3$, T_m: 858 °C)	Al series (Al–Si, T_m: 580 °C), Cu or Fe series (Cu–Si, T_m: 802 °C)
Performance					
Heat capacity	High	High	Middle	High	High
Thermal conductivity	Low	Low	Low	Low	Remarkably high
Volume expansion during phase change	~10%	~10%	~10%	~10–30%	~5%
Corrosion	Completely solved	Completely solved	Partially solved	Under development	Under development
Status of development	Commercial plant	Bench-scale heat-exchange study	Pilot-scale heat-exchange study	Laboratory-scale heat-exchange study	Material development

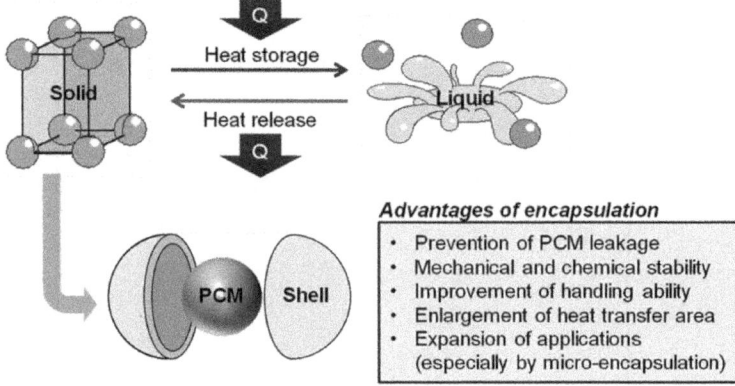

Figure 2   Concepts of encapsulating PCM.

advantages for a PCM, such as mechanical and chemical stability, improvement of handling ability and enlargement of its heat transfer area. Encapsulation technology can be classified in accordance with the size of the PCM capsule: macro-encapsulation (>1 mm) and micro–nano-encapsulation. Macro-encapsulated PCMs, in almost all cases, are applied to packed-bed LHS heat exchangers. Cylinders and spheres are typical shapes of PCM macro-capsules. In many cases of macro-encapsulation, the PCM is just packed in a shell and then the shell is sealed. Figure 3 shows examples of macro-encapsulation of metallic PCMs using cylindrical Al_2O_3 shells, which were prepared as in a previous study.[3] In this case, Al film was used as sealing material.

Micro–nano-encapsulation of PCMs is one of the main research and development fields for LHS because it can expand the applications of LHS, not only for LHS heat exchangers but also multifunctional materials for heat transportation (such as heat-transfer fluids) or temperature regulation (such as building materials and concretes). Figure 4 shows an example of a core–shell-type micro-encapsulated PCM that is composed of Al–Si microspheres as the PCM and Al_2O_3 as the shell, prepared as described in a previous study.[4] Recently, encapsulation by hard shells such as SiO_2 and Al_2O_3 has attracted much attention owing to their excellent corrosion stability and mechanical strength.

2.3.2   Composite Technology.   As shown in Table 1, common PCM candidates other than metals or alloys have low thermal conductivity. A low thermal conductivity of a PCM causes low heat-exchange rates during heat storage and release. To improve the thermal conductivity of common PCMs, PCM composites – mixtures of PCMs and high thermal conductivity materials – are one of the main research and development areas in LHS. Figure 5 shows a classification of the preparation methods for PCM composites. Cold or hot pressing of mixed PCMs and high thermal conductivity powders and impregnation of the PCM with high thermal

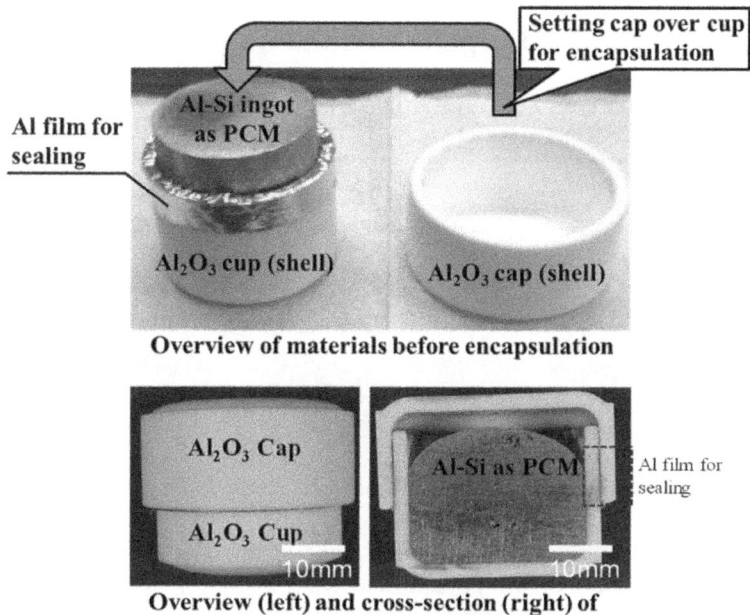

Overview of materials before encapsulation

Overview (left) and cross-section (right) of the PCM macro-capsule.

Figure 3 An example of macro-encapsulation of metallic PCMs using a cylinder-type Al$_2$O$_3$ shell.

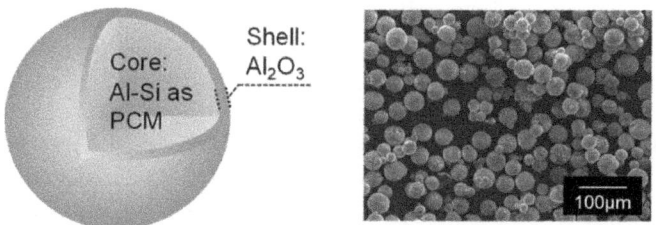

Figure 4 An example of a core–shell-type micro-encapsulated PCM that is composed of Al–Si microspheres as PCM and Al$_2$O$_3$ as shell.

conductivity porous materials are proven methods to improve the thermal conductivity of PCMs. Carbon materials, metals and ceramics (such as nitrides) have been investigated as high thermal conductivity materials. The development of high thermal conductivity composites with a low volume fraction of added materials is a goal of this technology.

2.4 Heat Exchangers for LHS

LHS heat exchangers can be classified according to indirect or direct contact between the PCM and the heat-transfer fluid (HTF), and passive or active dynamics of the PCM (see Figure 6). Indirect-passive types, which include

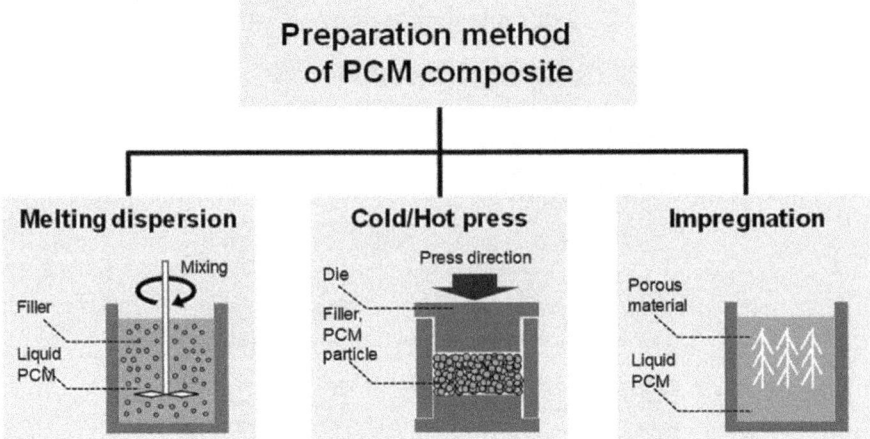

Figure 5 Classifications of preparation methods for PCM composites.

packed-bed and shell-and-tube types, are the most common LHS heat exchangers. In the case of the packed-bed type, PCM capsules are packed in a storage tank; the heat transfer area is easily enlarged simply by selection of smaller sized capsules or increasing the number of PCM capsules. In the case of shell-and-tube-type latent heat exchangers, large numbers of heat transfer tubes are set in the PCM layer. This type has been used for direct steam generation in a thermal energy storage system within concentrated solar power plants, where high-pressure steam can be used as the HTF.[5] A fluidized bed is classified as an indirect-active heat exchanger. In this system, PCM capsules (usually microcapsules) are circulated in order to transport heat from one process to another, giving greater efficiency than if only the HTF is circulated. A direct-contact heat exchanger has a very a simple, light structure and a high thermal storage density since the heat transfer proceeds between the PCM and the HTF without any barrier. To realize this system, insolubility and easy phase-separation characteristics are required for the PCM and HTF.

2.5   Applications of LHS

In the low-temperature range below 100 °C, various applications of LHS, such as ice thermal storage, building materials, *etc.*, have been commercialized. However, mid–high-temperature LHS at >100 °C is an underdeveloped technology. Here, recent advances in LHS in the mid–high-temperature range are described as a key technology for reducing fossil fuel consumption and utilizing renewable energy.

2.5.1   Latent Heat Transportation Systems.   A mobile latent heat transportation system using erythritol, a type of sugar alcohol, as a PCM has been commercialized in Japan. In this system, industrial waste heat at around

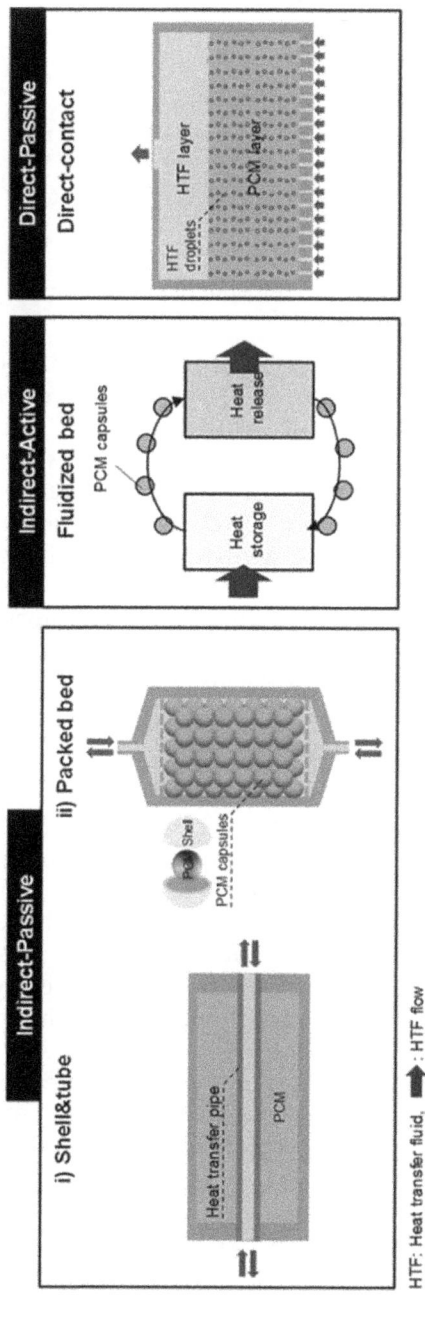

HTF: Heat transfer fluid,   : HTF flow

Figure 6   Classification of LHS heat exchangers.[2] Reprinted from ref. 2 with permission from John Wiley and Sons, Copyright © 2016 John Wiley & Sons, Ltd.

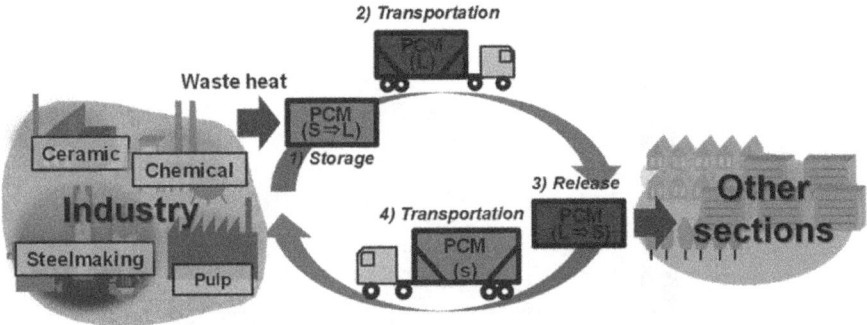

Figure 7   Overview of the latent heat transportation system.

200 °C is recovered and stored in the form of latent heat of the PCM, then transported by container truck to heat supply sites such as office buildings, hospitals, horticulture centres, spas and hotels.[6] This system enables heat to be transported up to about 35 km. Figure 7 shows an overview of this latent heat transportation system. A mobile latent heat storage container with heat exchanger is used to recover, transport and supply waste heat. Both direct and indirect contact latent heat exchangers can be used. Erythritol was used as the PCM for this storage system owing to its high latent heat (340 J g^{-1}) and its suitability (melting temperature 118 °C) for recovery of industrial exhaust heat. In addition, this melting temperature is high enough to operate a single-stage absorption chiller that supplies cold water at about 5–7 °C. Large (1.7–24 tonne) heat-transfer containers are used in the practical system.

2.5.2   LHS for Concentrated Solar Power.   An important research topic of high-temperature LHS is thermal energy storage systems for concentrated solar power (CSP) plants. Although sensible heat storage using liquid molten salts has been used in commercial CSP plants, a large amount of molten salt is required owing to the low heat capacity of sensible heat storage materials. Under these circumstances, the development of LHS using nitrates, especially sodium nitrate-based materials, with melting temperatures around 300 °C, as PCMs is accelerating. Although both packed-bed and shell-and-tube-type latent heat exchangers are being developed, the latter type, in particular, has attracted great attention since it can serve as a direct steam generator (DSG) by considering only the high-pressure design of the heat-transfer tubes. A DSG in which the heat capacity of the storage units, using a nitrate salt as the PCM, is around 100 kW$_{th}$ has been demonstrated.[7] However, new LHS technology using high melting temperature PCMs at >500 °C is strongly required, since innovations in CSP solar systems, from the troughs to the central towers, are raising the operating temperatures. Therefore, the development of an LHS system using a molten salt PCM with a melting temperature around 800 °C, such

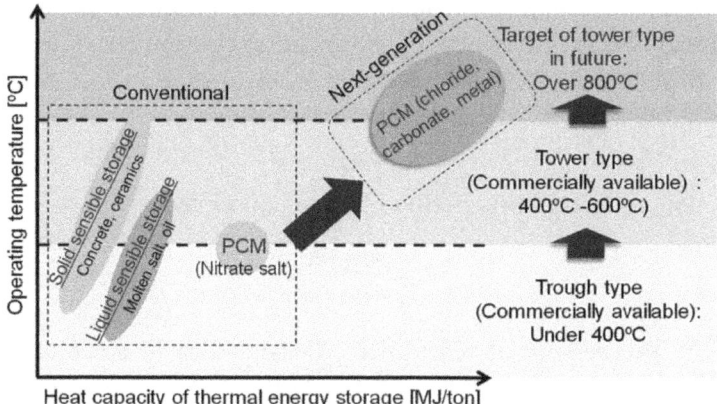

Figure 8 Target properties of next-generation PCMs for CSP.

as metal chlorides[8,9] and carbonates,[10,11] or metallic PCMs such as Al-based alloys,[3,4,12,13] is expected (Figure 8).

3 Thermochemical Energy Storage

Thermochemical energy storage (TCES) has the potential to be applied in high-temperature thermal storage. TCES is expected to have a great potential for high-temperature heat storage with high energy density. This section reviews the principle and application of TCES.

3.1 Principle of TCES

3.1.1 Chemical Reaction Equilibrium for TCES. When a heat source for TCES is targeted, an appropriate reaction system has to be selected for the conversion. The temperature range of the reaction should fit the temperature of the heat source. The turning temperature of a reaction, T_{turn} (K), can be a measure to find the reaction system. The Gibbs free energy change of reaction, ΔG (kJ mol^{-1}), is obtained from the reaction enthalpy change, ΔH (kJ mol^{-1}), and the enthalpy change, ΔS (kJ mol^{-1} K^{-1}):

$$\Delta G = \Delta H - T\Delta S \qquad (2)$$

ΔG has the following relationship with the reaction equilibrium constant, K:

$$lnK = -\frac{\Delta G}{RT} = -\frac{\Delta H}{R}\frac{1}{T} + \frac{\Delta S}{R} \qquad (3)$$

The reversible reaction condition is established around $K = 1$, that is, $\Delta G = 0$. Then, T_{turn} is given by

$$T_{turn} = \frac{\Delta H}{\Delta S} \qquad (4)$$

When a reversible chemical reaction has a temperature T_{turn}, the reaction can be applicable for thermal energy utilization at around that temperature.

Now, the following gas–solid reaction of magnesium oxide and water is used as an example:

$$MgO(s) + H_2O(g) \rightleftharpoons Mg(OH)_2, \quad \Delta H = -81.0 \text{ kJ mol}^{-1} \tag{5}$$

The reaction equilibrium is defined by the equation

$$K = \frac{a_{MgO} a_{H_2O}}{a_{Mg(OH)_2}} \tag{6}$$

where a_i is the activity of reactant i. The activities of solid and gaseous materials are expressed as unity and the partial pressure, respectively. Then eqn (6) is reduced to

$$K = P_{H_2O} \tag{7}$$

and

$$\ln P_{H_2O} = -\frac{\Delta H}{R}\frac{1}{T} + \frac{\Delta S}{R} \tag{8}$$

Eqn (8) shows that the reaction temperature corresponds to the reaction pressure in gas–solid reaction systems such as that shown as eqn (5). The relationship between $1/T$ and $\ln K$ ($= P_{H_2O}$) is linear over a range in which changes of ΔH and ΔS are negligible, and is called the reaction equilibrium line. The temperature at a reaction (vapour) pressure of 101.3 kPa is T_{turn} on the equilibrium line. It is recognized that nickel oxide (NiO) and cobalt oxide (CoO) have the potential to be reaction candidates for heat storage at <200 °C, and strontium oxide (SrO) and barium oxide (BaO) could be candidates for storage at >700 °C.

Figure 9a shows the equilibrium line and the change of the reaction state. At a constant temperature, an initial state (A) moves to an equilibrium state

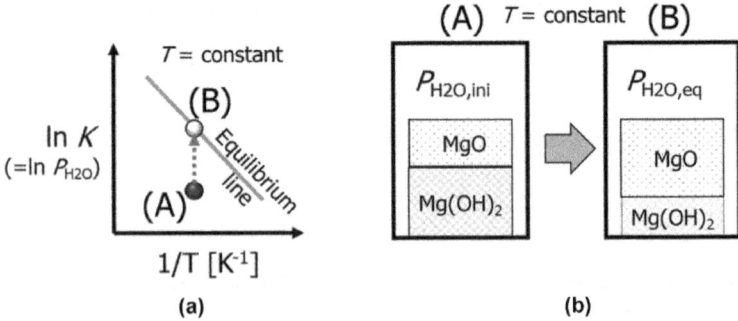

(a)   (b)

Figure 9   Reaction change in TCES operation: (a) reaction direction from a state A to an equilibrium state B; (b) reaction progress corresponding to the TCES operation of magnesium oxide/water system as an example.

(B) on the line. Figure 9b shows a schematic of the state change in a reactor from state A to the equilibrium state B for magnesium oxide–water as an example. The reaction proceeds to the equilibrium condition and the reaction pressure changes with the reaction progress; when the pressure attains an equilibrium state, the reaction is terminated.

3.1.2 Chemical Heat Pump Operation. A chemical heat pump uses a chemical reaction for thermal energy storage and conversion. The heat pump operation is based on a reaction equilibrium relationship and has two operational modes. Figure 10 shows the equilibrium relationship of a chemical heat pump cycle for the MgO/H$_2$O system in (a) the heat amplification and cooling mode and (b) the heat transformation mode.[14] Figure 11 shows the heat pump structure and operation in heat amplification and cooling modes. Numbers in parentheses on the equilibrium line in Figure 10a correspond to the state of the reactor at the same number in Figure 11. Generally, a chemical heat pump needs two reaction equilibria and two reactors to realize both reaction equilibria. In the heat

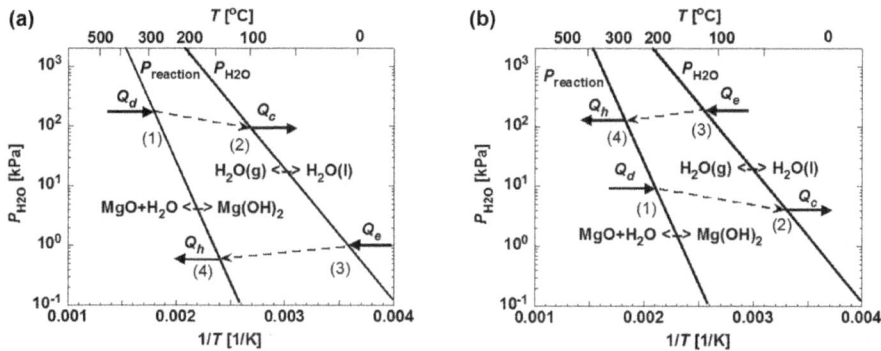

Figure 10 Equilibrium relationship of chemical heat pump cycle for the MgO/H$_2$O system: (a) heat amplification and cooling mode; (b) heat transformation mode.

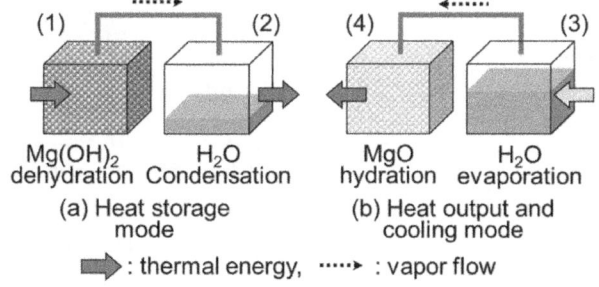

Figure 11 MgO/H$_2$O chemical heat pump operation in heat amplification mode with cooling operation.

amplification and cooling modes (see Figures 10a and 11), a high-temperature heat source (1) is stored in the system by endothermic dehydration and the condensation heat of the vapour is used on the heat demand side at (2) in the heat storage mode. In the heat output mode, hydration of magnesium oxide (4) proceeds, consuming the vapour, then evaporation of water (3) is enhanced and cooling output is generated at (3). Because the heat output produced $(Q_h + Q_c)$ is larger than heat source (Q_d), this mode is called the heat amplification mode. The heat transformation mode (see Figure 10b) consumes heat sources at mid-range temperatures (1, 3) to generate heat at higher temperature (4). The operation cycle resembles that of an adsorption heat pump. The advantages of the chemical heat pump over other heat pumps are a higher heat storage density, wider operating temperature range from near-ambient temperature to over 1000 °C, controllable operating temperature by choice of reaction conditions and thermal drivability with high efficiency.

3.2   Variety of TCES

Chemical reactions that are reversible can be used for chemical heat pumps. Figure 12 shows different types of chemical heat pump. The structure of a chemical heat pump depends on the reaction used in the pump. Generally, the heat pump consists of two reactors. A gas–solid reaction is applicable for batch operation. The driving force for heat pump operation comes from the pressure difference between the two reactors. Gas–liquid phase change or secondary gas–solid reactions are used to provide the driving force (see Figure 12a and b). Table 2 shows the classification of chemical heat pumps. Some driving force is required to operate a heat pump. Continuous operation as shown in Figure 12c is realized by gas–liquid reaction systems. Forward and reverse reactions proceed in different catalytic reactors. The driving force of the operation is concentration differences arising from the separation process using distillation, pressure differences generated by the compression process, or a combination of both. The reaction for a

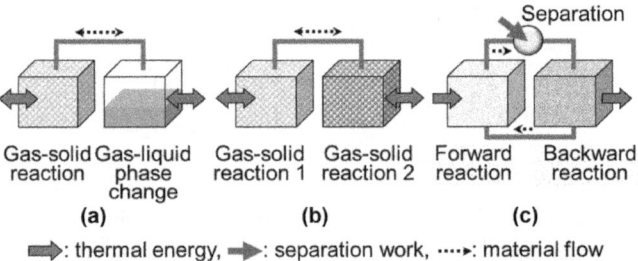

Gas-solid Gas-liquid   Gas-solid Gas-solid   Forward   Backward
reaction   phase   reaction 1 reaction 2 reaction   reaction
            change
   (a)   **(b)**   **(c)**

➡: thermal energy, ➤: separation work, ····▸: material flow

Figure 12   Types of chemical heat pump: (a) reaction and phase change batch type; (b) reaction and reaction batch type; (c) reaction and separation work continuous type.

Table 2 Classification of chemical heat pumps.

	Operation			
	Batch		Continuous	
Reaction phase	Gas–solid		Gas–liquid	
Driving operation	Gas–liquid phase change	Chemical reaction	Distillation, membrane separation	Compressor
Driving force	Reaction pressure	Reaction pressure	Concentration	Reaction pressure
Reactor system (Figure 12)	(a)	(b)	(c)	
Reaction examples	$CaCl_2/CH_3NH_2$ $CaCl_2/NH_3$ CaO/H_2O MgO/H_2O BaO/H_2O $MgCl_2/NH_3$ $MnCl_2/NH_3$ Adsorption TCES	MgO/CO_2 CaO/CO_2 $CaO/PbO/CO_2$ $Li_4SiO_4/CO_2/zeolite$ BaO/CO_2 Metal hydrate/H_2	Acetone/H_2 Absorption TCES	Isobutene/ H_2O Benzene/H_2

chemical heat pump needs to be reversible and, first, to have high reactivity and durability towards repetitive reactions and, second, to be safe in terms of materials and operation, of low-cost, compact and light weight.

There remains plenty of scope to find new reaction systems because there are thousands of reversible reactions and chemical material performance is continually developing.

3.3 Material and Reactor Technologies for TCES

As shown in Table 2, hydration is a key reaction type. Hydration TCES systems driven at low-temperatures of <100 °C are available for low-quality thermal heat recovery and also for cooling by thermal driving operation. The hydration reactions of calcium chloride $(CaCl_2/6H_2O)$,[15] calcium sulfate $(CaSO_4/0.5H_2O)$[16] and strontium bromide $(SrBr_2 \cdot H_2O/5H_2O)$[17] have been well studied for utilization of low-temperature heat. Magnesium oxide/water [eqn (5)] TCES has been studied for utilization of exhaust heat at medium temperatures of around 200–400 °C from internal combustion engines and industrial processes.[14] Material developments in the MgO/H_2O system are discussed with regard to enhancement of TCES performance. Lithium chloride-modified magnesium hydroxide $[Mg(OH)_2]$,[18] and also lithium bromide-modified $Mg(OH)_2$,[19] show higher TCES performance than pure magnesium hydroxide owing to their water-absorption enhancement effect on the surface of MgO particles. Although the $Mg(OH)_2$ pellet has high reactivity, thermal conductivity enhancement of the material is important for efficient heat exchange. A packed bed of $Mg(OH)_2$ pellets is characterized by a low effective thermal conductivity of around $0.2 \ W \ m^{-1} \ K^{-1}$.

Expanded graphite (EG) is a good candidate for thermal conductivity enhancement. An $Mg(OH)_2$ composite mixed with EG, named EM, was developed in previous work.[20,21] In comparison with pure $Mg(OH)_2$, EM has mouldability properties, which means that it is capable of being shaped easily into a specific shape by compaction in a mould. Measurements of thermal conductivity of EM tablets showed a higher effective thermal conductivity of up to $2.0~W~m^{-1}~K^{-1}$. A mixing mass ratio between $Mg(OH)_2$ and EG of $8:1$ showed optimal performance compared with others in a packed-bed reactor experiment.[22] Carbon nanotube-based hybrid materials for MgO/H_2O have also been developed for reactivity enhancement based on the heat transfer enhancement effect of carbon nanotubes.[23]

TCES material (see Figure 13a) is charged between heat-exchanging fins in a TCES reactor (see Figure 13b). The low thermal conductivity of $Mg(OH)_2$ powder, and any other powdered materials such as metal oxides, and the high contact thermal resistance between the material and the heat exchanger fins are problems for the TCES reactor (see Figure 14). This low thermal conductivity can cause poor performance of a TCES system, such as

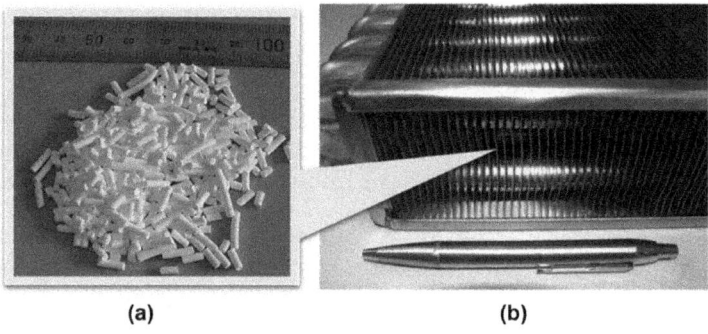

(a)   (b)

Figure 13   Components of a TCES reactor: (a) $Mg(OH)_2$ pellets (1.9×5–10 mm) as a TCES material; (b) packed-bed reactor with heat-exchanging function.

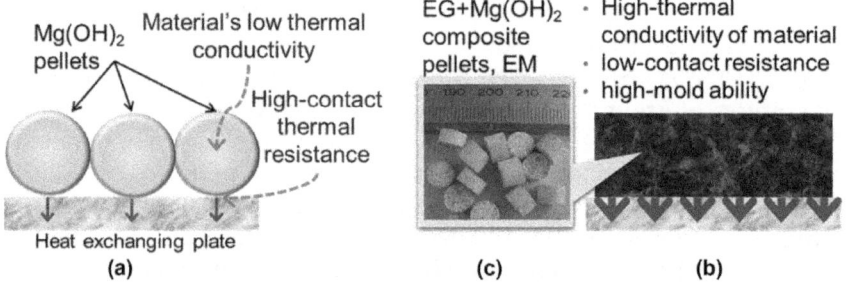

Figure 14   Heat transfer enhancement of a TCES reactor. (a) Conventional contact between TCES pellets and heat-exchanging plate; (b) composite of expanded graphite and $Mg(OH)_2$, EM; (c) heat transfer-enhanced packed-bed reactor introducing EM.

prolongation of the reaction time and lowering of the heat output rate. Methods for enhancement of the thermal conductivity of reaction materials and reactor design have been investigated for enhancement of TCES system performance. One promising method is the introduction of carbon-based materials, such as carbon nanofibers or nanotubes[23] and expanded graphite (EG), into heat-storage reactants.[15] EG has the advantages of high mass productivity and cost performance. Composites of EG and $Mg(OH)_2$ (EM) have been developed (see Figure 14b). The EM has high thermal conductivity and mouldability for making good contact with the heat exchanger plate, such that the reactor-introduced EM has a better thermal performance (see Figure 14c). It has been reported that the EM material has a thermal conductivity up to 10 times higher than that of pure $Mg(OH)_2$ material.[24] EM material having a mass mixing ratio $EM : Mg(OH)_2$ of $1 : 8$ (EM8) was found to be the optimum composite. A packed-bed packed EM8 in a random arrangement shows two times higher conductivity and a faster heat storage rate than an $Mg(OH)_2$ pellet bed.[22] Thermal conductivity enhancement and reactor-fitting design of TCES materials are important issues for practical TCES systems.

3.4 Applications of TCES

MgO/H_2O TCES has been studied for utilization of exhaust heat of medium temperature at around 200–400 °C from an internal combustion engine. The internal combustion engine in an automobile emits exhaust gas at over 500 °C and is a high-quality heat source. Stored heat from the exhaust gas combined with an on-board catalytic reactor operated at 300 °C is used for exhaust gas reduction. At the daily first start of the engine, the catalyst should first be heated from ambient temperature to the active temperature. The heating operation requires additional fuel. An MgO/H_2O TCES could assist in supplying heat to the reactor by storing surplus heat from the exhaust gas under normal running conditions on the day before, and releasing the stored heat to the catalyst by hydration.

Thermal energy storage is useful for solar thermal systems and other high-temperature industrial processes that require a stable thermal energy supply. A TCES system operating at 700 °C is needed for a concentrated solar power system.

A lithium orthosilicate/carbon dioxide (Li_4SiO_4/CO_2) reaction system has been proposed for use in TCES with chemical heat pump (CHP) systems at around 700 °C.[25] Carbonation of Li_4SiO_4 produces lithium carbonate (Li_2CO_3) and lithium metasilicate (Li_2SiO_3) exothermically:

$$Li_4SiO_4(s) + CO_2(g) \rightarrow Li_2CO_3 + Li_2SiO_3, \quad \Delta H = -94.0 \text{ kJ mol}^{-1} \quad (9)$$

CO_2 storage is a key issue. Zeolite is one of the candidates for a CO_2 adsorbent:

$$Zeolite(s) + CO_2(g) \rightarrow Zeolite\text{-}CO_2, \quad \Delta H = -60 \text{ kJ mol}^{-1} \quad (10)$$

Decarbonisation of these products is used for heat storage, and carbonisation is used for heat output in a TCES system. The Li_4SiO_4/CO_2/zeolite TCES system in a chemical heat pump operation is suitable for a concentrated solar power plant. By adjustment of the zeolite bed temperature, CO_2 can be moved between an Li_4SiO_4/CO_2 reactor and a zeolite/CO_2 reactor. When the zeolite bed temperature in heat output mode is higher than that in heat storage mode, the CO_2 pressure in the former is also higher than that in the latter; then the heat output temperature of Li_4SiO_4 carbonation is higher than that in heat storage operation. This heat pump function is a unique performance of TCES in comparison with conventional sensible and latent heat storage.

Generally, an electrical battery is still a popular candidate in energy storage for the stabilization of renewable energy output. However, thermal energy storage technologies cost less than one-tenth of electrical battery storage.[26] A TCES system could have potential for application to load and supply levelling in renewable energy storage. Recognition of the availability and benefits of TCES compared with electrical batteries is required for realization of its efficient utilization in renewable energy systems.

3.5 Challenges and Barriers to Implementation

Although low thermal conductivity and high corrosivity are major problems for mid–high-temperature PCM, these problems are being solved by PCM composites and encapsulation technology. This means that LHS have been evolving from the viewpoint of material development. The next step needed to implement mid–high-temperature LHS is an energy- and cost-effective LHS system using advanced PCM. The low thermal conductivity of powderstate materials, such as metal oxides, and high contact thermal resistance between the material and the fin are problems for TCES reactors. Enhancement of the thermal conductivity of materials and packed-bed reactors is a key issue for breaking the barrier for expansion of this new thermal storage market. High thermal storage density and high heat output of the reactor are important for the attractiveness of TCES systems in comparison with other thermal energy storage systems. The development of low-mass and low-cost heat exchanger equipment for packed-bed reactors will enhance the TCES system implementation.

References

1. T. Nomura, N. Okinaka and T. Akiyama, *ISIJ Int.*, 2010, **9**, 1229–1239.
2. T. Nomura and T. Akiyama, *Int. J. Energy Res.*, 2016, **41**, 240–251.
3. R. Fukahori, T. Nomura, C. Zhu, N. Sheng, N. Okinaka and T. Akiyama, *Appl. Energy*, 2016, **170**, 324–328.
4. T. Nomura, C. Zhu, N. Sheng, G. Saito and T. Akiyama, *Sci. Rep.*, 2015, **5**, 9117.

5. R. Tamme, T. Bauer, J. Buschle, D. Laing, H. Müller-Steinhagen and W. D. Steinmann, *Int. J. Energy Res.*, 2007, **32**, 264–271.
6. A. Kaizawa, H. Kamano, A. Kawai, T. Jozuka, T. Senda, N. Maruoka and T. Akiyama, *Energy Convers. Manag.*, 2008, **49**, 698–706.
7. R. Bayón, E. Rojas, L. Valenzuela, E. Zarza and J. León, *Appl. Therm. Eng.*, 2010, **30**, 2643–2651.
8. Z. Huang, X. Gao, T. Xu, Y. Fang and Z. Zhang, *Appl. Energy*, 2014, **115**, 265–271.
9. H. Tian, W. Wang, J. Ding, X. Wei, M. Song and J. Yang, *Appl. Energy*, 2015, **148**, 87–92.
10. N. Gokon, S. Nakamura, T. Hatamachi and T. Kodama, *Energy*, 2014, **68**, 773–782.
11. N. Gokon, S.-I. Inuta, S. Yamashita, T. Hatamachi and T. Kodama, *Int. J. Hydrogen Energy*, 2009, **34**, 7143–7154.
12. R. Fukahori, T. Nomura, C. Zhu, N. Sheng, N. Okinaka and T. Akiyama, *Appl. Energy*, 2016, **163**, 1–8.
13. T. Nomura, N. Sheng, C. Zhu, G. Saito, D. Hanzaki, T. Hiraki and T. Akiyama, *Appl. Energy*, 2017, **188**, 9–18.
14. Y. Kato, N. Yamashita, K. Kobayashi and Y. Yoshizawa, *Appl. Therm. Eng.*, 1996, **16**, 853–862.
15. Y. Hirata, K. Fujioka and S. Fujiki, *J. Chem. Eng. Jpn.*, 2003, **36**, 827–832.
16. H. Ogura, M. Kubota, H. Suzuki and T. Yamakawa, *Kagaku Kogaku Ronbunshu, (in Japanese)*, 2009, **35**, 506–510.
17. T. Esaki and N. Kobayashi, *J. Mater. Sci. Chem. Eng.*, 2016, **4**, 106–115.
18. H. Ishitobi, J. Ryu and Y. Kato, *Ind. Eng. Chem. Res.*, 2013, **52**, 5321–5325.
19. O. Myagmarjav, J. Ryu and Y. Kato, *Appl. Therm. Eng.*, 2014, **63**, 170–176.
20. S. T. Kim, J. Ryu and Y. Kato, *Progr. Nucl. Energy*, 2011, **53**, 1027–1033.
21. M. Zamengo, J. Ryu and Y. Kato, *Appl. Therm. Eng.*, 2013, **61**, 853–858.
22. M. Zamengo, J. Ryu and Y. Kato, *Appl. Therm. Eng.*, 2014, **64**, 399.
23. E. Mastronardo, L. Bonaccorsi, Y. Kato, E. Piperopoulosa, M. Lanza and C. Milone, *Appl. Energy*, 2016, **181**, 232–243.
24. M. Zamengo, J. Tomaškovič, J. Ryu and Y. Kato, *J. Chem. Eng. Jpn.*, 2016, **49**, 261–267.
25. H. Takasu, J. Ryu and Y. Kato, *Appl. Energy*, 2017, **193**, 74–83.
26. C. Forsberg, D. C. Stack, D. Curtis, G. Haratyk and N. A. Sepulveda, *Electri. J.*, 2017, **30**, 42–52.

Smart Energy Systems

SUSANA PAARDEKOOPER, RASMUS LUND* AND HENRIK LUND

ABSTRACT

Smart energy systems are a concept to support the design of coherent and sustainable energy supply strategies. A smart energy system is a combination of the currently isolated energy sectors, such as electricity, heating and transport, and it includes three smart energy grid infrastructures, namely the electricity, thermal and gas grids. These grids connect the energy resources with the demands, energy production, energy storage and interconnection points. When these grids and sectors are analysed in detail and assessed as one coherent energy system, a number of synergies can be identified across the sectors, which an approach that considers only one or two sectors could not have done. The two studies 'IDA's Energy Vision 2050' and 'Smart Energy Europe' are used as case studies to demonstrate and discuss the potential of the smart energy approach. In these studies, different storage options are considered and analysed, and generally it is found that hydroelectric storage, batteries in electric vehicles, thermal storage in district heating systems and storage of renewable electrofuels are important and provide a cost-efficient flexibility to the overall energy system. However, large-scale batteries on the grid level and stationary batteries in buildings are not found feasible in an energy system perspective.

*Corresponding author.

Issues in Environmental Science and Technology No. 46
Energy Storage Options and Their Environmental Impact
Edited by R.E. Hester and R.M. Harrison
© The Royal Society of Chemistry 2019
Published by the Royal Society of Chemistry, www.rsc.org

1 Smart Energy Systems

Energy systems will have to change as we move towards a more sustainable future. Smart energy systems are a concept that describes how the costs for energy supply can be reduced in monetary terms, but also in terms of fuel consumption and environmental impact, by integrating the different sectors of the energy supply (electricity, heating, transport, *etc.*).[1] This approach, particularly relevant in situations with a high penetration rate of fluctuating renewables, permits a number of synergies between the sectors, which result in a potential reduction in costs.[2] The concept of smart energy systems can be applied in analyses and planning to develop a more coherent and flexible energy supply. This approach, compared with a traditional single-sector approach, has implications for the feasibility of different types of energy storage and storage in different parts of the energy supply systems, which are elaborated further throughout this chapter.

1.1 General Objectives

The purpose of introducing the idea of smart energy systems is to improve the long-term sustainability of the energy supply by changing the perspective from looking at and analysing one sector at a time to considering all energy sectors in an integrated manner. This integration between the sectors is the underpinning factor between much of the primary energy reduction and flexibility that is built into the smart energy system within a 100% renewable energy context, where intermittent renewables play a central role.

The term 'sector' refers to the sum of all links in the supply chain, from resource to end-use demand. As an example, the electricity sector would include the demands, power distribution, power plants and other producers, and the required fuels. Additionally, sectors that are not traditionally considered to be energy sectors can be included here if they have implications for the energy supply. Fresh water supply, as an example, is not typically considered a part of the energy supply system, but as increasing demands for fresh water in some parts of the world start to require substantial capacities of desalination plants and corresponding consumption of electricity, this can be seen and included as an energy sector.[3] In such a situation, freshwater storage can have a significant influence on the rest of the energy system.

Smart energy systems are to a large extent an approach for planning, designing and comparing different overall energy supply strategies. This entails a perspective on a societal level, which means that the planning activity in this connection is on a national or regional level rather than just a business perspective.[2] By considering all energy sectors together, and also all parts of the energy supply chains, this approach seeks to uncover all of the relevant consequences for society through the comparison of different alternative supply strategies in a holistic way. Which consequences are relevant in this connection depends on the goals of society and the purpose of the planning of which the strategy is a part. These could be related to the total costs for society, greenhouse gas emissions, import dependency and security of supply.

In this perspective, analyses also need to include all costs that can be affected by the solution or scenarios in question, to be able to evaluate the economic consequences of different scenarios. The following is a list of cost items that usually are relevant to be included:

- fuel consumption;
- investments in infrastructure;
- operation and maintenance;
- electricity exchange;
- CO_2 emissions.

Many other costs can also be relevant in specific cases.[3,4] In studies with a focus on transport, it could be relevant to include, for example, the total costs for the national car fleet, since these are typically borne by private parties. In some cases, externalities in addition to carbon emissions might also be relevant for inclusion, alongside the direct economic costs. This could be damage or mitigation costs caused by the emission of particulate matter or the long-term storage and management of nuclear waste. The criteria and performance indicators of the smart energy system are contextualised by the planning objectives for that specific energy system.

1.2 Reducing the Need for Fuels

The high dependence on fuels is one of the key issues for which a smart energy system approach provides some solutions, through the identification of synergies in sector integration. The consumption of fossil and nuclear fuels has a range of negative economic, environmental and social consequences. Some of these negative consequences can be reduced by replacing the traditional fuel sources with bioenergy, *e.g.* net emission of greenhouse gases, or pollution connected to the extraction and storage of radioactive waste material. However, like fossil fuels, bioenergy is a limited resource, and the consumption of it in large quantities introduces a new set of challenges.

Bioenergy tends to play a large part in renewable energy systems, especially as energy systems transition to renewability. This is primarily because they are particularly effective at storing energy in a similar way to that in which fossil fuels store energy currently; *i.e.* easily, with relatively high density and few (if any) losses. In addition, many conversion technologies that currently exist can be easily adapted to combust the bioenergy counterpart of their current fossil fuel, which allows for fast reductions in emissions that contribute to climate change. This makes bioenergy an extremely valuable resource, especially in areas such as heavy transport and industrial processes, where energy savings and alternative energy sources are difficult to find.

However, overt reliance on bioenergy may contradict the objectives of a smart energy system. First, the import of biomass may contravene the energy-security objectives of energy planning and result in a continued reliance on other countries' resources. Second, over-use of bioenergy will certainly have sustainability implications. Bioenergy, in all its forms, continues to be a scarce resource that requires land and labour at a minimum. If the objective is to contain and sequester carbon within the natural environment to prevent it from entering and staying in the atmosphere, land-use changes need to be considered in order to understand the real contribution that bioenergy can make to reducing greenhouse gas emissions.[5] The interaction between bioenergy and food markets has attracted much attention,[6] specifically during the period of volatile global oil and grain prices in 2008. Later analysis showed that the correlation between the price variations was not fundamentally driven by an increased demand on biofuels; rather, the dependence of the global agricultural markets on global oil prices was mainly responsible for spikes in food prices and socially destabilising fears.[7,8] However, the concerns highlighted the complicated and delicate interaction and co-dependence between agricultural processes, carbon emissions and energy provision today, where bioenergy does not play a key role in many industrialised countries.

Connolly *et al.*, in the Smart Energy Europe scenario, compared a number of studies of the sustainable bioenergy potential in Europe (EU28), using a variety of assumptions and methods.[9] The results ranged from 9 to about 40 EJ per year, but with an average across the compared studies of around 20 EJ per year. Some of the main differences lie in the assumptions of what can be considered as sustainable bioenergy. If only the residual biomass that is left without any other use or value can be considered sustainable as a fuel in the energy supply, the potential is much more limited. On the other hand, if biomass from many different origins and types of biomass can be used for energy purposes, including some energy crops, before reaching a sustainable limit, the potential is higher.

These numbers should be compared with the current fossil fuel consumption in the region, which is about 60 EJ per year. This clearly indicates the necessity to reduce the current fuel consumption and integrate intermittent renewable energy sources in the transition towards an energy supply based on renewable and sustainable energy. For example, Mathiesen *et al.*[10] found that biomass can be reduced to a minimum in the heating sector, because there are good, efficient alternatives that do not consume biomass. Therefore, biomass resources should be use efficiently and prioritised for the demands that are hard or very expensive to meet without biomass, limiting the consumption in other sectors.

A smart energy system can contribute to a reduction in the need for fuel to a certain required level, no matter whether it is fossil, nuclear or renewable fuel. The approach facilitates comparison of different strategies and provides a catalogue of technological solutions for the reduction of fuel consumption and introducing a capacity for producing renewable fuels.

The reduction in fuel consumption is mainly generated through improving the efficiency of the general fuel utilisation in the system and through enabling a replacement of fuel consumption with fluctuating renewable energy sources. This is elaborated in the following sections.

1.3   Smart Electric, Thermal and Gas Grids

The term 'smart grid' is traditionally used to describe how the electricity sector can be developed in the future, increasing the use of fluctuating renewable energy sources. Typically, the notion 'smart' refers to the ability of a particular unit to react to external signals, using information and communication technology (ICT) systems.[11] An example of a smart consumer in this sense could be an electric vehicle that is connected to the electricity supply in the evening for charging overnight. If the owner of the car does not need it until next morning, it might not necessarily be fully charged immediately if the decision regarding when exactly during the night to charge the battery is left to a charger with access to an electricity price signal. This allows the battery to be charged when electricity is most abundant, *e.g.* due to high levels of (intermittent) renewable energy during periods of otherwise low demand. This may reduce the electricity costs for the consumer and at the same time improve the energy and cost efficiency of the electricity supply.

Within the smart energy system concept, this idea is expanded from a focus on the electricity grid to include three core grid infrastructures for energy transmission and distribution, making the connections between supply, storage and demand:

- *Smart electricity grids* to connect flexible electricity demands, such as heat pumps and electric vehicles, to intermittent renewable resources, such as wind and solar power.
- *Smart thermal grids (district heating and cooling)* to connect the electricity and heating sectors. This allows the utilisation of thermal storage for creating additional flexibility and the recycling of heat losses in the energy system.
- *Smart gas grids* to connect the electricity, heating and transport sectors. This permits the utilisation of gas storage for creating additional flexibility. If the gas is refined to a liquid fuel, then liquid fuel storages can also be utilised.[2]

These three grids form the backbone of a smart energy system and allow the integration of the operation across the different energy sectors through ICT. These grids to some extent overlap the energy sectors, as described above, but not completely however, and this is the reason why it is important to consider all these types of grids so as to identify the best overall strategy. The electricity grid, of course, covers the electricity sector, but it also feeds into all the other sectors. To mention a few examples, electric heat pumps producing heating or cooling, process energy can be supplied electrically, and electric vehicles are increasingly common. The thermal grids, apart from heating, can also cover low-temperature process energy, but can additionally play a role in utilising excess heat from power production or industrial processes. The gas grids are not associated with one particular sector, but as in the case of the electricity grids, they are connected to more or less all energy sectors, *e.g.* for power production, heat supply and industrial process energy.

1.4 Coupling of Energy Sectors

In the preceding section, the general and overall ideas associated with a smart energy system were presented. In this section, this approach is illustrated and compared with alternative large-scale approaches.

Figure 1 illustrates the principles of a smart energy system. The resources of the energy system, or the primary energy supply, are located in the left-most column and the demands are in the right-most column. In between are placed the different categories of conversion and storage technology, which link the resources to the demands. In the figure the three smart energy grids can been seen, and crucially it can be observed how each of them has several interaction points, connecting the sectors with each other. The smart electric grid is represented in the top-most horizontal connection arrow and its branches. The smart thermal grid is the horizontal arrow between 'Boiler' and 'Heat demand' and the branches to this and the smart gas grid are represented in the L-shaped arrow from "Fuel" to "Industry" and the branches to this.

The sector coupling is materialised in technologies in the system that allow an interaction between two or more sectors. One example is a combined heat and power (CHP) plant, which couples the electricity sector with a district heating system. The coupling can be more or less strong, depending on the flexibility of the applied CHP technology to regulate up or down its production of either electricity or heating. In a situation with a low production from fluctuating renewable electricity sources or high electricity prices, it can be feasible to start or increase the production of electricity. This will also increase the potential utilisation of excess thermal energy from the plant for district heating (DH) production, replacing the alternative fuel consumption for the DH supply. If the DH system also has a large-scale thermal storage capacity connected, the excess heat from the power production can be utilised, even if there is no current heat demand, by storing it for later use. In the opposite situation, where the production of fluctuating renewables is high and the electricity prices might be low, the CHP plant can be reduced or shut down. In this situation, the district heating system could cover the heating demands by using a cheaper alternative heat source, producing heat with a (large- or small-scale) heat pump, or possibly drawing heat from a thermal storage. In this way, both the electricity and the DH system benefit from the CHP plant as a coupling point between the two.

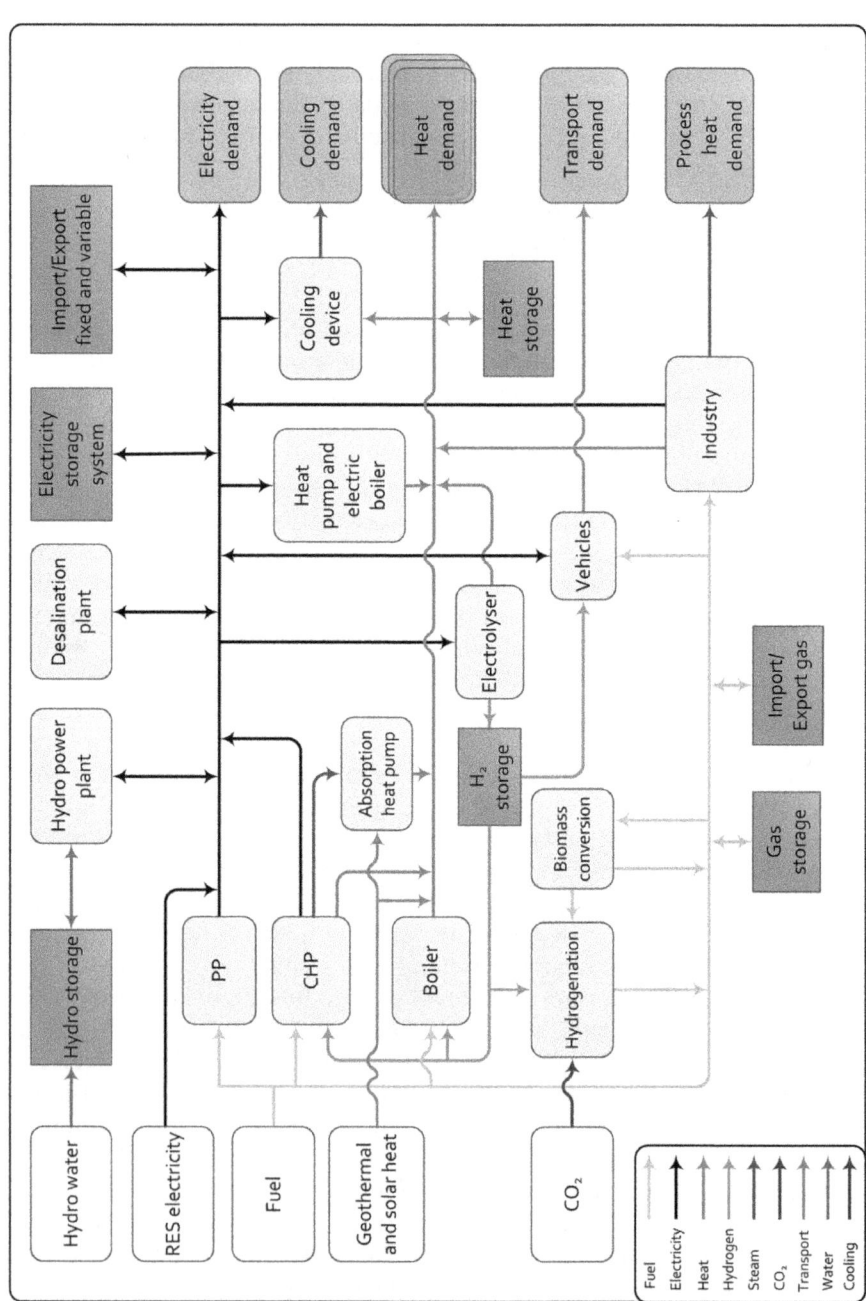

Figure 1 Energy system diagram illustrating resources (left column), demands (right column) and conversion and storage in between connecting the resources with the demands.[23]

This sector coupling represents the key source of flexibility in the smart energy system, because it maximises the different conversion technologies that can be used to provide a final energy demand. For example, heat can be retrieved from storage, produced with renewables, produced electrically, as a by-product of industrial processes, as a by-product of electricity generation or simply from a boiler. The smartness of the system is based on being able to use the right kinds of energy when they are available – and being able to transfer that energy between sectors. As a result, the need for storage is already diminished, since the energy can be redirected. However, this also means that energy can be directed towards the cheapest storages. Section 3 and 4 consider these operations, with particular relevance to storage technologies, using two scenarios based on sector integration and the smart energy system approach.

The smart energy systems approach can be compared with several other general concepts on how to achieve a largely fossil-free and sustainable energy system. One alternative strategy is a general electrification of all energy sectors. Electricity can be transmitted over long distances with few losses and electricity grids are already established in most places. There are also available technologies for covering most types of demand using electricity, so this may intuitively seem like a good solution. The challenge to this type of solution is that it is generally not a very energy-efficient way of supplying all demands. Particularly for heating and cooling, there are some very efficient alternatives, in both monetary and energy terms.[2,12] Thermal networks can also accommodate electrification through the use of large-scale heat pumps and direct electric boilers. However, as seen in Figure 1, thermal networks can also use excess heat from power generation and industry in contribution to the overall energy efficiency of the system, and integrate renewables such as geothermal energy – which would otherwise go unrecovered. In this way, the use of thermal networks in a smart energy system contributes to a better performance of the system in terms of both primary energy reductions and cost.

Another alternative approach is to solve the sustainability issues regarding energy supply at building level and small-scale community level, in a fully distributed way. This can be through insulating the buildings to zero energy and installing heating and electricity production capacities locally, *e.g.* solar thermal, solar photovoltaic or small wind turbines together with battery packs to balance the fluctuations in supply. This approach will generally be very expensive, because the small-scale production capacities are more expensive per unit energy produced than large-scale ones, which could alternatively have been connected to the electricity grid elsewhere.[13] At the same time, this kind of solution will hardly remove the need for the electricity grid completely, so the costs of maintaining this grid can at best be kept the same.[14]

Compared with the above-described approaches, the smart energy system approach seeks to identify a balanced solution with a focus on how to reach an energy supply with a consumption of fuel resources at a sustainable level in the most cost-effective way possible. This is done by implementing different types of smart grids and making use of the strengths of each of the sectors to utilise the potential synergies *via* coupling of the sectors.

2 Potential of Smart Energy Systems and Sector Coupling

In this section, two studies applying the smart energy system approach are presented to bring the principles described in Section 1 into actual use, and cases illustrating what kind of implications this approach can have are illustrated. The first study is called 'IDA Energy Vision 2050', where the case of Denmark is analysed and a 100% renewable energy scenario is proposed. The second study is called 'Smart Energy Europe', where Europe (EU28) is analysed and a scenario for a decarbonisation of the total energy supply in the region is suggested. These cases will also be referred to in Section 4, where the use of several types of energy storage in these studies is presented and discussed in detail. In both of the two studies, the energy system analysis tool EnergyPLAN was used. The tool is presented at the end of this section.

2.1 IDA Energy Vision 2050

IDA Energy Vision 2050 was a project initiated by the labour union Danish Society of Engineers (in Danish abbreviated IDA), which was completed in 2015, with the purpose of creating an input for public debate and policy. IDA Energy Vision 2050 builds on two previous studies with similar purposes: IDA Climate Plan 2050 from 2009 and IDA Energy Plan 2030 from 2006.[15,16] The aim of the project is to create a scenario for how the Danish energy supply can be provided without fossil fuel in 2050, as this is the official national goal. The Danish Energy Agency (DEA) created a number of scenarios of how to reach the national goal of fossil-free energy supply in 2050 one year earlier. In contrast, IDA Energy Vision explicitly makes use of a smart energy system approach. The objective was to create a cost-effective energy supply that requires a lower amount of biomass for the energy system than the scenario from DEA, based on system integration and the smart energy system perspective.

The overall method in IDA Energy Vision was to use the newest scientific findings about technological development, renewable energy potentials, demand reduction possibilities and similar considerations to create a sound reference for what can be done in the energy system – both technically and economically. This reference forms the basis for the deliberate design of a new energy system scenario in the context of the Danish energy system, using the energy system analysis tool EnergyPLAN. By simulating and analysing a variety of different combinations of technologies, energy sources and other initiatives, and their corresponding costs in an energy system perspective, it is found which technologies and combinations are most feasible and why. This is continued in an iterative process until the desired performance (biomass, primary energy and costs) is achieved and the relevant alternatives are assessed. The components used for the analysis include a wide range of initiatives and technologies. The main elements of the design of the energy system in IDA Energy Vision are as follows:

- demand reduction and flexible demands in electricity;
- changes in transport patterns and modal shifts to reduce energy consumption in transport;
- electrification of light road vehicle and rail transport;
- reduction of demands in existing and new buildings;
- improved efficiency in industry including utilisation of excess heat;
- replacement of boilers with heat pumps in buildings not connected to district heating;
- introduction of large-scale electric heat pumps in district heating;
- introduction of large-scale renewable electrofuel production infrastructure;
- implementation of district cooling for certain cooling demands.

The production of electrofuels is a process in which electricity is used to produce hydrogen, which is stored or directly used in a hydrogenation process to upgrade a carbon source, *e.g.* CO_2, biogas or gasified biomass, to methane. This can be converted into a number of other fuel products, *e.g.* methanol or dimethyl ether. Using this process, electrofuels can be produced based solely on renewable energy.

Using this simulation and design approach fits well with the objective of IDA Energy Vision, which was to contribute to the public dialogue, discuss alternatives and bring more detail to how a future scenario for Denmark could operate.[17]

As a reference point for the results, two of the DEA scenarios are shown here together with the statistical reference for 2015 in Figure 2. The DEA Fossil scenario is a business-as-usual scenario, which represents no significant change in policy, whereas the DEA Wind scenario is a 100% renewable strategy including large capacities of wind power. Figure 2 shows the primary energy supply for the compared scenarios. It can be seen that the DEA Wind 2050 scenario requires the same amount of total primary energy as but one-third more biomass than IDA Energy Vision 2050 scenario. The lower level in IDA is achieved with a focus on creating flexibility in the energy system to integrate the fluctuating renewables, mainly wind power, by introducing sector-coupling technologies in the supply in line with the smart energy system principle. The total energy system costs, in monetary terms, are slightly higher in the IDA 2050 scenario than what the costs are today. On the other hand, it is about 10% lower compared with the DEA Wind scenario. At the same time, the cost of the energy system in IDA Energy Vision is more robust to external changes because it is less reliant on importation of electricity and fuel. This means that the IDA 2050 scenario uses the same amount of primary energy less biomass, at a lower cost than the DEA

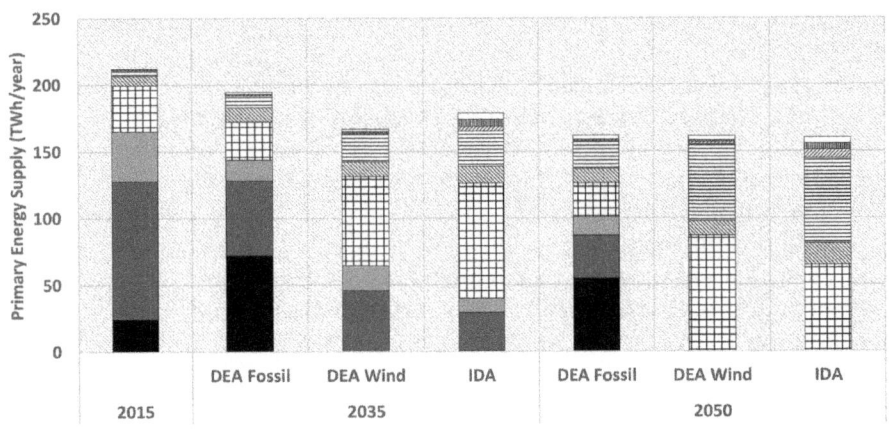

Figure 2   Primary energy supply in 2035 and 2050 in IDA Energy Vision, in 2015 and in the DEA scenarios for the case of Denmark. All scenarios using a medium fuel-price assumption corresponding to US$105 per barrel of crude oil.

scenario, and is more robust, using all of the same assumptions. The identification of this better scenario is possible because of the application of the smart energy systems approach.

2.2   Smart Energy Europe

In Denmark, as the case was for IDA Energy Vision, there is already a tradition in the energy sectors of having some sector-coupling conversion technologies in the form of CHP plants with thermal storage, which contribute to balancing the fluctuations in wind power. In addition, the practice of using excess heat for district heating supply is relatively common. On a European scale, sector coupling is much less common, so here the potential for this approach is much greater. European Union policy has traditionally been very heavily focused on the power sector, but increasingly the subject and, for example, the potential of integrating district heating with the electricity and industry sectors is becoming a significant topic.[12,18] The project Smart Energy Europe was undertaken for the European Commission to shed light on the possibilities of covering the energy demands in the EU with renewable energy to meet the decarbonisation targets.[19,20]

The method applied in this project uses a stepwise structure, where each step accounts for a defined part of the transition towards the decarbonised energy system. Since the simulation approach towards energy system design involves making all the alternatives (including the suboptimal ones) explicit, this stepwise approach represents a way of breaking down the analysis of the many alternative scenarios modelled. The steps illustrate the impact of the conversion of each part of the energy system of renewable energy. The steps are as follows:

1. As a starting point, the EU28 reference scenario for 2050 is used, based on the EU Energy Roadmap.[21]
2. All nuclear production is replaced with natural gas-based production, to accommodate a nuclear phase-out.
3. The heating demands in buildings are reduced by 35%.
4. Most passenger cars are converted from fossil fuel to electricity.
5. Individual heat pumps are installed to cover all space heating and hot water demands.
6. District heating systems are installed in all dense urban areas instead of heat pumps.
7. The remaining fossil fuels in transport are replaced with renewable electrofuels.
8. Coal and oil for various purposes are replaced with renewable electrofuels.
9. Natural gas is replaced with renewable electrofuel.

These steps are not ordered chronologically, but according to their political and scientific certainty. Hence the first three steps from the starting point (2, 3 and 4), are approached as having some sort of consensus in the public discourse, concluding that they need to happen and that it is technically possible today. The next two steps, related to the heating sector, are technically possible today but are further from a political consensus, whereas the last three steps concerning electrofuels include technologies which are still under development.[22] As for IDA Energy Vision, a simulation approach was used and the EnergyPLAN tool was applied for the quantification of the potentials in the scenario.

The results, in terms of primary energy supply, of the study are shown in Figure 3 in a stepwise way, along with the corresponding carbon emissions. It can be seen that up until step 6 the primary energy supply decreases, mainly because of the improved efficiency that the different measures imply. From step 2 to step 5, the carbon emissions decrease even more rapidly than the primary energy supply. This is caused by these measures improving the overall energy system's ability to integrate more fluctuating renewables sources as it becomes more interconnected. From step 6 to step 9, the primary energy supply has an increasing tendency, due to the conversion losses in the production of electrofuels compared with the fossil alternatives. On the other hand, the electrofuel production consumes electricity and thereby allows for a significant increase in the integration of fluctuating renewable electricity, which can also be seen in the figure. The electrofuel production process also allows for flexibility through large-scale storage of the components used to produce the electrofuel, *e.g.* hydrogen or syngas, but also the final product.

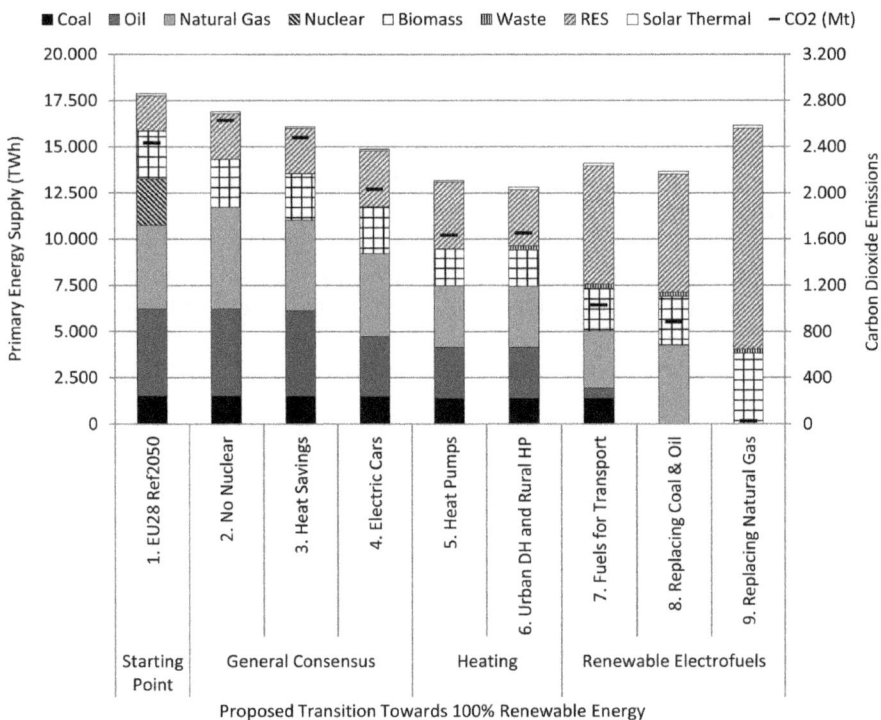

Figure 3   Primary energy supply by fuel and carbon dioxide emissions for all steps in the transition to a smart energy system for Europe.

The total energy system costs will be of the same order of magnitude as the fossil-based EU28 Energy Roadmap 2050, with an increase of about 10%. However, the economics will be based on infrastructure investments within the EU rather than on imports of fossil fuels. It is concluded that the energy system in EU28 can be converted to being covered with renewable energy, at similar cost levels to those of the alternatives.

2.3   The Energy System Analysis Tool EnergyPLAN

EnergyPLAN, which is used in both the IDA Energy Vision 2050 and Smart Energy Europe, is an advanced energy system analysis tool, designed to analyse the integration of fluctuating renewable energy into an energy system and includes all parts of the energy system relevant to a smart energy system approach (see Figure 1). The tool simulates the operation of the entire energy system for every hour of one year and the dynamics between technologies and sectors from hour to hour. The hourly simulation is important for the temporal distribution between demands and resources. The tool can provide information on when a demand can be directly covered with, for example, wind power, when it can be redirected to another sector and when it makes sense to consider storage of energy. When different technologies, resources, storage types or capacity are assessed, the tool provides a detailed numerical description of how the different components are used, which can be used for further analysis or assessment. The costs for each scenario are evaluated when a simulation is performed, in addition to a variety of other relevant parameters.[3] More information and documentation about the EnergyPLAN tool can be found at www.energyplan.eu.[23]

3 The Need for Storage in a Smart Energy Systems Perspective

Energy storage has a number of roles to play in a smart energy perspective: mechanical, electrochemical, hot and cold thermal, hydrogen and hydrocarbon fuel storage. Each plays an important role. However, the needs and relevance of energy storage in a smart energy system perspective might be different from results based on single-sector approaches or local case studies or business economic assessments. The analysis and conclusions presented here are based on studies with a societal perspective and, as far as possible, including all sectors and all links in the supply chains.

3.1 Assessment of Storage Needs: A Function of the Demands

Smart energy systems represent an approach where the energy system moves from a fuel-based energy system to a system based on (fluctuating) renewable energy. In addition to any other aims, the underlying objective is to design an energy system that can meet the energy demands of its citizens, at all times. Especially if there is a simultaneous ambition to constrain the use of biomass, this represents an inherent need to create flexible flows of energy and storages – not just in the electricity system, but also in the thermal and transport sectors. The different needs and solutions for these types of storage have to be assessed in a coherent, comprehensive way.

The smart energy system approach aims to design energy systems that are economically feasible and socioeconomically affordable; in many cases, this means embracing the fact that the cheapest type of storage is the storage that is not necessary. Energy savings represent a key approach in reducing the need for energy storage, based on reduced energy demands. Electricity savings have traditionally been difficult; however, over the past few decades, Denmark has managed to stabilise its final electricity demand.[24] In IDA Energy Vision, a substantial increase in appliance use is assumed, but a reduction in the conventional electricity demand of 25% is foreseen, when excluding demands for heat pumps and electric vehicles, for example. This is partially due to technical improvements of electricity-using devices, but mostly to behavioural changes and changes in people's practices regarding their electricity consumption. Changes in transport demand are both difficult to envision and difficult to predict; however, here the potential for the conversion technologies to become more efficient is much higher.

The thermal sector has a much longer history of implementing energy savings, typically through higher standards in building codes and the renovation of existing buildings. Heating is an area where savings have, traditionally, been much more embraced. A significant body of research exists, using Smart Energy System-inspired approaches, investigating the potential and cost optimality of heat savings in a European context.[25,26] Typically, a level between 30 and 50% delivered heat savings, compared with today, is found to be feasible and economically preferable in these studies. This corresponds to, or is conservative in relation to, others' perspectives and scenario development.[27] In IDA Energy Vision, (delivered) heat demand to buildings is reduced by 40% compared with 2015; in Smart Energy Europe, a reduction of 50% is found to be cost optimal compared with 2010. These savings are the first step in reducing the need for energy demands overall, by proxy also reducing the overall need for energy storage. In addition to the overall reduction in energy needed, savings (especially in the thermal sectors) reduce the amplitude of the temporal variations, also contributing to a reduced need for storage. This is an important first step to consider in a smart energy system perspective, to avoid over-investment in storage when energy savings could bring increased benefits.

The interconnection between the sectors that characterises smart energy systems also reduces the need for energy storage by allowing more energy to be used directly, and creating flexibility in the system. As described in Section 2, the ability also to use electricity in the thermal and transport sectors in a flexible way reduces the need to store electricity in high-producing hours. For example, the ability to use excess wind in heat pumps or electrolysers if more electricity is being produced than is demanded at that moment both reduces primary energy supply and cost but also reduces the need for electricity storage. In this way, storage in smart energy systems is only used where necessary and cost-effective.

3.2   Comparison of Costs for Different Storage Types

A coherent approach towards all sectors of the energy system also necessitates a comprehensive assessment of different types of energy storage. Electricity storage, as evidenced in the first four chapters, is not self-evident and typically involves using an electricity input, transforming electrical energy for storage and then retrieving it as electricity again. In comparison, the storage of both fuel and thermal energy in (insulated) tanks or spaces is relatively straightforward. Comparing the relative costs and efficiencies of these storage types can give an insight into how smart energy systems can best be designed to prevent the need for over-investment in storage capacity.

A variety of different technologies for energy storage can be considered in EnergyPLAN and have been considered in the smart energy systems scenarios exemplified by IDA Energy Vision and the Smart Energy Europe scenario. These include the following:

- pumped hydroelectric storage;
- dammed hydroelectric storage;

- compressed air energy storage;
- (flow) battery energy storage;
- flywheel energy storage;
- supercapacitor energy storage;
- superconducting magnetic storage;
- hydrogen and (electro)fuel-based energy storage;
- thermal energy storage;
- electric vehicles.

Of these, not all are considered in depth in smart energy systems scenarios, or described in Section 4. Hydroelectric energy storage, battery energy storage, thermal energy storage, hydrogen and (electro)fuel-based energy storage and electric vehicles have been identified as the most attractive energy storage technologies in a smart energy system.[28–30] Within smart energy system settings, batteries are typically not found to be viable at a building level, mainly owing to their cost.[30] The other storage technologies are likely to continue to play a role in the energy system – particularly the power grid – but are not expected to contribute in large amounts to the flexibility of the energy system. However, this still allows for a variety of storage technologies that could be deployed in IDA Energy Vision and the Smart Energy Europe scenario.

Between the different types, the main drivers differentiating their costs and use in a smart energy system scenario are their scale and in what form they store energy. When comparing the different storage technologies directly (Figures 4–6), two main tendencies emerge: larger storages tend to be cheaper (per MWh) than small ones, and electricity storage tends to be more expensive than other forms of energy storage. However, in order to be able to use the more cost-efficient types of storage, such as large-scale thermal and fuel storages, infrastructures such as thermal and gas grids need to be present. These considerations play a central part in determining how the technologies are used in a smart energy system and how the design of the smart energy system interacts with them.

The form in which energy is stored greatly affects the cost of energy storage. Figure 4 illustrates several representative annualised costs (on a non-linear scale) and efficiencies for the storage of different types of energy. Of these, the costs for fuel, gas and thermal energy storage are the more reliable; the costs of hydroelectric storage are more variable, given the specificity and variability of the geographies and contexts of their application.[30] However, the cost of electricity storage is more than 100 times higher than for the other types of storage, indicating that there is a comparative advantage to avoiding it. The connection between the sectors in the smart energy system perspective allow this – both by reducing the impetus for electricity storage and by facilitating interaction between the power and other sectors.

Similarly, the scale at which energy storages are deployed impacts the price of energy storage. Figures 5 and 6 show the relative (annualised) costs

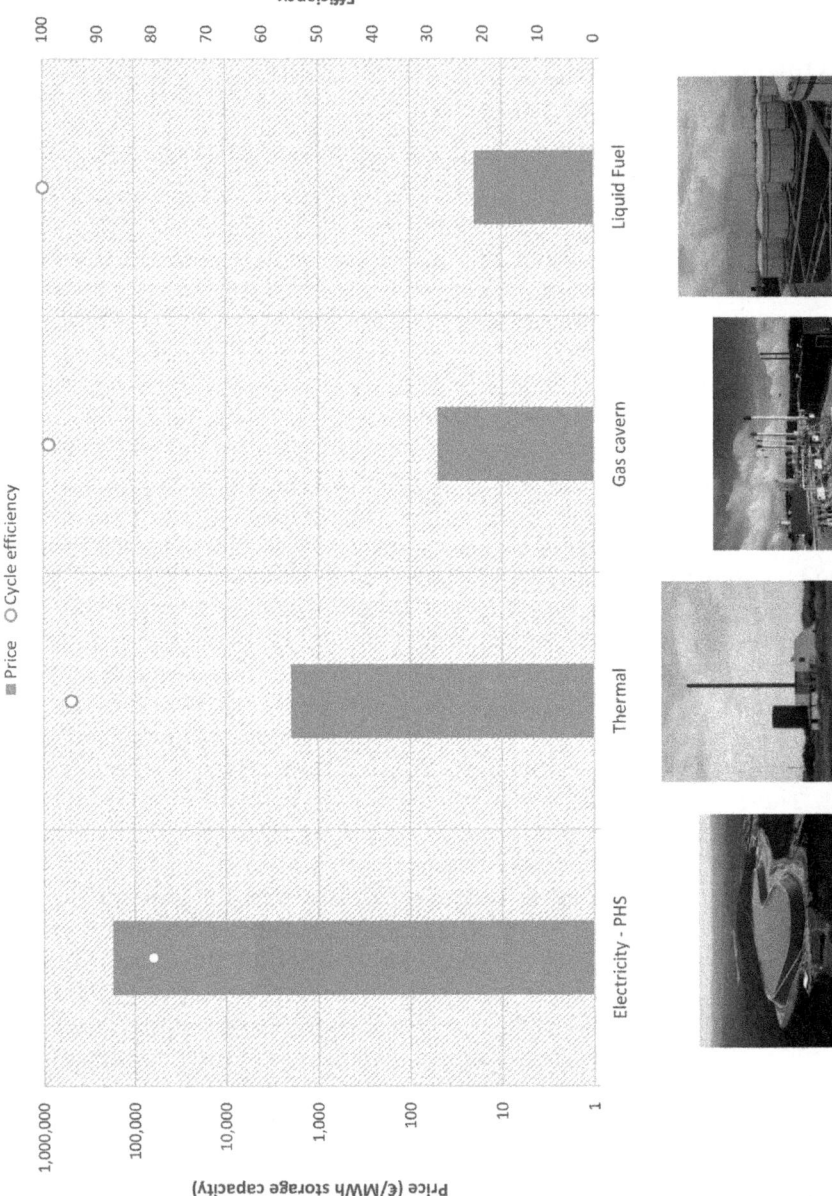

Figure 4 Investment cost and cycle efficiency of electricity, thermal, gas and liquid-fuel storage technologies. For data and assumptions, see ref. 30. Reproduced from ref. 30 with permission from Aalborg University Press.

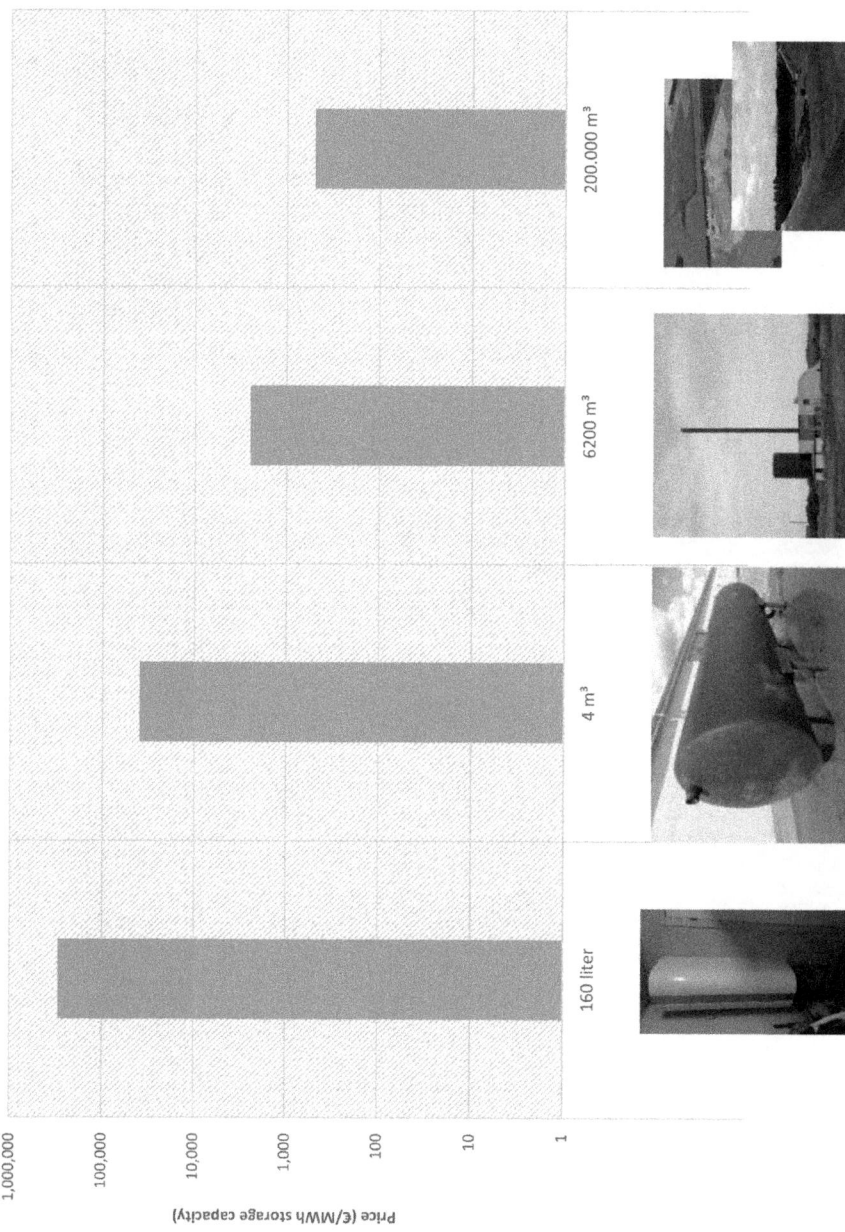

Figure 5 Investment cost comparison of different types of thermal energy storage technologies. For data and assumptions, see ref. 30. Reproduced from ref. 30 with permission from Aalborg University Press.

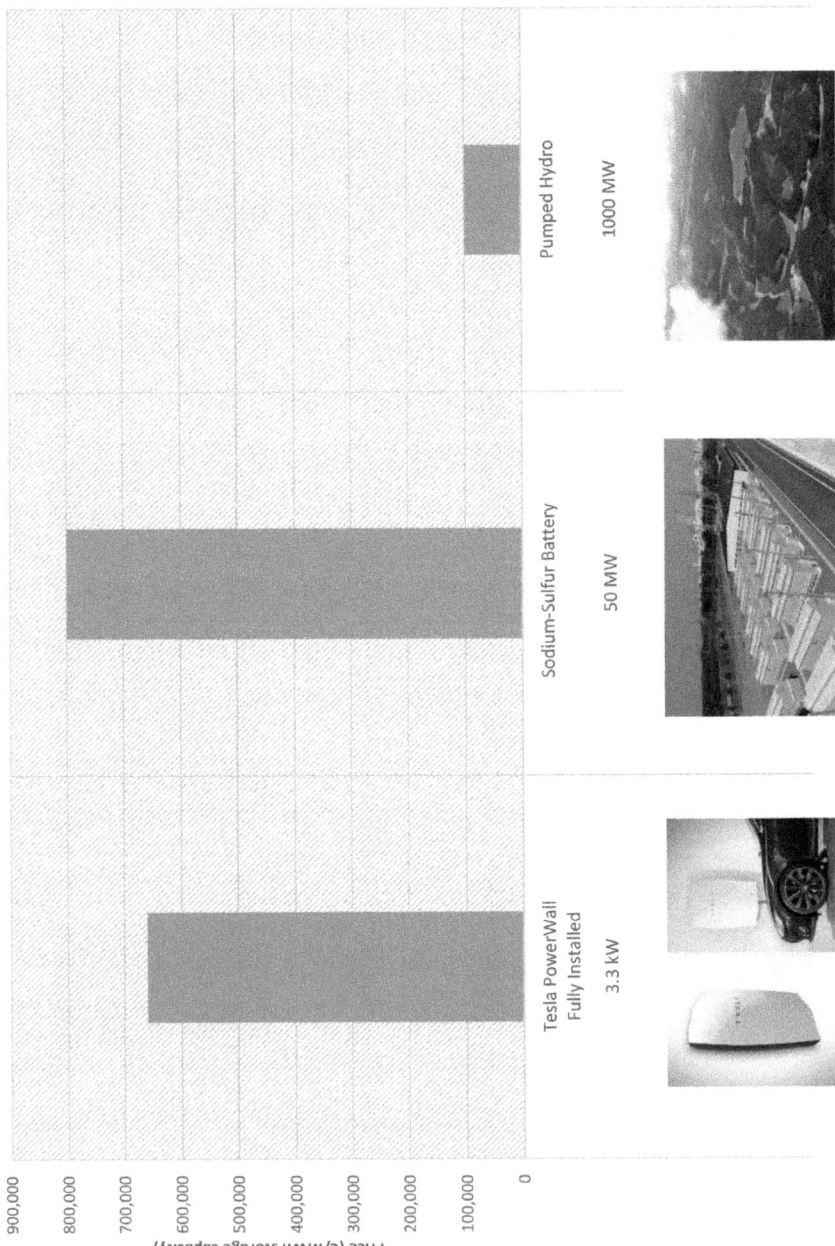

Figure 6 Investment cost comparison of different sizes of electricity energy storage technologies. For data and assumptions see ref. 30. Reproduced from ref. 30 with permission from Aalborg University Press.

of different sizes of thermal and electricity storages, respectively. In both cases, larger scale storages have a significantly lower cost than smaller scale storages.[30] This means that, typically, it is more advantageous to store energy in a decentralised way than in a fully distributed way. However, using these larger scale storages centrally implicitly requires having the infrastructure to be able to transport the energy to the end user, highlighting the importance that smart grids have in a smart energy system. Without these infrastructures, it is not possible to access large-scale thermal storages, for example, which could increase the energy system costs. In a smart energy system, which seeks to reduce overall cost, energy storage is prioritised in the fuel and thermal sectors, and at the largest scale feasible.

4 The Relevance of Storage in a Smart Energy System

This section details the use of the way in which different storage options are operationalised in IDA Energy Vision and the Smart Energy Europe scenario.[9,31] As described, these scenarios are developed as technically possible, economically feasible scenarios for the future, based on the smart energy system concept. By interconnecting the electricity, heating and cooling and transport sectors through smart grids, the overall need and cost for storage overall are reduced. Then, where storage is necessary, it can be concentrated in the larger scale applications (such as large thermal storage or hydro-electric storage rather than individual batteries) and in the cheapest form of energy available.

Since these examples of a smart energy system approach were simulated in the EnergyPLAN tool, results are available for how the systems operate and how they balance the demands and supply options at an hourly level.[32] This allows for not only an analysis of what the storage capacity and charge/discharge capacity are, but also how much energy is stored and discharged throughout the year, and in what contexts the storages are being emptied and replenished. Table 1 presents summarised results of the key types of storages that are used in IDA Energy Vision and the Smart Energy Europe scenario.

An important note here is that although both of these scenario-building exercises use a smart energy system perspective, and are designed using the EnergyPLAN tool, they are not necessarily directly comparable to each other. Cost, weather and technology data have not been aligned in a consistent manner between the scenarios, and the results presented here use EnergyPLAN version 11.3 for the Smart Energy Europe scenario but version 13.2 for IDA Energy Vision. For this reason, they both represent the simulation of a smart energy system but may not be directly comparable in the exact operation of the different technologies and storages.

4.1 Renewable Fuels

In the 100% renewable energy system analyses done in IDA Energy Vision and Smart Energy Europe, bioenergy is largely limited so the production of

Table 1 Key figures for the operation of storages in the Smart Energy Europe scenario and IDA Energy Vision (PES, primary energy supply).

Study	Parameter	Units	Fuel storage	Dammed hydroelectric storage	Pumped hydroelectric storage	V2G	Thermal storage
Smart Energy Europe	Storage capacity	GWh	721 000	19 867	286	6635	580
	Charge and discharge capacity	GW	1496	90	53	1327	117
	Annually discharged energy	TWh year^{-1}	4454	24	28	283	42
	Percentage of PES	%	28	0.15	0.17	1.75	0.26
IDA Energy Vision	Storage capacity	GWh	6000	0	0	15	112
	Charge and discharge capacity	GW	11.3	0	0	5.7	3.2
	Annually discharged energy	TWh year^{-1}	15.1	0	0	1.5	3.2
	Percentage of PES	%	10.2	0.00	0.00	1.01	2.16

hydrogen and electrofuels plays a particularly large role. This is the final, and most expensive, step towards the full decarbonisation of the scenarios.[22] Given the scarcity and expense of creating these fuels, it is important to ensure that they are used only where they are most needed, based on a cross-sectoral and integrated approach.[30] For this reason, electrification takes place in many sectors where fuels could be used (such as passenger transport and heating and cooling options).[33] This ensures both that bioenergy is used sustainably and that the need for (expensive) upgrading and production of electrofuels is minimised to those situations where there are no better or cheaper alternatives for decarbonisation.

In terms of fuel uses, in IDA Energy Vision, around 40% of the bioenergy is used directly for transport; mostly towards heavy transport modes, aviation and shipping. Small shares of bioenergy are used directly in heat-only boilers (in IDA Energy Vision, less than 2% of the available bioenergy). Given the scarcity of bioenergy, the resulting share of bioenergy in these smart energy systems is upgraded and used as an electrofuel. Hydrogen is produced using electricity in an electrolyser and then is combined either with biogas from a gasifier or with other forms of carbon, *via* a chemical synthesiser. Based on the cross-sectoral approach, this energy is then used in the sectors that are difficult to decarbonise (such as industry and the residual transport) and to power the remaining plants in the energy system.

The continued use of fuels is threefold. First, the 100% renewable energy system cannot use only intermittent and thermal renewable sources, particularly in sectors such as heavy-goods transport, aviation, shipping and industrial processes that require either extremely high temperatures or combustion processes. This prioritisation is made possible by considering all the sectors, not only from a perspective of integration, but also in terms of distributing resources in a cost-effective and comparatively advantageous way.[2] Where electrification is not possible, fuels are still needed and must become renewable.[34] Electrification of these fuels then both reduces the bioenergy need and functions as a proxy electrification of these otherwise difficult-to-decarbonise sectors.

While the integration of the smart thermal, electricity and gas grids allows for a much greater use of intermittent electricity sources, there is still a need for some electricity production with electricity-only plants [mostly combined cycle gas turbines (CCGTs)] and CHP plants.[35] In IDA Energy Vision, only 12% of electricity production comes from power or CHP plants, with the remainder being provided by a variety of (intermittent) renewable sources. Using bioenergy and electrofuels allows for the decarbonisation of these processes, while still ensuring that the energy systems in both IDA Energy Vision and the Smart Energy Europe scenario are balanced in terms of supplying the demanded energy reliably at all times.

Lastly, the use of fuel and specifically the production of electrofuels can contribute to the overall flexibility and efficiency of the system. The deployment of the electrolysers and hydrogenation plants is not very flexible, meaning that they operate effectively at a baseload level. This reduces the overall investment needed in production capacity, but means that they cannot respond to fluctuations in the system very well. However, chemical synthesis plants operate partially in response to the availability of (intermittent) electricity, and can then easily and cheaply store the produced energy in fuel storages. Overall, in the Smart Energy Europe scenario, chemical synthesis plants go down to 4% of total installed capacity during their lowest periods, with a yearly average of 44%. Given the size of the fuel storages, this flexibility occurs primarily as a function of the availability of intermittent renewables, particularly in hours of high wind power. As these fuels are electrified, this operation allows for an indirect storage of (intermittent) renewable electricity options and, through their use of otherwise excess electricity, also maximises the use of installed wind, solar and thermal capacity. Then, as with traditional fossil fuels, bio- and electrofuels can very easily be stored for extended periods of time, at a very low cost and with high efficiency.

This ease of storing fuels, specifically electrofuels, becomes clear from Table 1. The largest storages in the system are fuel storages, both in terms of storage capacity and charge/discharge capacity and in terms of volumes of energy passing through them yearly. Since the storage of fuels is so cheap and does not represent a significant loss in energy, this means that the production of fuels can be based on the availability of the production inputs as much as possible. Around half of the fuels produced in the smart energy system scenarios are made for storage and later use, 47% in IDA Energy Vision and 55% in the Smart Energy Europe scenario, respectively. This also translates in terms of primary energy. Although the production of these fuels is expensive and not as flexible as other technologies, their storage (especially at a large scale) is relatively cheap. This means that this plays a large role in the design of energy systems such as IDA Energy Vision and Smart Energy Europe, which are 100% renewable and must therefore also decarbonise the more challenging sectors such as industry and non-vehicle transport. The cheap storage of fuels, especially at a larger scale, thus plays an important role in ensuring that the sectors which are most difficult to decarbonise can be addressed and that energy provision is possible at all times, even in a highly renewable smart energy system.

4.2   Large-scale Hydroelectric Storage

Hydroelectric storage represents a large-scale, relatively cheap way of storing electric energy mechanically. The analysis of hydroelectric power and storage has not typically stood central in the development of the smart energy system concept since Denmark does not have the natural geography to develop significant hydroelectric power. In reality, Denmark is connected to several countries that have resources for hydroelectric power, in particular Norway. In practice, this represents a very valuable source of highly flexible electricity in the Nordpool region, which can avoid the use of peak electricity-generating capacity for periods of time when there are few (intermittent) renewables available, as has been discussed elsewhere.[31,36] However, in its conceptualisation, one of the overarching objectives of a smart energy system has been not only to have sustainable and cost-effective energy systems but also to provide energy security. There is then an incongruity in designing and developing an energy system scenario that is largely dependent on on-going international electricity trade and foreign resources. For this reason, IDA Energy Vision for 2050 is modelled as a closed system and does not consider the widespread application of hydro-storage. However, European resources are available (albeit limited), and, since the Smart Energy Europe scenario does consider Europe as an open system with continued electricity trade, this section discusses the operation of hydroelectric storage in the context of the Smart Energy Europe scenario.

Hydroelectric storage is simulated in two ways in the Smart Energy Europe scenario. It can be considered as dammed hydro, falling naturally with precipitation and collecting in rivers and lakes behind dams throughout the year – but not being used at the same time as when it falls and collects. Second, pumped hydro options are available that can use electricity, mechanically store it through water displacement behind a dam, and then discharge it when necessary. Table 1 also gives an overview of the storage capacity, discharge capacity, and total amount of energy discharged throughout the year for both types of energy.

The first relevant point is that within the context of the EU28, hydro-electric storage does not play an overly significant role, with the stored energy throughout the year representing only 1.1% of the total electricity demand. This corroborates findings by others who have used smart energy system principles to design 100% renewable energy system scenarios.[37] Although hydroelectric power is currently one of the most prevalent sources of electricity storage worldwide, as systems become more flexible and inte-grated it is likely that hydro will be overtaken by other forms, notably elec-tricity storage in distributed electric vehicles. This need not be general for all smart energy systems: the inclusion of Norway and Switzerland, and po-tentially the Balkan regions, would likely increase the role of hydroelectric storage for Europe.[38] Similarly, 100% renewable energy scenarios based on smart energy system principles for these countries (and others such as Canada, Brazil and China) could also look different.

The second is that where technically possible, all hydroelectric storage potential is economically feasible in the Smart Energy Europe scenario. This also is not uncommon in scenarios using smart energy system approaches where hydroelectric power has been applicable.[37,38] This underlines the superior cost-effectiveness of large-scale electricity storage in an energy system compared with individualised options such as batteries, and the assumption that much higher levels of hydroelectric storage are viable in countries where the potential is higher, since it is such a cheap form of storing electricity. If so, it would stand to reason that in a smart energy system scenario for such countries, the need for peak electricity generation capacity in the form of power and CHP plants would be significantly reduced, allowing for lower levels of (expensive) electrofuels and scarce bioenergy consumption.

4.3 Local Electric Storage in Electric Vehicles

In this section, we discuss the use of local electric storage in electric vehicles (EVs) using battery and vehicle-to-grid (V2G) principles, with a focus on private passenger transport. In a smart energy system, EVs perform two very important roles: linking mobility demands with the electricity sector in a very efficient way (thus reducing the demand for fuels) and functioning as a sink for intermittent renewable electricity where necessary. The provision of mobility through electricity, be that through EVs, plug-in hybrids or electrified buses and rail transport, is actually very valuable in a smart energy system in order to decarbonise the transport sector. Savings are not as straightforward as, for example, in heating and cooling; although Denmark has managed to keep and even reduce its energy demands in the built environment over the last 40 years, demand for mobility, fuels in transport and CO_2 emissions from transport have continued to increase.[24,31] In addition, the transport sector is not as interconnected as some other sectors, *e.g.* heat, where there are many different types of conversion technologies that could fulfil that demand. Lastly, some of the alternatives to electrification are not necessarily as desirable, partially owing to the discussion of fuel prioritisation in a cross-sectoral approach towards heavy transport and industry, but also because the combustion of the fuel (especially in cities) still contributes to local air pollution. For these reasons, the cheapest and most energy efficient solution for the transport sector in both IDA Energy Vision and the Smart Energy Europe scenario is to electrify what is possible in the transport sector, which in turn results in a significant battery storage capacity in the transport sector.

In both IDA Energy Vision and the Smart Energy Europe scenario, the main drivers for the level of EVs is the cost of the vehicles and the amount of biomass and electrofuels that are being consumed, not the amount of storage that the EVs can contribute *via* V2G operations. This is primarily because the cost, relative to other types of storage, is so high. In practice, there are some differences between the way in which transport is modelled in these two smart energy system scenarios. IDA Energy Vision assumed an extremely ambitious level of savings in terms of mobility demand and also assumed a relatively high level of modal shifts (which results in overall lower costs and energy consumption); this was not the case to the same extent in the Smart Energy Europe scenario. In both of these scenarios, there is an assumption that the majority of cars will become EVs, but that there will also be some plug-in hybrid vehicles, biofuel-based and fuel-cell vehicles using hydrogen. Further increasing the level of electrification in the passenger vehicle fleet does not lead to a lower primary energy supply but does drive up costs, underlining the fact that the value of electrification is due to reductions in fuel use, and only secondarily due to contributions towards the flexibility of the energy system.

In the Smart Energy Europe scenario, 80% of the car transport is electrified, with 25 kWh storage simulated per car. As seen in Table 1, this results in a relatively large storage capacity, although representing less than 1% of the fuel storage capacity available. In IDA Energy Vision, EVs represent only 75% of an already largely reduced personal vehicle fleet. This results in a comparatively significantly lower storage capacity. This also results in a different contribution to the use of the storage in the personal vehicle fleet; while the proportion of energy stored and then returned from the vehicles in IDA Energy Vision (representing around 12% of the energy used by the EVs in total), the exchange of energy *via* V2G represents 42% of the energy finally used by the EVs in the Smart Energy Europe scenario. It is evident that the different ways in which the transport sector is decarbonised have a large effect on the way that storage in the transport sector can be utilised.

4.4 Thermal Storage

The heating sectors in both IDA Energy Vision and the Smart Energy Europe scenarios are fully decarbonised using a variety of conversion technologies and grids. Electric heat pumps play an important role, particularly in areas where district heating is not viable. District heating in a smart thermal network is modelled in urban and suburban areas, using a combination of large-scale heat pumps, excess heat from industry and cogeneration, solar thermal and geothermal energy, and direct heat use. This is one of the primary ways in which thermal networks can contribute to smart energy systems.[26] In IDA Energy Vision, 66% of the (residential and service) heat demand is covered by district heating; in the Smart Energy Europe scenario this is 50%.

Thermal storage is defined here as sensible thermal storage, at temperature levels suitable for district heating production (70–100 °C). This means that the sole purpose of the stored energy is to supply heat at a later time. In terms of thermal energy storage, the storage uses in the district heating sector are the more interesting to consider, since they are more variable in size and can store thermal energy from a mix of sources. In terms of a district heating system, this includes the following:

- Heat from stable sources, which needs to be stored in order to satisfy (intermittent) demands. Examples of this are geothermal and heat from waste, and certain types of industrial excess heat. Both of these provide significant amounts of energy towards the district heating system; in the Smart Energy Europe scenario, 12 and 11% of the district heat demand, respectively. By being able to store these types of heat when heat demands are low and there is no immediate heat demand, these types of renewable and sustainable heat can still be used later and contribute to the renewability and efficiency of the energy system. Particularly for the heat that is produced as a by-product of industrial processes, which was initially produced through biomass or electrofuels, this can provide a good opportunity to re-use and re-store energy from valuable and scarce sources.

- Heat that is produced intermittently and needs to be stored for when it can meet demand. This is the case, for example, with solar thermal installations, which provide more than 4% of the district heat in the Smart Energy Europe scenario. Solar thermal plants produce most of their heat during summer days; however, there are already today examples of district heating systems that combine with seasonal storage and can fulfil their demand through to February on solar heat alone. This is especially important, since the viability of solar district heating is largely dependent on the overall system and storage availabilities.[39]

These types of storage play an important role in ensuring that the system is balanced in terms of temporal demand and supply of heat, and preventing the need for peak boilers. In IDA Energy Vision, 5% of the heat in the district heating system is provided by (biomass and direct electric) peak boilers; in the Smart Energy Europe scenario, this is around 10%. This reduces both the amount of fuel necessary in the system and the need to invest in peak production capacity.

The second way in which thermal storages can contribute to the overall performance and flexibility in the district heating system is by allowing for the conversion of abundant or excess intermittent electricity to be converted to heat *via* heat pumps or direct electric heating, then to be used during a later period where there might not be as much intermittent electricity. The simulation of the district heating system in EnergyPLAN depends on a number of hourly variables: the energy demands (both heating and electricity) in that hour, the weather patterns and the other technologies simulated in the system. The system operates to balance both the heating system and the electricity market, resulting in differing operations of the system (see Table 2, using an example from the Smart Energy Europe scenario). In hours where there are many intermittent renewables, it is more beneficial to use those directly or indirectly *via* a heat pump. If extra heat is required and there is an abundance of intermittent electricity, a direct electric boiler may be preferable to a methane or electrofuel boiler. This connection allows not only for the use of intermittent renewables such as wind and sun in the thermal sector but also for the cheap storage thereof.

As shown in Table 1, the relative impact of the sizes of the thermal storages in the district heating systems is not very large, especially when compared with gas. In IDA Energy Vision, the higher level of district heating allows for some more flexibility in using the storages in the energy system, with overall 9% of the energy distributed to the district heating system having come through the large thermal storages. In Smart Energy Europe, this is only 3.5%. In both cases, around one-third of this energy comes out of the 'inflexible' heat supply sources such as geothermal and (constant) industrial excess heat; the remaining two-thirds comes either from the storage of heat from a CHP plant (thereby maximising the efficiency of the electricity

Table 2 Operation of (large) district heating systems for various hours during the year, showing the interactions between hourly weather patterns, electricity demands and generation and heat demand, using an example from the Smart Energy Europe scenario. (RES, renewable energy supply.)

Parameter	Units	Large CHP	Large HPs	Large boilers	Large (direct)	Large solar thermal	Large thermal storage
Installed capacity	MW	150 000	45 000	101 482	25 000	43 101	350 000
Hour with low intermittent RES, storage filled	MWh	88 858	849	0	0	0	350 000
Hour with intermittent RES, storage emptied, boiler backup required	MWh	0	45 000	78 882	0	0	0
Hour with high intermittent RES, direct electric heating possible, storage filling with intermittent RES	MWh	0	45 000	0	24 598	11 243	151 199
Hour with very high intermittent RES and heat demands, storage empty, both direct electric and boilers required	MWh	0	45 000	16 046	25 000	0	0

system, when it requires power from plants) or from intermittent renewable sources – either directly from solar thermal or from intermittent RES through large heat pumps. In these ways, the low cost of the thermal storages – especially at a large scale – both contributes to the overall efficiency of the system and allows for a greater integration of renewable energy.

5 Conclusion

The smart energy system concept has been developed to design 100% renewable energy systems that are technically possible, socioeconomically feasible and do not rely on the over-exploitation of bioenergy. In order to achieve this, high levels of (intermittent) renewable energy sources must be integrated in the energy system. To do so, the development of smart electricity, thermal and gas grids is proposed, with the objective of coupling the different energy sectors in order to exploit the synergies that occur between them and utilising energy storage in the most cost- and energy-efficient manner. This allows for increased flexibility, a reduction in primary energy supply and a decrease in costs compared with other proposed decarbonisation approaches.

The development of IDA Energy Vision and the Smart Energy Europe scenario have been presented to exemplify the smart energy system approach. The system designs rely on energy savings on the demand side, using thermal grids in order to unlock the potential of renewable and excess energy sources, electrifying the remainder of the heating and cooling sector, electrifying transport where possible and replacing the remaining fossil fuels with bio- and electrofuels. These scenarios are developed and assessed in EnergyPLAN, an hourly energy system simulation model that has been explicitly developed to analyse the operation and performance of 100% renewable energy systems. Through this, two energy systems are designed that can integrate higher levels of intermittent energy sources, present cost- and energy-efficiency improvements and can fully decarbonise the respective energy systems without imposing undue burdens on the available bioenergy potentials.

By coupling sectors, (excess) renewable energy can more easily be dispersed through the energy system. This reduces the need for energy storage, since there are more options for directly using high levels of (intermittent) renewable energy. In addition, this interconnection allows for the specific deployment of energy storage where it can contribute most. Energy storages of all kinds are more expensive at a small scale than at a large scale, and electricity storage at all scales is significantly more expensive than thermal energy and fuel storages. By connecting the thermal, transport and power sectors, large-scale and cheaper forms of storage can be used better.

In terms of the ways in which energy storages are used in the two smart energy system scenarios presented, the production and storage of (electro)-fuels represents the most significant. This can be observed in terms of storage capacity, charge/discharge capacity and amount of energy stored and discharged throughout the year. Since this is the cheapest form of energy storage, this represents a valuable service in terms both of allowing for the decarbonisation of otherwise difficult energy demands such as industry, heavy transport and peak electricity demands, and of allowing for chemical synthesis plants to run in the most flexible way they can.

Hydroelectric storages, in terms of both dammed and pumped hydroelectric storages, are not present in IDA Energy Vision of flat Denmark, since the natural geography does not naturally allow for the development of any potential. However, in the Smart Energy Europe scenario, all available hydroelectric storage potential is found to be cost-effective owing to its large scale and value in providing electricity in a highly flexible way. Even then, the natural constraints mean that hydroelectric storage in Europe is still too limited to represent a significant part of the storage capacity of the energy system.

The storage capacity afforded through EVs plays a larger role in the smart energy system scenarios analysed, particularly in the Smart Energy Europe scenario. However, since the cost of smaller scale batteries is high, the storage role of electric vehicles is secondary to their contribution towards decarbonising the passenger vehicle fleet without increasing biomass use. Reducing transport and vehicle demand is typically cost-effective, since the contribution that the batteries in EVs make in terms of storage is outweighed by the cost and energy use of road transport.

Thermal storages, specifically connected to district heating, play an important role in the operation of the smart thermal grids by allowing the integration of both intermittent renewables and stable, baseload-renewable thermal sources. The manifestation of the coupling between the thermal and power sectors, and the role that large-scale thermal storages can play in this, are shown in the context of different (hourly) weather and energy-demand situations throughout the year.

Energy storages play a key role in the smart energy system concept, since they contribute to the integration of (intermittent) renewable energy sources and flexibility of the system. In this chapter, the role and operation of different types of energy storage are presented, exemplified by IDA Energy Vision and the Smart Energy Europe scenario. By coupling the different energy sectors and using smart electricity, thermal and gas grids, the need for energy storage can be reduced, and energy storage can be deployed there where it is most cost-effective. In doing so, two technically possible and socio-economically feasible energy systems are designed, without using unsustainable amounts of bioenergy. Looking forward, the design of these scenarios and the development of the smart energy system concept contribute towards the planning and design of an energy transition that can help counteract the climate and environmental pressures that our current energy systems have created.

References

1. D. Connolly, H. Lund, B. V. Mathiesen, P. A. Østergaard, B. Möller, S. Nielsen, I. Ridjan, F. Hvelplund, K. Sperling, P. Karnøe, A. M. Carlson, P. S. Kwon, S. M. Bryant and P. Sorknæs, *Smart Energy Systems: Holistic and Integrated Energy Systems for the Era of 100% Renewable Energy*, Denmark: Sustainable Energy Planning Research Group, Aalborg University, 2013. http://vbn.aau.dk/files/78422810/Smart_Energy_Systems_Aalborg_University.pdf.
2. H. Lund, *Renewable Energy Systems: A Smart Energy Systems Approach to the Choice and Modeling of 100% Renewable Solutions*, Academic Press, Burlington, USA, 2014, vol. 2.
3. P. A. Østergaard, Reviewing EnergyPLAN simulations and performance indicator applications in EnergyPLAN simulations, *Appl. Energy*, 2015, **154**, 921–933. DOI: 10.1016/j.apenergy.2015.05.086.
4. P. A. Østergaard, Reviewing optimisation criteria for energy systems analyses of renewable energy integration, *Energy*, 2009, **34**(9), 1236–1245. DOI: 10.1016/j.energy.2009.05.004.
5. T. Searchinger, R. Heimlich, R. A. Houghton, F. Dong, A. Elobeid, J. Fabiosa, S. Tokgoz, D. Hayes and T. hsiang Yu, Emissions from land-use change, *Science*, 2008, **423**, 1238–1240.
6. M. Rosegrant, Biofuels and grain prices: impacts and policy responses, *Int. Food Policy Res. Inst.*, 2008, 1–4.

7. A. Ajanovic, Biofuels versus food production: Does biofuels production increase food prices?, *Energy*, 2011, **36**(4), 2070–2076. DOI: 10.1016/j.energy.2010.05.019.

8. D. Zilberman, G. Hochman, D. Rajagopal, L. Angeles, S. Sexton and G. Timilsina. The Impact of Biofuels on Commodity Food Prices: Assessment of Findings. *Science (80-)*. 2011. https://s3.amazonaws.com/academia.edu.documents/44272884/The_Impact_of_Biofuels_on_Commodity_Food20160331-16479-1qo4r7w.pdf?AWSAccessKeyId=AKIAIWOWYYGZ2Y53UL3A&Expires=1522450952&Signature=yQ4K1GObSPJEsKzevYosb9RLnYQ%3D&response-content-disposition=inline.

9. D. Connolly, H. Lund and B. V. Mathiesen, Smart Energy Europe: The technical and economic impact of one potential 100% renewable energy scenario for the European Union, *Renewable Sustainable Energy Rev.*, 2016, **60**, 1634–1653. Accessed April 3, 2016.

10. B. V. Mathiesen, H. Lund and D. Connolly, Limiting biomass consumption for heating in 100% renewable energy systems, *Energy*, 2012. DOI: 10.1016/j.energy.2012.07.063.

11. B. V. Mathiesen, H. Lund, D. Connolly, H. Wenzel, P. A. Østergaard, B. Möller, S. Nielsen, I. Ridjan, P. Karnøe, K. Sperling and F. K. Hvelplund, Smart Energy Systems for coherent 100% renewable energy and transport solutions, *Appl. Energy*, 2015, **145**, 139–154. DOI: 10.1016/j.apenergy.2015.01.075.

12. D. Connolly, K. Hansen, D. Drysdale, H. Lund, B. V. Mathiesen, S. Werner, U. Persson, B. Möller, O. G. Wilke, K. Bettgenhäuser, W. Pouwels, T. Boermans, T. Novosel, G. Krajačić, N. Duić, D. Trier, D. Møller, A. M. Odgaard and L. L. Jensen. *Enhanced Heating and Cooling Plans to Quantify the Impact of Increased Energy Efficiency in EU Member States (Heat Roadmap Europe 3)*. Copenhagen; 2015. www.heatroadmap.eu.

13. S. Paardekooper, The implications of an increasingly decentralised energy system, 2015. http://projekter.aau.dk/projekter/files/213566904/2015_Susci4_Susana_Paardekooper.pdf.

14. H. Lund, A. Marszal and P. Heiselberg, Zero energy buildings and mismatch compensation factors, *Energy Build.*, 2011, **43**(7), 1646–1654. DOI: 10.1016/j.enbuild.2011.03.006.

15. The Danish Society of Engineers. *The Danish Society of Engineers' Energy Plan 2030*. The Danish Society of Engineers; 2006. http://ida.dk/sites/climate/introduction/Documents/Energyplan2030.pdf.

16. B. V. Mathiesen, H. Lund and K. Karlsson. *The IDA Climate Plan 2050 - Background Report. Technical Energy System Analysis, Effects on Fuel Consumption and Emissions of Greenhouse Gases, Socio-Economic Consequences, Commercial Potentials, Employment Effects and Health Costs*, The Danish Society of Engineers, IDA, Copenhagen, Denmark, 2009. http://vbn.aau.dk/files/38595718/The_IDA_Climate_Plan_2050.pdf.

17. H. Lund, F. Arler, P. A. Østergaard, F. Hvelplund, D. Connolly, B. V. Mathiesen and P. Karnøe, Simulation versus optimisation: Theoretical

positions in energy system modelling, *Energies*, 2017, **10**(7), 1–17. DOI:10.3390/en10070840.

18. D. Connolly, B. V. Mathiesen, P. A. Østergaard, B. Möller, S. Nielsen, H. Lund, D. Trier, U. Persson, D. Nilsson and S. Werner, Heat Roadmap Europe: First pre-study for EU27, 2012. http://vbn.aau.dk/files/77244240/Heat_Roadmap_Europe_Pre_Study_1.pdf.

19. D. Connolly, B. V. Mathiesen and H. Lund. Smart Energy Europe: A 100% renewable energy scenario for the European Union. In: Dubrovnik, Croatia: Proceedings from 10th Dubrovnik Conference on Sustainable Development of Energy, Water and Environment Systems, 2016. www.vbn.aau.dk.

20. D. Connolly, H. Lund and B. V. Mathiesen, Smart Energy Europe: The technical and economic impact of one potential 100% renewable energy scenario for the European Union, *Renewable Sustainable Energy Rev.*, 2016, **60**, 1634–1653. DOI:10.1016/j.rser.2016.02.025.

21. European Commission, Roadmap 2050, *Policy*, 2012, 1–9. DOI:10.2833/10759.

22. I. Ridjan, *Integrated Electrofuels and Renewable Energy Systems*, Department of Development and Planning, Aalborg University, Copenhagen, Danmark, 2015.

23. H. Lund, Department of Development and Planning Aalborg University. EnergyPLAN: Advanced Energy System Analysis Computer Model. http://www.energyplan.eu/. Published 2015.

24. L. Kørnøv, M. Thrane, A. Remmen, H. Lund, *Tools for Sustainable Development*, Aalborg University Press, Aalborg, Denmark, 2007.

25. T. Boermans, K. Bettgenhäuser, M. Offermann and S. Schimschar, Renovation Tracks for Europe up to 2050: Building Renovation in Europe - What Are the Choices?, *Ecofys*, 2012.

26. H. Lund, S. Werner, R. Wiltshire, S. Svendsen, J. E. Thorsen, F. Hvelplund and B. V. Mathiesen 4th, Generation District Heating (4GDH): Integrating smart thermal grids into future sustainable energy systems, *Energy*, 2014, **68**(0), 1–11. DOI:10.1016/j.energy.2014.02.089.

27. D. Connolly, B. V. Mathiesen, P. A. Østergaard, B. Möller, S. Nielsen, H. Lund, U. Persson, S. Werner, J. Grözinger, T. Boermans, M. Bosquet and D. Trier, Heat Roadmap Europe: Second pre-study, 2013, http://vbn.aau.dk/da/publications/heat-roadmap-europe-2050(306a5052-a882-4af9-a5da-87efa36efeaa).html.

28. D. Connolly and M. Leahy, *A Review of Energy Storage Technologies: For the Integration of Fluctuating Renewable Energy*, University of Limerick, Limerick, 2010, http://vbn.aau.dk/files/100570335/Energy_Storage_Techniques_v4.1. pdf.

29. H. Lund and G. Salgi, The role of compressed air energy storage (CAES) in future sustainable energy systems, *Energy Convers. Manage.*, 2009, **50**(5), 1172–1179.

30. H. Lund, P. A. Østergaard, D. Connolly, I. Ridjan, B. V. Mathiesen, F. Hvelplund, J. Z. Thellufsen and P. Sorknæs, Energy Storage and Smart

Energy Systems, *Int. J. Sustainable Energy Plan Manage.*, 2016, **11**, 3–14 DOI: 10.5278.

31. B. V. Mathiesen, H. Lund, K. Hansen, I. Ridjan, S. Djørup, S. Nielsen, P. Sorknæs, J. Z. Thellufsen, L. Grundahl, R. Lund, D. Drysdale, D. Connolly and P. A. Østergaard, *IDA's Energy Vision 2050*, Aalborg University, Copenhagen, 2015.

32. H. Lund, EnergyPLAN - Advanced Energy Systems Analysis Computer Model, Documentation Version 12, http://www.energyplan.eu/, Published 2015, Accessed October 17, 2016.

33. I. Ridjan, B. V. Mathiesen, D. Connolly and N. Duić, The feasibility of synthetic fuels in renewable energy systems, *Energy*, 2013, **57**, 76–84. DOI:10.1016/j.energy.2013.01.046.

34. D. Connolly, B. V. Mathiesen and I. Ridjan, A comparison between renewable transport fuels that can supplement or replace biofuels in a 100% renewable energy system, *Energy*, 2014, **73**, 110–125. DOI:10.1016/j.energy.2014.05.104.

35. R. Lund and B. V. Mathiesen, Large combined heat and power plants in sustainable energy systems, *Appl. Energy*, 2015, **142**(0), 389–395. DOI:10.1016/j.apenergy.2015.01.013.

36. J. Z. Thellufsen and H. Lund, Cross-border versus cross-sector interconnectivity in renewable energy systems, *Energy*, 2017, **124**, 492–501. DOI:10.1016/j.energy.2017.02.112.

37. D. Connolly and B. V. Mathiesen, A technical and economic analysis of one potential pathway to a 100% renewable energy system, *Int. J. Sustainable Energy Plan Manage.*, 2014, **1**, 7–28. DOI:10.5278/ijsepm.2014.1.2.

38. B. Ćosić, G. Krajačić and N. Duić, A 100% renewable energy system in the year 2050: The case of Macedonia. *6th Dubrovnik Conf Sustain Dev Energy Water Environ Syst SDEWES 2011*, 2012, **48**(1), 80–87. DOI: 10.1016/j.energy.2012.06.078.

39. B. V. Mathiesen, K. Hansen. *The Role of Solar Thermal in Future Energy Systems*. Paris, 2017. http://vbn.aau.dk/files/265304574/IEA_SHC_Task_52_STA_AAU_report_20170914.pdf.

Life-cycle Analysis for Assessing Environmental Impact

HEIDI HOTTENROTH,* JENS PETERS, MANUEL BAUMANN, TOBIAS VIERE
AND INGELA TIETZE

ABSTRACT

In this chapter, stationary energy storage systems are assessed concerning their environmental impacts *via* life-cycle assessment (LCA).
The considered storage technologies are pumped hydroelectric storage,
different types of batteries and heat storage. After a general introduction to the method of LCA, some methodological implications for
energy storage systems and the selection of impact indicators are
outlined. Subsequently, the environmental impacts of different energy
storage options are assessed in three case studies. The first case study
compares pumped hydroelectric storage and utility-scale battery
storage applying a screening LCA. Both of the two following case
studies are based on an island micro grid application and follow a
stepwise approach. The starting point is a pair of cost-optimal energy
scenarios – one with and the other without use of stationary battery
storage. First, based on the given operational parameters, the environmental performance of different lithium-ion batteries is assessed.
This allows the identification of the most appropriate battery chemistry
for this specific application (case study 2). Applying these results, the
battery-using energy system scenario is compared in terms of environmental performance with an alternative scenario without battery
use in order to determine the contribution of energy storage within the
whole energy system (case study 3). Under the given modelling

---

*Corresponding author.

Issues in Environmental Science and Technology No. 46
Energy Storage Options and Their Environmental Impact
Edited by R.E. Hester and R.M. Harrison
© The Royal Society of Chemistry 2019
Published by the Royal Society of Chemistry, www.rsc.org

assumptions, the use of battery storage results in increased environmental impacts in the majority of the assessed categories, both in comparison with pumped hydroelectric storage (case study 1) and in comparison with the standard small-scale energy system without battery storage (case study 3). Regarding heat storage, the underlying case study shows a low relevance of environmental impacts within the energy system.

1   Introduction to Life-cycle Assessment

Life-cycle assessment (LCA) is a standardized process for assessing the environmental impacts of product and service systems over their whole life cycle from raw material extraction to final disposal or recycling.[1,2] It is a methodology commonly accepted and widely applied for sustainability assessment on various levels. For instance, LCA is used to provide product comparisons for marketing purposes, assessments of new technologies for product and process design and macro-scale studies on the consequences of environmental policy options. Starting decades ago with studies on packaging and beverage systems, LCA has developed into a widespread methodology with huge academic and practitioners' networks globally.

LCA follows a well-established procedure defined in ISO standards.[3,4] Four major process steps can be distinguished:

1. Goal and Scope define the LCA framework, where the goal and the intended audience of the study are defined, the boundaries of the assessed system and the functional unit, *i.e.* the basis on which product and service systems are assessed and compared. For instance, to compare various options for European electricity systems of the future, goal and scope need to define baselines, target years, included and excluded technologies, technical and quality requirements and so forth. Typical distinctions in terms of system boundaries are cradle to gate, *i.e.* the analysis of a system from raw material extraction to the final product or service, and cradle to grave, *i.e.* also including distribution, use and end-of-life of the product or service.

2. Within a Life-cycle Inventory (LCI), the system under investigation is modelled, including all physical flows (usually material and energy flows) that enter or leave the system (raw materials, emissions, intermediates, *etc.*). This phase is usually time consuming because an entire product or service system involves a multitude of upstream and downstream processes at the various stages of the life cycle. Whereas inventory data for major processes are gathered and measured directly (foreground system), large LCI databases are used for processes further up- or downstream for which data collection is not feasible or possible (background system).[5]

3. The Life-cycle Impact Assessment (LCIA) 'translates' the LCI into environmental impacts of the system. There is a large range of potential environmental impacts, *e.g.* global warming, ionizing radiation, resource depletion and human toxicity. Each impact has a common unit of measurement; global warming, for instance, is usually measured in kilograms of CO_2-equivalents for a time span of 100 years, called Global Warming Potential 100 (GWP100). All LCI entries that contribute to global warming are classified as contributors to that particular impact and then converted into their particular GWP100 by scientific characterization factors. These factors consider the substances' different potential damage, such as 1 kg of dinitrogen monoxide being roughly as damaging as 300 kg of carbon dioxide.
4. Within Interpretation, results are discussed and analysed. The interpretation part also includes the identification of errors and insufficiencies of the system model and might lead to a refinement of goal and scope, LCI and/or LCIA. The whole process of LCA is therefore an iterative process and is repeated until the desired quality of the results is achieved.

These procedures form the core of LCA; but life-cycle thinking and methods have been diversified and developed further. Carbon footprints, water footprints and global footprint follow the main LCA procedures, but deliver a single-score result in terms of global warming potential or cubic metres water consumed. Environmental Product Declarations and also Product Environmental Footprints proposed by the European Union (EU) aggregate and standardize LCA results much further than the ISO standards following defined rules in order to increase comparability of results within a product group. Corporate carbon footprints, organizational LCA and organizations' environmental footprints extend the idea of LCA to whole organisations and their performance over time.

2 Life-cycle Assessment of Energy Storage Systems

From financial and technical perspectives, it is not always easy to answer the question of which energy storage system is the best option in a specific setting. The same is true when it comes to environmental assessments. To provide some insights into the outcomes but also challenges of assessing energy storage systems by means of LCA, we will present three different case studies. Beforehand, we explore some general challenges for LCA studies on energy storage systems.

Storage systems only provide their benefits in combination with other power- or heat-generating technologies. For this reason, the definition of the provided service for the derivation of the functional unit needs thorough elaboration, particularly when comparing different technologies. Even if individual storage systems are assessed, the larger energy system is of interest, for instance, to understand the benefits of the storage system in comparison with increasing capacities for providing the required energy without storage. The exact quantity of product or service required to fulfil the functionality constitutes each system's reference flow. Depending on lifetime, efficiency, sizing and other features, each system under comparison might have different requirements in terms of physical storage (*i.e.* different reference flows) to fulfil the same overall functional unit. For this reason, comparisons of battery systems on a per-kilogram basis are often not meaningful because of different energy densities. Within an LCA study on energy storage systems, the use phase determines the physical storage required and hence environmental impacts in raw material extraction, production and disposal/recycling of storages. The use phase is usually a significant contributor to the overall results mainly due to charge–discharge efficiencies, but particularly for utility-scale batteries also the demand for external ventilation and cooling systems causes impacts that cannot be ignored.

As end-of-life treatment of battery storages such as recycling and remanufacturing options and impacts of disposal are not well known, this stage of the life cycle is often disregarded and left out of the system boundary. This is a potential flaw in studies as end of life might have a significant impact on the outcome of a study.[6]

3   Selection of Impact Indicators

For the LCIA of the following case studies, ILCD,[7,8] a specific predefined method recommended by the EU, was chosen. The ILCD method recommends a comprehensive list of environmental impacts, which is supported by the ecoinvent database that we made use of. We followed the ILCD recommendations with two exceptions. As mainly renewable energy systems are assessed, ionizing radiation, *e.g.* from nuclear power, does not need to be included. Second, we included an impact indicator for use of water resources from another LCIA method (Ecological Scarcity 2013[9]) in order also to include water resource use. The following description of impact indicators is partly taken from the Product Environmental Footprint Initiative by the European Commission.[10]

- Resource depletion – metals, minerals, fossils and renewables
  The Earth contains a finite amount of non-renewable resources such as metals, minerals or fossil fuels like coal, oil and gas. The basic idea behind this impact category is that extracting a high concentration of resources today will force future generations to extract lower concentration or lower-value resources.
  Unit of measurement: kilograms of antimony equivalent (kg Sb-eq.).
- Resource depletion – water
  The withdrawal of water from lakes, rivers or groundwater can contribute to 'depletion' of available water on-site, while water itself is seen as a renewable resource. This impact category considers the scarcity of

water in the regions where the activity takes place if this information is known.

Unit of measurement: eco-points (EP).

- Land use

The impact category land use attempts to estimate the damage to ecosystems and also the limitations on land as a resource due to the effects of occupation and transformation of land. Examples of land use are agricultural production, mining and human settlement. Transformation is the conversion of land from one use to another. Impacts are diverse; for instance, loss of species, organic matter content in soil, reduced primary production or even loss of the soil itself (erosion).

Unit of measurement: kilograms of carbon deficit [kg soil organic carbon (SOC)].

- Climate change

The indicator climate change refers to the changes induced to the World's climate as a consequence of the emissions to the atmosphere of the so-called greenhouse gases, such as CO_2, N_2O, CH_4 or hydrocarbons. The most important human contribution to the emissions of greenhouse gases is attributed to the combustion of fossil fuels. The consequences include increased global average temperatures and sudden regional climatic changes, which can directly and indirectly negatively affect the natural environment, human health and the availability of natural resources.

Unit of measurement: kilograms of carbon dioxide equivalent (kg CO_2-eq.).

- Acidification

Acidification has contributed to a decline of coniferous forests and increased fish mortality. Acidification can be caused by emissions into the air, water and soil. For instance, when gaseous SO_2 is released and reaches a water body, it reacts with water to form sulfuric acid. When acids (and compounds that can be converted to acids) are emitted to the atmosphere and deposited in water and soil, the addition of hydrogen ions may result in an increase of the water body's acid content. The most significant anthropogenic sources of acidification are combustion processes in electricity, heating production and transport.

Unit of measurement: moles of hydron equivalent (mol H^+-eq.).

- Ecotoxicity – freshwater aquatic

A substance contributing to ecotoxicity affects the function and structure of the ecosystem by exerting toxic effects on the organisms that live in it. Toxic effects can occur as soon as the substances are released (acute ecotoxicity), or may appear after repeated or long-term exposure to the substances (chronic ecotoxicity). The chronic ecotoxicity of a compound is determined by its toxic effects, its biodegradability and its ability to accumulate in living organisms.

Unit of measurement: comparative toxic unit for ecosystems (CTUe).

- Eutrophication
  Eutrophication is an impact on the ecosystems mainly from substances containing nitrogen or phosphorus. As a rule, the availability of one of these nutrients will be a limiting factor for growth in the ecosystem, and if this nutrient is added, the growth of algae or specific plants will be increased.
  - Freshwater
    In lakes and rivers, eutrophication will be mainly due to the increase of phosphorus. Too rapid growth of algae can lead to a lack of oxygen in the water for fish to survive once the algae die and are degraded (which consumes oxygen). Emissions of nitrogen to the aquatic environment are caused largely by the agricultural use of fertilizers, but oxides of nitrogen from combustion processes are also of significance for aquatic ecosystems. The most significant sources of emissions of phosphorus are sewage treatment plants for urban and industrial effluents and leaching from agricultural land.
    Unit of measurement: kilograms of phosphorus equivalent (kg P-eq.).
  - Terrestrial
    On land, ecosystems that need an environment with only few nutrients are gradually disappearing, mainly as a result of the addition of nitrogen. Oxides of nitrogen (NO_x) from combustion processes are of significance to terrestrial ecosystems.
    Unit of measurement: moles of nitrogen equivalent (mol N-eq.).
  - Marine
    For the marine environment, eutrophication will be mainly due to the increase in nitrogen. Emissions of nitrogen are caused largely by the agricultural use of fertilizers, but oxides of nitrogen from combustion processes are also of significance for marine ecosystems.
    Unit of measurement: kilograms of nitrogen equivalent (kg N-eq.).
- Ozone depletion
  The stratospheric ozone (O_3) layer protects us from hazardous ultraviolet radiation (UV-B). Its depletion can have dangerous consequences in the form of an increased frequency of skin cancer in humans and damage to plants. Stratospheric O_3 is broken down as a consequence of anthropogenic emissions of halocarbons [as chlorofluorocarbons (CFCs) and hydrochlorofluorocarbons (HCFCs)], halons and other long-lived gases containing chloride and bromine.
  Unit of measurement: kilograms of CFC-11 equivalent (kg CFC-11-eq.).
- Photochemical ozone formation
  Whereas stratospheric ozone protects us, ozone on the ground (in the troposphere) attacks organic compounds and increases the frequency of respiratory problems when photochemical smog ('summer smog') is present in cities. When solvents and other volatile organic compounds (VOCs) are released to the atmosphere (*e.g.* by emissions from

combustion processes), they can be degraded within a few days. The reaction occurs under the influence of light from the sun. In the presence of NO_x, ozone can be formed.

Unit of measurement: kilograms of ethylene equivalent (kg ethylene-eq.).

- Human toxicity – cancer effects
  Chemicals emitted as a consequence of human activities can contribute to cancer in humans *via* exposure to the environment. The behaviour of the cancer-causing substance has to be considered. The most important routes of exposure are *via* the air breathed in or *via* other materials ingested orally, *e.g.* food or water.
  Unit of measurement: comparative toxic unit for humans (CTUh).
- Human toxicity – non-cancer effects
  Chemicals emitted as a consequence of human activities can be poisonous to humans *via* exposure to the environment. The behaviour of the poisonous substance has to be considered. The most important routes of exposure looked at in those categories are *via* the air breathed in or *via* other materials ingested orally, *e.g.* food or water.
  Unit of measurement: comparative toxic unit for humans (CTUh).
- Particulate matter/respiratory inorganics
  Particulate matter (PM) and its precursors (*e.g.* NO_x, SO_2) cause adverse effects on human health. The mechanism for the creation of secondary emissions involves emissions of SO_2 and NO_x that create sulfate and nitrate aerosols. Usually, the smaller the particles, the more dangerous they are as they can penetrate deeper into the lungs.
  Unit of measurement: kilograms of particulate matter 2.5 equivalent (kg PM 2.5-eq.).

4 Case Study 1: Life-cycle Assessment of Pumped Hydroelectric Storage and Battery Storage

4.1 Goal and Scope

Two electricity storage options will be compared, a pumped hydroelectric storage and a large-scale lithium-ion battery storage, in order to answer the question: which storage technology performs better environmentally over the entire life cycle? The target audience may be policymakers where the LCA could provide decision support for future energy system planning. The functional unit is the use of the energy storage over 80 years to provide 2600 GWh per year of electricity in Germany.

The database ecoinvent v3.3[11] (system model: allocation, cut-off) is used for calculating the LCA. The ecoinvent database is relied on for a substantial share of the required data for the life-cycle inventory and upstream processes although it is supplemented with real data – mostly for the pumped hydroelectric storage. The LCA presented here can be considered a screening assessment.

For upstream processes, the system boundaries include components of the storage medium, the built structure and technical components up to the point of hand-over to the grid. The use phase accounts for the energy losses or the operation electricity. For the pumped hydroelectric storage, use of lubricating oil and methane emissions from basins are also included. For downstream processes, the disposal of the built structure is assessed. The batteries are treated with hydrometallurgical and pyrometallurgical processes at the end-of-life stage.

4.2   Description of Compared Systems and Functional Equivalency

To compare the impacts of the two systems, they have to be sized in a way that allows for comparable functionality so as to define the reference flows for both technologies. In this case, the pumped hydroelectric storage is the reference system, providing 1.4 GW of power with a storage capacity of 13.4 GWh. The sizing of the battery has to be comparable to meet functional equivalency.

Utility-scale batteries, which have emerged only recently, consist of a large number of battery units on racks filling large halls.[12] The particular lithium-ion cell chemistry considered here is a lithium manganese oxide cathode in combination with a graphite anode.

The extent to which the technical properties of pumped hydroelectric storages and utility-scale battery storages permit their employment in comparable applications has to be established. This question has been discussed by Tietze *et al.*[13] in greater detail. In summary, it can be said that the suitability of both storage technologies is similar enough to allow for a comparison. It must be remembered, however, that they differ in the extent to which they can provide the services. Batteries are particularly well suited for fast-response, short-term balancing requirements.[14] Larger storage capacities for longer term services are not currently common.[15] Pumped hydroelectric energy storages, on the other hand, tend to hold large volumes, have far higher energy-to-power (ETP) ratios, and thus are able to provide longer term services, even bridging prolonged periods of low renewable energy output at times of low sun and low wind (in addition to the provision of short-term balancing services). It is these longer term services that are expected to be in greater demand as the share of renewable electricity grows.[16]

There are also differences in their preferred running modes. On the one hand, modern batteries will last longer if charging and discharging are done incrementally, avoiding maximum charge and depletion. On the other hand, if pumped hydroelectric power is running on part load, its efficiency is being compromised. However, any storage technology will have to weigh up its technically preferred running mode against grid requirements and related economic impacts. Hence a trade-off has to be made between maximum operating hours and optimum operational loads.

After elaboration of the options in Tietze *et al.*,[13] storage capacity is chosen as the sizing criterion. Choosing the capacity of the battery as the determining factor takes into account the pumped hydroelectric storage's ability to deliver long-term balancing services. As it is this longer term service that will see the principal increase in demand, this option will be pursued.

4.3   Underlying Data

4.3.1   Pumped Hydroelectric Storage.   For pumped hydroelectric storage, data could be obtained from a commercial operator in an aggregated form. These data are complemented by data from ecoinvent and from the literature. Technical and operating characteristics are based on real-life data from the operator.

The system boundaries include the reservoirs and the water contained within them as storage medium, other built structures such as the underground turbine hall, all necessary services for the turbine hall, such as lighting, ventilation, *etc.*, the tunnel penstock and the surge tank. Furthermore, the material use for technical components of the pumps, turbines and cabling is included. Diesel burned in building machines and electricity for construction processes are based on the ecoinvent database. Methane generation in basins is assumed according to ecoinvent data for hydro-power.

In the literature, the life span of pumped hydroelectric storage ranges from 50 to 150 years.[17,18] A period of 80 years is chosen for this case study. Further input data are given in Table 1.

4.3.2   Utility-scale Battery Storage.   For the utility-scale battery, the battery cells and battery casing make up the storage medium. The built structures include the industrial hall with shelving racks and trays, which house the battery units, and also building services for heating, cooling, ventilation and lighting.

For the utility-scale battery, performance data were obtained from an existing storage in Schwerin, Germany, as found in the literature.[19,20] In particular, the efficiency, including all operational losses, is based on these sources.

Since the assumed storage system has a capacity of 13.4 GWh, the 5 MWh battery storage from the reference literature has to be scaled up by a factor of 2680. It is assumed that the battery may lose 20% of its storage capacity within 20 years (*e.g.* Wolfs[21]) due to ageing and degradation processes (reflecting its 20 year warranty[22]). It therefore has to be over-dimensioned by 10%, over-producing in the beginning by 10% and under-producing towards the end-of-life stage by 10% of the originally needed storage capacity – also bearing in mind that the individual battery cells would be replaced gradually when necessary. Hence, in order to provide comparable average output over the course of its life span, the scaling factor is 2978. It is unlikely that a utility-scale battery 2978 times the size of the installation in Schwerin would

Table 1 Data and assumptions for pumped hydroelectric storage and utility-scale battery (GLO, global; RoW, rest of world).

	Pumped hydroelectric storage	Utility-scale battery
Reference flow	One pumped hydroelectric storage generating 2.08E11 kWh in 80 years	2980 utility-scale battery storages generating 2.08 E11 kWh in 80 years
Main raw materials	Steel: 43.6 Mt Concrete: 2 966 Mt Copper: 0.5 Mt	Battery production, Li-ion, rechargeable, prismatic (GLO) Building construction, hall, steel construction (RoW) Racks: copper, polyethylene, steel
Direct use of land	98 ha	400 m^2 (estimated) per unit
Life span	80 years	Building: 80 years Batteries: 20 years (= current best practice)
Maintenance and replacement cycles	Continuous use of lubricating oil Major overhaul of pumps, turbines and generators every 25 years	Replacement of battery units every 20 years
Efficiency	75.0%	72.5%
Total losses per MWh generated	0.350 MWh MWh$_{\text{generated}}^{-1}$ (electricity use for pumps, building services, control and management systems)	0.379 MWh MWh$_{\text{generated}}^{-1}$ (electricity use for ventilation and cooling, building services, control and management systems)
Other data	Methane generation in basins as per ecoinvent data for hydro-power	Energy density of 114 Wh kg^{-1}, low self-discharge rate
Electricity mix	Future German electricity mix used over the whole life cycle (see Section 3.2)	

be installed in a single location – more likely it would be spread over a number of locations. This allows for the scaling up of a suitable building using the same factor as with the battery components.

There is no long-term evidence yet for life spans of utility-scale batteries, as this is a recent and continuously evolving technology. However, a life span of 20 years can be found in the literature[23] and is in line with the warranty for the existing storage in Schwerin. Hence replacement requirements for the battery units every 20 years are assumed.

For the battery cells, inventory data are taken from the ecoinvent database, which itself is based on work by Notter *et al.*[24] The corresponding battery chemistry is a spinel-type lithium manganese oxide cathode and a graphite anode (LMO-C). The energy density is 114 Wh kg^{-1} on a battery pack level, and the overall efficiency of the system is 72.5%, including internal inefficiencies, but also auxiliary electricity consumption of the storage facility. Further input data are given in Table 1.

Table 2 Future energy generation scenario according to the German Federal Ministry for the Environment, Nature Conservation, Building and Nuclear Safety.[25]

Energy carrier	Share in 2035/%
Nuclear power	0
Lignite	11.3
Hard coal	9.8
Natural gas	13.4
Others	2.6
Renewables	
Water	4.5
Wind	41.9
PV	10.9
Biogas	1.1
Biomass	2.6
Geothermal	1.1
Pumped hydro	0.7
Total	100

4.3.3 Stored Electricity. The construction of pumped hydroelectric storage has a long lead time for engineering planning, approval procedures and consultation and the construction itself can last for several years. For future storages, the use phase is further in the future. It is to be expected that by then the transformation of the (German) energy system to a lower carbon system will be well advanced. Therefore, an electricity mix for 2035 is applied and the environmental impacts of the use phase are calculated on this basis. Manufacturing and construction processes are modelled with the current electricity mix. Table 2 shows assumed shares per energy carrier. This energy mix results in a carbon footprint of 0.41 kg CO_2-eq. kWh^{-1} electricity.

4.4 Results

Comparing the two technologies, pumped hydroelectric storage shows more than 20% lower impacts in all but two categories (water resources and climate change). The resource use is only one-fifth of the battery store and shows the greatest difference. The overall efficiencies of pumped hydroelectric storage (75%) and utility-scale battery storage (72.5%) is nearly identical and impacts of the use stage are nearly the same for both technologies. In Figure 1 the impacts of both technologies are juxtaposed (utility-scale battery = 100%). Table 3 shows absolute values per functional unit.

For the pumped hydroelectric storage, impacts of the operational stage ('use stage') dominate those of the production stage in all categories. The use stage is largely made up of the impacts of operational energy losses, *i.e.* the difference between stored energy and released energy. These losses depend on efficiency losses and internal energy demands of the installations.

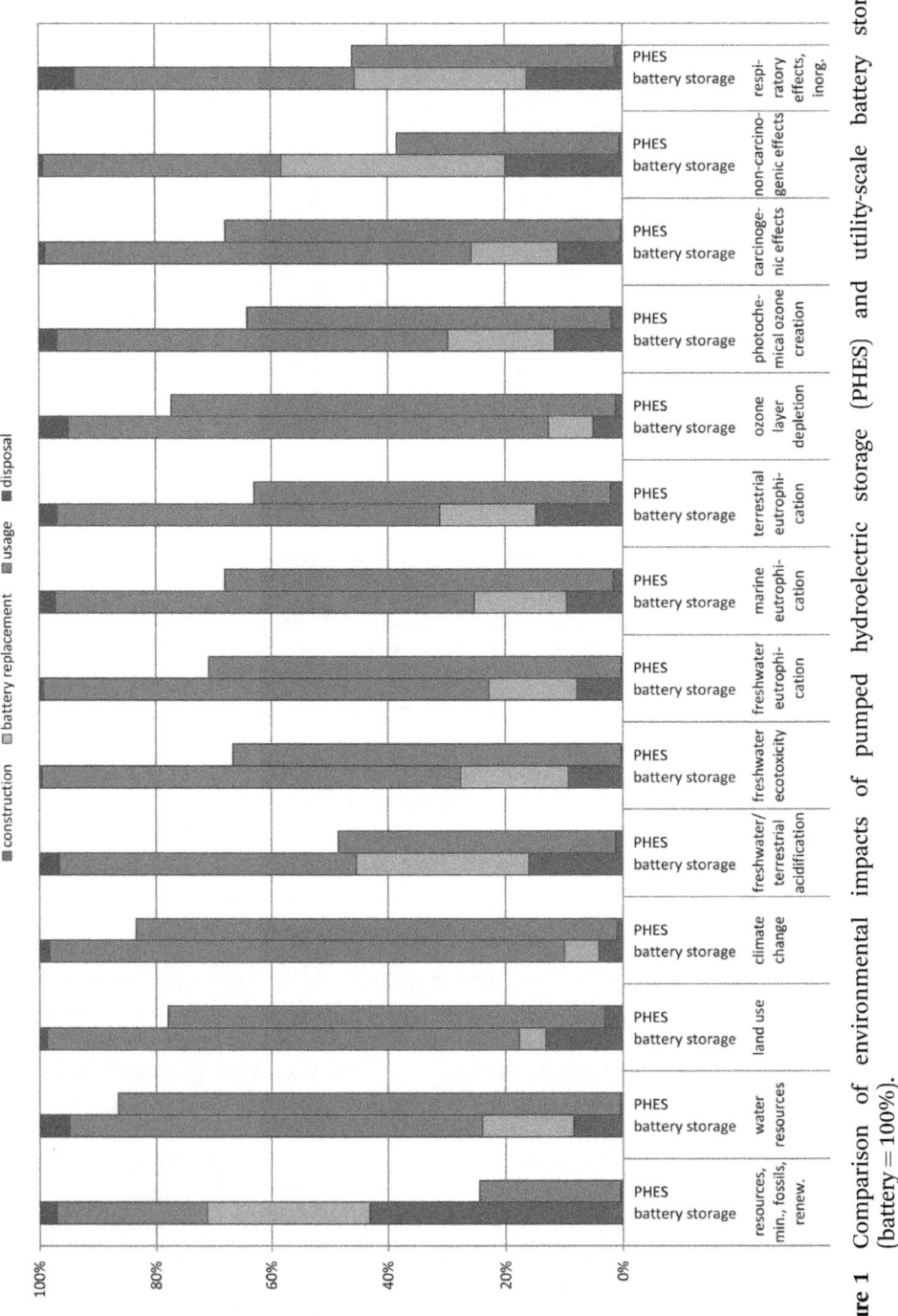

Figure 1 Comparison of environmental impacts of pumped hydroelectric storage (PHES) and utility-scale battery storage (battery = 100%).

Table 3   Impact indicator results per functional unit of 80 years' use of the energy storage providing 2600 GWh per year of electricity in Germany.

Impact indicator	Battery storage	Pumped hydroelectric storage	Units
Resources, minerals, fossils, renewables	3.37×10^6	8.24×10^5	kg Sb-eq.
Water resources	9.94×10^{10}	8.60×10^{10}	EP
Land use	7.00×10^{10}	5.46×10^{10}	kg SOC
Climate change	3.63×10^{10}	3.02×10^{10}	kg CO_2-eq.
Freshwater/terrestrial acidification	1.42×10^8	6.88×10^7	mol H^+-eq.
Freshwater ecotoxicity	1.41×10^{12}	9.41×10^{11}	CTUe
Freshwater eutrophication	4.32×10^7	3.06×10^7	kg P-eq.
Terrestrial eutrophication	2.16×10^8	1.36×10^8	mol N-eq.
Marine eutrophication	2.58×10^7	1.75×10^7	kg N-eq.
Ozone layer depletion	1.96×10^3	1.51×10^3	kg CFC-11-eq.
Photochemical ozone creation	6.03×10^7	3.86×10^7	kg ethylene-eq.
Carcinogenic effects	4.10×10^3	2.78×10^3	CTUh
Non-carcinogenic effects	2.95×10^4	1.13×10^4	CTUh
Respiratory effects, inorganics	1.29×10^7	5.95×10^6	kg PM 2.5-eq.

Impacts of decommissioning and disposal for the pumped hydroelectric storage are barely visible.

In contrast to the pumped hydroelectric storage, for the utility-scale battery storage impacts from the construction in combination with replacement are larger. Since the use stage is nearly the same for both technologies, this stage is responsible for higher overall impacts of the utility-scale battery. Particularly resource use is dominated by battery cell production and the steel construction of the buildings. In this category, the use of metals determines the result. Impacts from decommissioning and disposal are discernible.

4.5   Sensitivity Analysis

Alternative approaches to scaling the battery are also possible, as explained previously. The chosen approach of scaling to equal capacity ensures that the battery can store the same amount of energy as the pumped hydroelectric storage, thus allowing for longer term balancing and ancillary service provision. However, this is at odds with the typical sizing of battery storage. A utility-scale battery with a storage capacity of 13.4 GWh would, according to common ETP rules, have a much higher power rating than the pumped hydroelectric storage. It would therefore be able to provide short-term balancing services to a far greater extent than assumed for the pumped hydroelectric storage. It was therefore decided to investigate a utility-scale battery sized to generate merely the same annual output (MWh per year) as the pumped hydroelectric storage. In this case, the number of annual full charging cycles for the battery is the decisive parameter.

The assumptions made regarding the running of both technologies in the reference scenario would result in 194 charging cycles per year for the utility-scale battery. However, in order for such a battery to be economically viable, 300–360 cycles per year would need to be achieved.[15,19] In other words, sizing the battery based on equal capacity in GWh, while allowing for the same long-term services, results in the utility-scale battery being under-used or over-sized for the assumed use expressed in GWh per year.

The cycle life of the best available technology (lithium titanate anode with iron phosphate-based cathode; LFP-LTO batteries) is currently already >10 000 cycles. Scientists assume that batteries with a cycle life of up to 20 000 cycles are feasible.[26] This would equate to 1000 full cycles per year over 20 years. A sensitivity scenario assumes that the battery would complete 1000 cycles per year, still generating the same 2600 GWh of short-term balancing services per year. For comparability purposes, the same lithium manganese composition as used in the base case is assumed, although such long cycle lives would probably be achieved by alternative battery chemistries. Putting it another way, it is assumed that the same batteries would be improved in order to be much more resilient to withstand the additional charging cycles. The example installation in Schwerin needs to be scaled up by only 577- rather than 2680-fold. This implies, however, that longer term balancing and ancillary services would have to be excluded from the comparison in this case.

The results of the comparison presented in Figure 2 show a considerably smaller difference between the two storage technologies. Only four indicators are significantly lower for the pumped hydroelectric storage (resources, acidification, non-carcinogenic effects and respiratory effects). Since the assessment has a screening character implying a high uncertainty, all other indicator differences are deemed not to be significant.

4.6 Discussion

Pumped hydroelectric storage is typically designed to serve both short- and long-term requirements, including the bridging of longer periods of low sun and simultaneously low wind, whereas utility-scale batteries are particularly well suited to fulfil short-term incremental balancing requirements. Hence the two technologies are not unconditionally comparable, nor are they interchangeable, even though they are capable in principle of providing largely similar balancing and ancillary services. The demand for balancing and ancillary services is expected to increase in line with the increase in intermittent renewables. Rather than one technology substituting the other, they should be employed to complement each other in providing for this increased demand.

The overall results are dominated by the environmental impacts related to the use stage. These are caused by the impacts of the share of electricity that is lost due to internal inefficiencies of the battery and the installations.

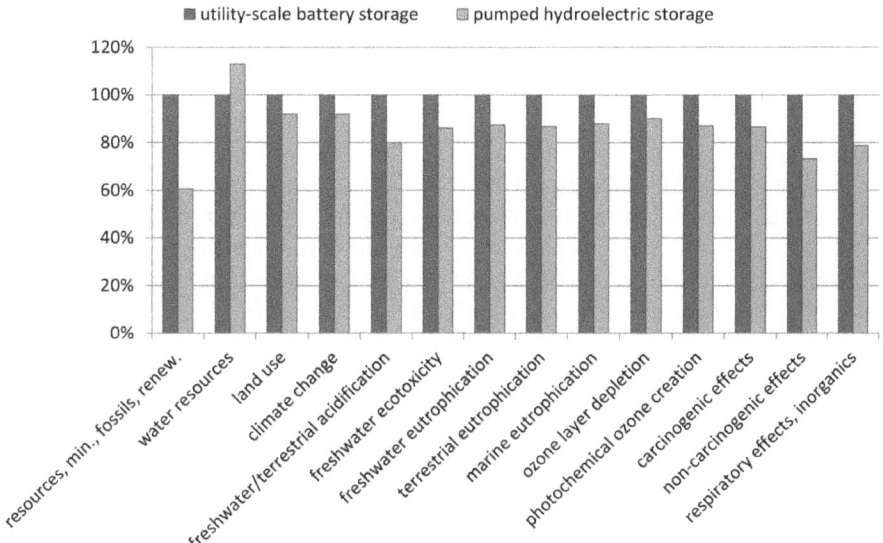

Figure 2 Comparison of environmental impacts of pumped hydroelectric and utility-scale battery storage (battery = 100%) for the sensitivity analysis.

This means that the system efficiency and internal energy requirement of the examined technologies are crucial for the overall result, as they define electricity 'lost' in the 80 years' use stage. This aspect is of great significance, since for the battery storage great uncertainty exists concerning the internal energy requirements stemming from only one literature source.

Concerning the sensitivity analysis, it should be remembered that achieving the assumed amount of full cycles within 1 year is not only dependent on battery technology but also requires an electricity grid that often requires services of the battery. This scenario makes very favourable assumptions for the battery and is taken to be representative of future developments. However, the results of the sensitivity analysis make clear the dependency of battery storage's impacts on scaling of the battery.

The limitations of this study are mainly the underlying data for batteries, since the data set used reflects not a stationary battery but a battery for mechanically driving an electric vehicle. Developments in the field of stationary batteries are currently moving at a fast pace and are certain to be considerable within the very long time span of 80 years considered here.

With the use of utility-scale batteries being an emerging field, developments can only partially be anticipated. The assumptions of this study would therefore have to be reconsidered when battery technology evolves and more data are available.

5   Case Study 2: Life-cycle Assessment of Different Lithium-ion Battery Chemistries for a Small-scale Energy System

5.1   Goal and Scope

Very different battery technologies exist for stationary energy storage. Among these, lithium-ion batteries (LIBs) have become the dominant technology owing to their superior technical performance in terms of efficiency, energy density and cycle life. In fact, they are now the leading technology for stationary energy storage after pumped hydroelectric in terms of storage capacity.[27] However, there are different LIB chemistries available, each with its specific properties, advantages and disadvantages, and the choice of the most appropriate one depends on the requirements defined by the application.

The previous case study considered a lithium manganese battery, applying a simplified ageing model. However, owing to their different properties, no battery is suitable for all applications, and trade-offs between cell costs, cycle life and efficiency need to be considered when choosing the most promising battery in terms of environmental aspects. For this purpose, an assessment of the environmental performance of the different battery chemistries within the application is needed, taking into account also the impact of the operation profile on the life-cycle expectancy (and thus replacements) of the battery. In this case study, the effect of these operational parameters on the environmental performance of the batteries is assessed in a small-scale energy grid application, allowing the identification of the most appropriate battery chemistry for this specific application and pointing out the key parameters that determine the choice. The unit of comparison (functional unit) is set as 1 kWh of electricity provided by the battery storage system over its projected use life of 20 years. This allows also the operation of the battery system and therefore all performance parameters such as cycle life and internal efficiency to be considered. The use phase considers the losses through ac–dc and dc–ac electricity conversion, *i.e.* all environmental impacts caused by the internal inefficiencies of the battery storage system. The stored electricity is generated by a photovoltaic (PV) installation.

The battery storage system is modelled based on the existing literature,[28–32] with the inventory data from previous studies being recompiled, unified and modularised.[33] This allows the minimisation of the differences due to varying assumptions (*e.g.* regarding cell housing, electrolyte modelling, *etc.*) made in studies of different origins. Since no specific data for the disposal for every single battery chemistry exists, the end-of-life stage is beyond the scope of this study.

5.1.1   Description of Compared Systems and Functional Equivalency.   A typical application for battery storage is a small-scale energy grid (island solution), where the battery storage system helps in balancing the grid and maximises the use of renewable energy sources. A typical representative of

such a small-scale grid is the island Mainau, situated in Lake Constance in southern Germany, where the conversion of the heat and electricity supply into 100% renewables is being investigated as a case study and subject to an energy system optimisation model.[34] The cost-optimised electricity system is based on several roof-mounted, multi-silicon PV installations, supported by two combined heat and power (CHP) plants that ramp up when the demand cannot be covered by the PV panels or the battery. One of the CHP plants is powered by biomass (wood gas) and the other by natural gas. A 2.15 MWh battery charged by the PV panels provides grid balancing services and peak-load electricity. Detailed load and generation profiles are available for the case study with an hourly resolution, allowing an application-specific optimisation of the battery configurations.

The energy storage service is provided by a lithium-ion battery installation. However, as different LIB chemistries exist; the assessment aims at comparing the different LIB chemistries and identifying the most promising one for the chosen application. Currently, the most relevant LIB chemistries are the following:

- LMO-C: lithium–manganese oxide cathode/graphite anode – comparatively cheap battery type with good energy density, but poor cycle life.
- NCM-C: lithium–nickel–cobalt–manganese oxide cathode/graphite anode – high energy density, good cycle life. Currently the battery of choice for automotive applications.
- NCA-C: lithium–nickel–cobalt–aluminium oxide cathode/graphite anode – high energy density, good cycle life, slightly higher safety risk.
- LFP-C: lithium–iron phosphate cathode/graphite anode – average energy density but very safe. Good power rating and longer cycle life than NCM or NCA.
- LFP-LTO: lithium–iron phosphate cathode/lithium titanate anode – the very stable LTO anode provides an extraordinarily long cycle life and high power ratings, but at the price of a low energy density.
- LCO-C: lithium–cobalt oxide cathode/graphite anode – very high energy density, expensive, comparable safety risks. Used almost exclusively for hand-held applications (*e.g.* mobile phones, laptops).

Since LCO-type batteries, owing to the high energy density, high costs and short cycle lifetime, are used only in portable devices and not in larger storage systems, only the first five chemistries are considered for this case study because of their favourable properties. Furthermore, all named systems are already commercially available.[35]

5.2 Underlying Data

Time series with detailed data about load and generation within the island grid system are provided on an hourly basis from the Mainau case study.

From these time series, the charge–discharge cycles provided by the battery per day and the discharge depth can be derived for the battery optimisation model (explained in the following). The operation profile (probability distribution of charge–discharge cycles per day) as the basis for the optimisation is provided in Figure 3. An average of 0.66 cycles for the entire year is used for calculation (mean values).

The assessment is based on technoeconomic data on battery technologies from a continuously updated battery database (Batt-DB) in combination with environmental performance data (LCA input) from the literature for the given application case. Batt-DB is a database containing up-to-date technoeconomic data from industry, the literature and scientific reports for all types of secondary batteries (Table 4).[36,37] With these inputs, the battery storage system's nominal capacity is optimised with respect to economic aspects, identifying the lowest life-cycle costs (LCCs) as a trade-off between battery over-sizing (increasing the cycle life by reducing the average depth of discharge), battery replacements (due to the limited cycle life and/or calendric lifetime) and costs of electricity dissipated due to internal inefficiencies. The optimisation follows economic criteria and varies the minimum state of charge (SoC), which itself influences the battery cycle lifetime.[36,38,39] A large depth of discharge (DoD), *i.e.* deep cycling, generally reduces the battery cycle life, which is why batteries are often over-sized in order to extend the operation time. The optimisation aims to minimize the overall LCC by finding an optimal equilibrium between initial investment cost (battery over-sizing) *versus* replacement costs (reduced battery life) under given conditions for the different application cases. The minimum SoC is obtained as a function of the estimated number of cycles and the average discharge depth

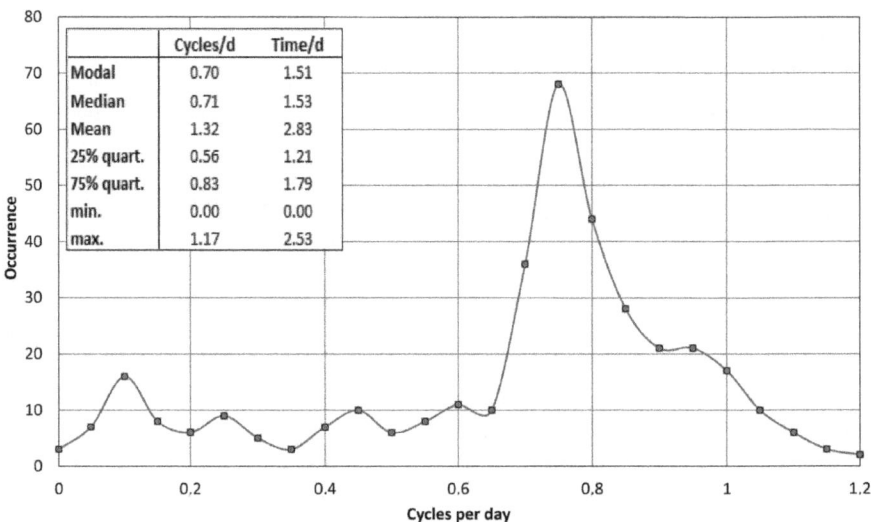

	Cycles/d	Time/d
Modal	0.70	1.51
Median	0.71	1.53
Mean	1.32	2.83
25% quart.	0.56	1.21
75% quart.	0.83	1.79
min.	0.00	0.00
max.	1.17	2.53

Figure 3   Operational profile (probability distribution of charge–discharge cycles per day) of the LIB within the island grid application.

Table 4 Key performance parameters of the assessed batteries using upper, median and lower quartile values based on available recent literature and industry data sources from the Batt-DB using the most recent (complete) datasets as indicated in the second row for each technology.[6,37] [a]

Component	Units	Range	LTO	LFP	LMO	NCM	NCA
Oldest source used from Batt-DB			2013	2013	2013	2013	2013
Cost	€ kWh^{-1}	Min. 75 q	600	289	153	192	172
		Median	900	309	238	318	213
		Max. 25 q	1200	315	564	554	355
Cycle life @ DoD 80%	—	Min. 75 q	4500	1750	1000	1000	1250
		Median	8000	5000	1500	4000	3000
		Max. 25 q	9750	5325	5000	4875	5125
Efficiency	% dc–dc	Min. 75 q	81	83	85	83	90
		Median	90	96	94	93.8	91.55
		Max. 25 q	94.5	96.5	98.25	97.275	93.1
Calendric lifetime	Years	Min. 75 q	10	7.5	5	5	10
		Median	17.5	15.0	10.0	10.0	10.0
		Max. 25 q	25.0	20.0	15.0	15.0	15.0
O&M cost	€ kW-year^{-1}	Min. 75 q	11.0	17.0	20.0	20.0	20.0
		Median	25.0	25.0	25.0	20.0	25.0
		Max. 25 q	33.8	31.3	30.0	30.0	30.0

[a]DoD, depth of discharge; O&M, operation and maintenance; q, quartile.

as determined by application over the 20 year time horizon, and the calendric lifetime of the batteries. Typical SoC ranges are assumed to be between 10 and 95%, because most battery types are usually not charged to 100% or discharged to 0% in order to avoid over-charge and -discharge.[40,41]

The optimised battery configuration then provides the base for the following LCA. The desired operation period for the entire energy storage system within the application is assumed to be 20 years.[39,42] Owing to the rack configuration, batteries can be scaled easily to suit the required gross storage capacity; for a nominal 2.15 MWh storage system, 83 racks would be required when allowing 100% DoD (no oversizing), and 104 racks for a maximum 80% DoD (equivalent to 2.7 MWh gross capacity).

Based on the previously described modular battery inventories, a rack-mounted stationary storage system with a storage capacity of 26 kWh is modelled,[43] in which round cells (type 18 650) are placed in a steel tray for modular insertion in a standard 48×43 cm rack. The tray housing is assumed to be made of steel with a polypropylene insulation. A single tray (48×43 cm) can host 570 round cells in a rectangular packing. The battery rack structure is assumed to be made of powder-coated steel and hosts, apart from the battery trays, a battery management system and a fan tray for providing forced-air cooling. Whereas the storage capacity of a single tray is determined by the capacity of the 18 650 battery cells, the number of trays mounted in a rack is varied so that the final storage capacity of a battery rack equals roughly 26 kWh for all battery chemistries considered. The capacity of a single cell is obtained from the underlying publications, scaled according to the different contents of active material in the assumed 18 650 cells. The LFP-C type cells show a capacity of 4.19, the LFP-LTO 2.5, the LMO-C 5.6, the NCM-C 6.3 and the NCA-C 6.4 Wh per cell. For a tray with 570 cells, 11 trays with LFP-C cells are required for a 26 kWh battery module, 18 trays for LFP-LTO, eight for LMO-C and seven for NCM-C and NCA-C. The rack is dimensioned accordingly, with 11 trays fitting in a 190 cm rack, eight in a 125 cm rack and 18 in a 220 cm rack. For the use phase, data for roof-mounted, multi-silicon PV plant from the ecoinvent database are applied.

5.3   Results

The optimization results show that, owing to the small average discharge depth and moderate use intensity within the application (average daily cycles of 0.66), the battery lifetime is determined mainly by the calendric lifetime and no battery over-sizing is required to extend the cycle life. Therefore, for all batteries except LMO, a minimum DoD of 10% is calculated (the minimum possible value, since a DoD of 0% is not recommended for any LIB). Only for the LMO, with a very short average cycle life, a slight over-sizing is calculated to extend the lifetime, and thus a minimum SoC of 22%.

For the optimised battery systems, the environmental impacts per functional unit, *i.e.* per kWh of electricity provided by the battery storage installation over the lifetime, can then be calculated. Figure 4 provides the relative and absolute characterisation results for the different batteries within the application, including the contribution of the different life-cycle stages (construction, use and replacement). Table 5 gives absolute values. The individual impact categories and the key drivers for the impacts obtained are discussed briefly in the following. As can be observed, the contribution of the different life-cycle stages varies strongly between categories.

5.3.1   Resource Depletion – Mineral, Fossils and Renewables.   Regarding resource depletion aspects, the LFP and LTO batteries score best, in spite of their lower energy densities (requiring a higher battery mass and hence more resources for providing the same battery capacity). However, they show long lifetimes and use mainly abundant materials such as iron and phosphorus. The high impact of the LMO battery is due to the battery model used in the source publication,[24] where the amount of copper for the anode current collectors is significantly higher than those for the other battery models. This is further aggravated by the shorter lifetime and more frequent battery replacements for the LMO battery.

5.3.2   Depletion of Water Resources.   For water use, the drivers are numerous and vary fundamentally between the different batteries. In addition, not all of the source publications from which the inventory data are derived account thoroughly for water consumption, which is why

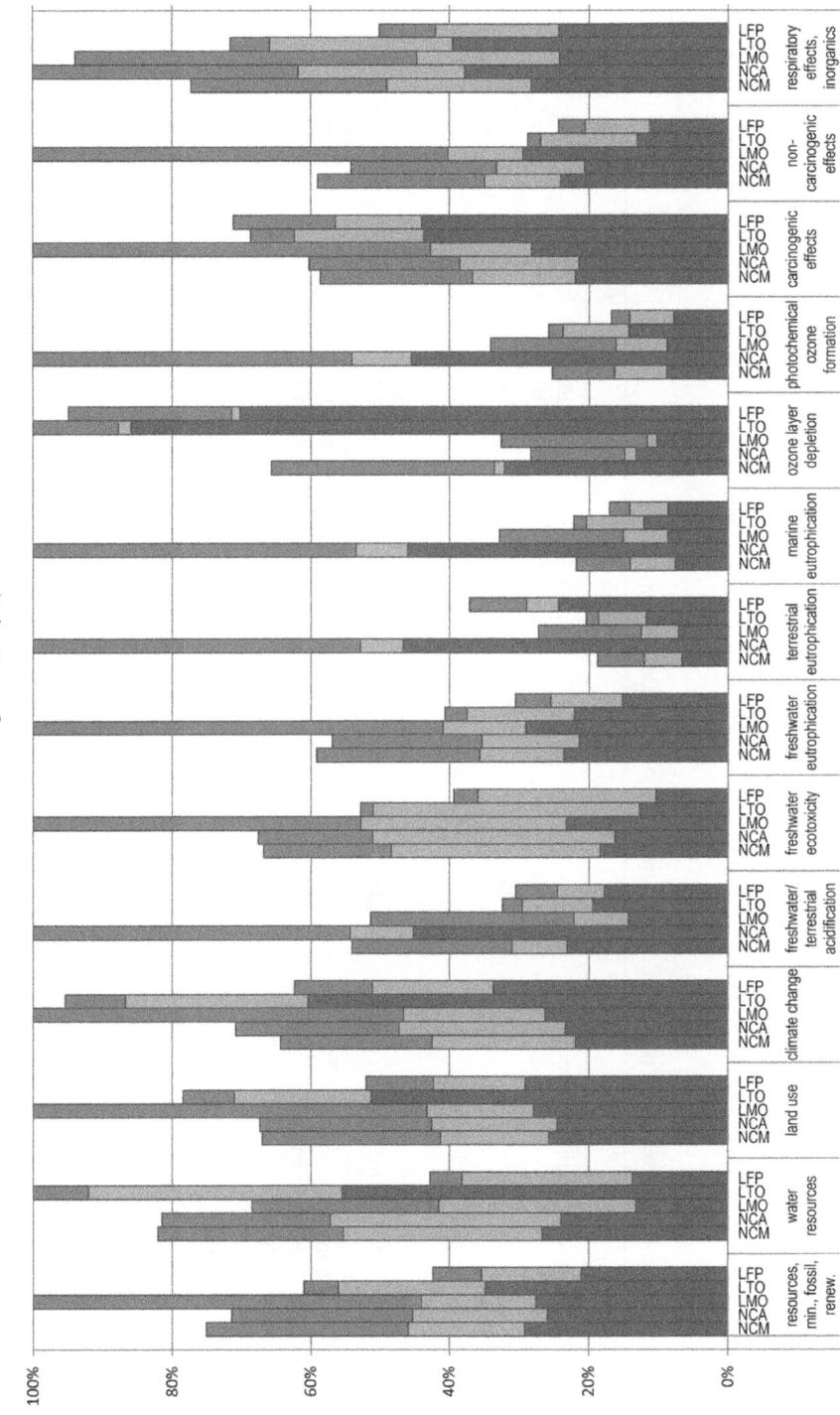

Figure 4 Results for different LIBs considering construction, use and replacement of cells.

Table 5 Impact indicator results per functional unit (provision of 1 kWh electricity).

Impact indicator	LFP	LTO	LMO	NCA	NCM	Units
Resources, minerals, fossil, renewables	1.56×10^{-5}	2.25×10^{-5}	3.69×10^{-5}	2.64×10^{-5}	2.77×10^{-5}	kg Sb-eq.
Water resources	1.49×10^{-4}	3.48×10^{-4}	2.38×10^{-4}	2.83×10^{-4}	2.85×10^{-4}	m^3
Land use	7.97×10^{-2}	1.20×10^{-1}	1.53×10^{-1}	1.03×10^{-1}	1.03×10^{-1}	kg SOC
Climate change	5.41×10^{-2}	8.27×10^{-2}	8.68×10^{-2}	6.15×10^{-2}	5.59×10^{-2}	kg CO_2-eq.
Freshwater/terrestrial acidification	5.66×10^{-4}	6.01×10^{-4}	9.53×10^{-4}	1.86×10^{-3}	1.00×10^{-3}	mol H^+-eq.
Freshwater ecotoxicity	2.51	3.37	6.37	4.30	4.26	CTUe
Freshwater eutrophication	4.35×10^{-5}	5.78×10^{-5}	1.42×10^{-4}	8.10×10^{-5}	8.42×10^{-5}	kg P-eq.
Terrestrial eutrophication	1.35×10^{-3}	7.44×10^{-4}	9.93×10^{-4}	3.64×10^{-3}	6.86×10^{-4}	mol N-eq.
Marine eutrophication	5.73×10^{-5}	7.44×10^{-5}	1.10×10^{-4}	3.35×10^{-4}	7.32×10^{-5}	kg N-eq.
Ozone layer depletion	1.89×10^{-7}	1.99×10^{-7}	6.48×10^{-8}	5.64×10^{-8}	1.31×10^{-7}	kg CFC-11-eq.
Photochemical ozone formation	1.60×10^{-4}	2.45×10^{-4}	3.24×10^{-4}	9.51×10^{-4}	2.40×10^{-4}	kg ethylene-eq.
Carcinogenic effects	1.34×10^{-8}	1.29×10^{-8}	1.88×10^{-8}	1.13×10^{-8}	1.10×10^{-8}	CTUh
Non-carcinogenic effects	5.25×10^{-8}	6.20×10^{-8}	2.16×10^{-7}	1.17×10^{-7}	1.27×10^{-7}	CTUh
Respiratory effects, inorganics	5.26×10^{-5}	7.51×10^{-5}	9.86×10^{-5}	1.05×10^{-4}	8.11×10^{-5}	kg PM 2.5-eq.

comparisons are difficult. However, a major driver is the electricity demand for cell manufacture.

5.3.3 Land Use. The land use impact for producing a certain mass of battery is similar for the different batteries, and differences are mainly due to the different energy densities (giving high impacts for the LTO battery). For the use phase, the land use depends on the charge–discharge efficiency, which is why the LTO batteries (with slightly lower efficiency) show higher land use impacts from this phase. However, the long lifetime of the LTO batteries leads to very small impacts from battery replacement, reducing the total impacts. In contrast, the LMO batteries, with comparably low impacts from manufacturing and operation, show the highest total impacts in this category owing to their short lifetime and frequent battery replacements. In total, the LFP battery gives the best results regarding land use aspects owing to a combination of high efficiency and long lifetime expectancy.

5.3.4 Climate Change. Greenhouse gas (GHG) emissions are (on a mass basis) similar for the different battery types. However, their energy density varies, which is why, per kWh of storage capacity, the batteries with higher energy density (NCM and NCA) show advantages. LFP, in spite of higher impacts from battery production, give similar results owing to their longer lifetime (fewer impacts from battery replacements).

5.3.5 Freshwater Terrestrial Acidification. Freshwater terrestrial acidification impacts are dominated by the mining of nickel and cobalt from sulfidic ores, where significant emissions of SO_2 occur. Hence nickel- and cobalt-containing batteries (NCA and NCM) show comparably high impacts in this category, followed by LMO, mainly from battery replacements owing to their short lifetime.

5.3.6 Freshwater Ecotoxicity and Eutrophication. Very similar results are obtained for these categories. Here, the mining and production of copper and thus current collectors and electronic parts are the main drivers for impacts in these categories. The high impact of the LMO battery is again (as for resource depletion) a combination of the high amount of copper used for the anode current collector and the more frequent battery replacements.

5.3.7 Terrestrial Eutrophication. Here, the NCA battery shows by far the highest impact. Again, this is an artefact of the modelling approach used by the source publication,[29] which assumes that the NO_2 released during the synthesis of the NCA cathode material is emitted entirely into the atmosphere. A modern industrial plant would most probably recover the NO_2 from the exhaust air and show drastically lower impacts in this category. The same applies (although to a lesser extent) to the LFP battery, where ammonia emissions from the LFP synthesis are accounted for. When these effects are disregarded, the differences between the battery types would be relatively small.

5.3.8 Marine Eutrophication. The profile obtained in this category is very similar to terrestrial eutrophication, except for the LFP battery, where the ammonia emissions are less relevant, obtaining impacts similar to those of the other battery types.

5.3.9 Ozone Depletion. Ozone depletion is driven mainly by the emissions of hydrofluorocarbons that are released during the production of binder components and thus the amount of poly(vinylidene fluoride) (PVdF) required for the binder. Here, the different battery models assume varying amounts of PVdF, which is why the large differences are mainly due to different modelling approaches and less to the different electrochemistries.

5.3.10 Photochemical Ozone Formation. The (previously discussed) NO_2 emissions from NCA synthesis are responsible for the high impacts obtained for this battery and, for the LTO battery, also the ammonia emissions from the LTO synthesis process. The uncertainties of the underlying assumptions require that these results be treated with caution. For the remaining batteries, the impacts from production and use are very similar, which is why the differences in the final results are basically due to the varying lifetimes between NCM, LMO and LFP batteries.

5.3.11 Human Toxicity (Carcinogenic and Non-carcinogenic Effects). In both categories, the LMO battery shows the highest impacts, again due to the high amounts of copper required for its anode in combination with frequent replacements. For carcinogenic impacts, the other battery types show comparable impacts; the LFP and LTO score significantly better regarding non-carcinogenic effects.

5.3.12 Respiratory Effects. Inorganics stem from the electricity for battery cell manufacture (due to the share of coal power in the electricity mix) and the mining of metals such as copper and aluminium. For the LTO and NCA batteries, the previously mentioned direct emissions of ammonia and NO_2 contribute an additional share, increasing the burdens from battery production. However, the LTO battery can compensate for this by requiring less frequent battery replacements, obtaining the second best result in this category after the LFP battery.

5.3.13 Summary. In summary, the LFP battery seems to be the most recommendable technology among those considered for the chosen application, showing the lowest contributions in the majority of the impact categories and comparably high impacts in only one category, ozone depletion. However, for several categories, uncertainties in the battery models provided by the source publications are responsible for major impacts. Especially the previously mentioned high NO_2 and ammonia emissions cause high impacts, which in reality would be significantly lower. These aspects have to be considered when interpreting the results, reducing the significance of the results obtained in the affected impact categories.

5.4 Discussion

Although the results seem to indicate the LFP battery as the most promising candidate for the given application case, some limitations of the study need to be considered.

Owing to a lack of reliable data, the end-of-life phase/recycling of the batteries is not considered. While the same recycling process would most probably be applied to the different LIBs, the corresponding environmental impacts would also be comparable. However, the metals contained in the batteries have very different economic values. Expensive metals such as nickel and cobalt are usually recovered from the spent batteries because of simple economic considerations. These metals are often also those that show the highest environmental impact from mining and processing, which is why their recovery can reduce significantly the total impacts associated with the battery. Cheaper metals such as iron, lithium and phosphorus are of little economic interest and therefore often disregarded in recycling processes or left as residue in the slag. Hence these batteries (although with lower production impacts) would probably also show lower recycling benefits. This could lead to a disadvantage for the LFP batteries and better results for cobalt- and nickel-containing batteries such as NCM and NCA. Further work on assessing recycling impacts and benefits, specifically for the different lithium-ion battery types, would therefore be of high interest in future studies.

A high contribution of the use phase is shown in many categories, in spite of the already high efficiency of the assessed lithium-ion batteries (>90%, significantly higher than for other competing battery technologies, *e.g.* lead–acid or redox-flow batteries). This indicates the relevance of a high efficiency, where lithium-ion batteries are highly competitive. However, this also depends strongly on the use case, *i.e.* the origin of the charged electricity.

The dimensions of the battery are set by the application and therefore fixed for the given use case. However, when looking at the results of the battery optimisation, the battery seems to be over-sized. For almost all batteries, the lifetime is determined by calendric ageing, *i.e.* the batteries do not age because of their use. This indicates that maybe a smaller, more intensively used battery would score better, reducing the impacts from battery production due to better capacity usage. The optimisation of the battery size also in this direction would be another task and could confer advantages on batteries with a longer cycle life, such as the LTO battery. Apart from that, the optimization is based on hourly values (as provided by the use case's operational data), and consequently short-term fluctuations of PV generation and related battery operation are not considered. Furthermore, an optimal power flow optimisation is used. A security-constrained power flow optimisation (SCOPF) may lead to a different operation profile of the storage unit. These factors would affect the operating hours of the battery storage unit and thus also the final results.

In summary, the case study assessment gives highly valuable insights into the key drivers of environmental impacts in the different categories and allows the identification of the most promising battery type for this application with regard to environmental considerations. However, it is based on a comparatively simple battery and operational models that include numerous assumptions and estimations, which is why a high level of uncertainty is associated with the outcomes. These should therefore be considered only as indicative. Apart from that, the results are case specific and cannot be generalised, since the choice of a certain battery chemistry is always a trade-off between different battery performance parameters (energy density, efficiency, cycle and calendric lifetime), which need to fit to the specific requirements of the application. Thus, for every application an optimisation of the battery is needed and calculating the environmental impacts can yield completely different results in another application.

6   Case Study 3: Life-cycle Assessment of Energy Scenarios with Various Uses of Heat and Battery Storage for a Small-scale Energy System

6.1   *Goal and Scope*

The previous case study compared different battery technologies for the selected application, but did not provide further details about the share that the energy storage system contributes to the total environmental impacts of the energy system. Therefore, this section presents a contribution assessment of battery and thermal storage in a small-scale energy system with sector coupling.

The functional unit of the assessed system is the provision of 2275 MWh of electricity and 5960 MWh of heat (space heating and hot water) per year for a tourist company in southern Germany, located on an island in Lake Constance. The system boundary is cradle to grave. The target audience is the managers of the company when making decisions about a future energy system. While different scenarios for energy systems are modelled on a cost optimal basis, these scenarios will now be assessed in relation to their environmental impacts.

Two energy systems are modelled, mainly using the database ecoinvent v3.3[11] (system model: allocation, cut-off). The datasets for energy generation are adapted to site-specific conditions. Data for battery storage come from the previous section (LFP-type lithium-ion battery, as identified there as the most recommendable technology for this application).

6.2 Description of Compared Systems and Functional Equivalency

Two scenarios for a future small-scale energy system generated by an optimisation model[44,45] are compared. They are based on site-specific energy potentials for different energy technologies and demand profiles for electricity and heat with an hourly resolution. The target function of the optimisation is cost minimisation. The two scenarios are as follows:

- *Scenario 1:* The energy system consists of two combined heat and power (CHP) plants, one using natural gas and the other wood gas from the gasification of wood chips. Apart from these, electricity is also generated by PV roof installations and partly purchased from the grid. Heat is also generated by a boiler using wood chips. Heat is partly stored in heat storages.
- *Scenario 2:* There is no purchase of electricity from the grid, but a larger PV installation, leading to a decreased use of CHP from wood. Instead, a battery is implemented that stores excess electricity from the PV installation. Owing to the lower CHP generation, more heat needs to be generated by the boiler (unchanged heat demand), which is why the size of the heat storage also increases.

The output of the natural gas-based CHP is the same for both scenarios. The energy generation shares of each energy technology for both scenarios are shown in Figure 5. Both energy systems generate the same yearly amount of heat and electricity according to the demand.

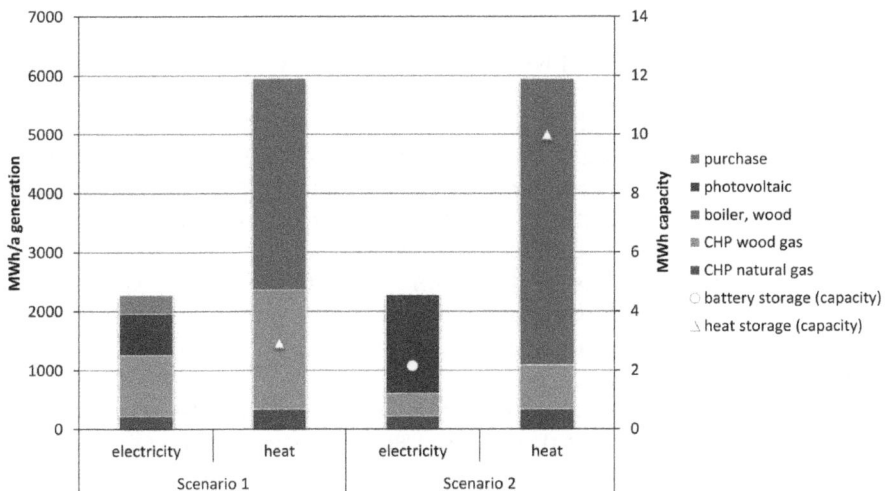

Figure 5   Shares of each energy technology for both scenarios.

6.3   Underlying Data

Each heat- or power-generating technology is assessed using a specific dataset as implemented in the ecoinvent database[11] adapted to the site-specific conditions. The data sets of CHP plants and the multi-silicon PV panels are adapted to the site-specific conditions concerning efficiency and annual electricity generation, respectively. The allocation of efforts between electricity and heat for CHP plants is done based on exergy content. Wood for CHP and boiler is assumed to be spruce, pine, beech and oak in equal proportions stemming from sustainable forest management.

Since the energy scenarios show a future energy system, the purchase of electricity in scenario 1 is modelled with the future energy mix laid down in Section 4.3.3.

Battery production data are based on the previous case study, applying LFP-C technology. The capacity is 2.15 MWh and the lifetime of the installation is assumed to be 15 years (see Table 4). Since no specific data for end-of-life processes for this technology exist, data for recycling of LMO batteries are applied representatively. These data assume a 1:1 share of hydro- and pyrometallurgical processes.

The data set for heat storage is based on for a 2000 L steel tank insulated with mineral wool, including heat exchanger and boiler. Since only the storage capacity is given by the energy model, the volume needed is calculated according to the heat capacity of water and the data set was linearly scaled to larger volumes.

6.4   Results

From an environmental point of view, there is no clear advantage for one or the other scenario (see Figure 6 and Table 6 for absolute values). Scenario 1 shows lower impacts in the categories resource use, acidification, freshwater ecotoxicity, ozone layer depletion, carcinogenic effects and respiratory effects, whereas scenario 2 scores better in the categories land use and eutrophication and, although only very marginally, also marine eutrophication and photochemical oxidant creation. For the categories where scenario 2 scores worse, the battery storage system is mainly responsible for these higher impacts. It contributes 77% to the total ozone layer depletion impacts, 21% to carcinogenic effect and 15% to resource use impacts, whereas for all other categories its contribution is less than 10%. In contrast, the impact of heat storage is far below 10% in all categories and therefore negligible. Carcinogenic effects might become more relevant with increasing size of the storage unit.

Most indicator results are dominated by either the PV installation or the wood chip furnace for heat production. For scenario 1, electricity purchase also contributes a significant proportion to several impact categories, especially freshwater eutrophication and carcinogenic effects. This is mainly

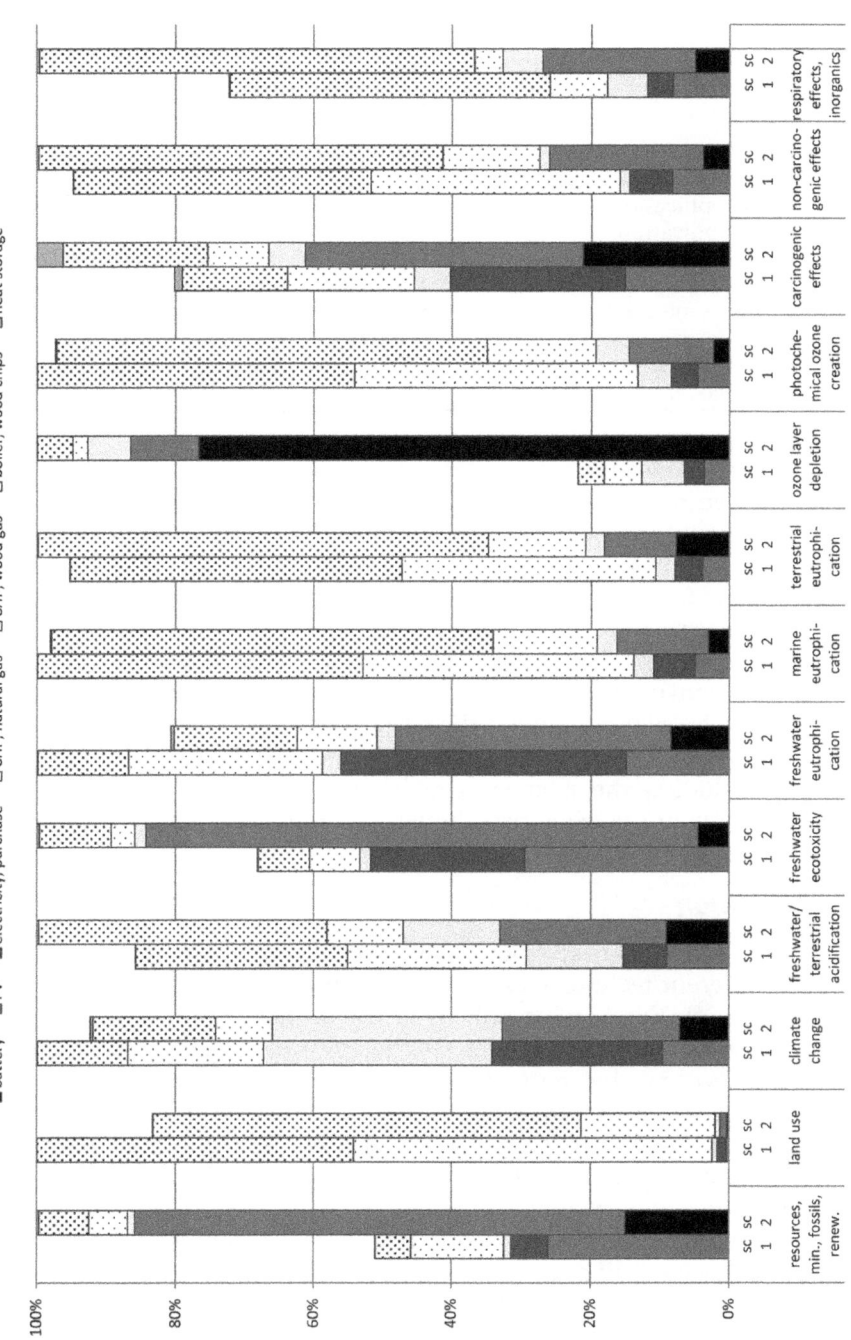

Figure 6 Comparison of environmental impacts of scenarios 1 and 2 with contributions of each technology (highest impact per scenario = 100%).

Table 6   Impact indicator results per functional unit [provision of 2275 MWh electricity and 5960 MWh heat (space heating and hot water) per year].

Impact indicator	Scenario 1	Scenario 2	Units
Resources, minerals, fossils, renewables	33.0	64.6	kg Sb-eq.
Land use	1.63×10^7	1.36×10^7	kg SOC
Climate change	5.21×10^5	4.81×10^5	kg CO_2-eq.
Freshwater/terrestrial acidification	3.91×10^3	4.55×10^3	mol H^+-eq.
Freshwater ecotoxicity	1.22×10^7	1.79×10^7	CTUe
Freshwater eutrophication	3.19×10^2	2.57×10^2	kg P-eq.
Terrestrial eutrophication	1.33×10^4	1.40×10^4	mol N-eq.
Marine eutrophication	1.24×10^3	1.22×10^3	kg N-eq.
Ozone layer depletion	4.75×10^{-2}	2.17×10^{-1}	kg CFC-11-eq.
Photochemical ozone creation	4.14×10^3	4.03×10^3	kg ethylene-eq.
Carcinogenic effects	3.79×10^{-2}	4.73×10^{-2}	CTUh
Non-carcinogenic effects	7.42×10^{-1}	7.83×10^{-1}	CTUh
Respiratory effects, inorganics	4.90×10^2	6.78×10^2	kg PM 2.5-eq.

due to the coal-based electricity generation that is also still used in the future electricity grid mix. However, a direct comparison of technologies is not possible since each technology provides different amounts of energy.

The carbon footprint of the electricity generated within the two scenarios is nearly the same: 0.17 kg CO_2-eq. kWh^{-1} for scenario 1 and 0.15 kg CO_2-eq. kWh^{-1} for scenario 2, *i.e.* the use of the battery and thus the increased self-consumption of PV electricity (or reduced purchase from the grid, respectively) slightly reduces the GHG emissions of the whole system. The carbon footprint of the produced heat is 0.02 kg CO_2-eq. kWh^{-1} for both scenarios (including the heat storage system), basically because heat is produced mainly in the wood-fired boiler, *i.e.* from biogenic sources.

6.5   Discussion

The uncertainty in this case study is lower than that in case study 1, which compared different technologies. This is because both scenarios consist of the same technologies, varying only in their contributions.

Concerning heat storage, only one technology is considered. Nevertheless, the results show that the share of heat storage in all scenarios is less than 10% and therefore much lower than that of the battery storage. Additionally, its impacts are probably still overestimated, since in reality for this application the storage volumes will be significantly higher than the 2000 L assumed for this study, which would reduce the environmental impacts. Owing to a better surface area-to-volume ratio, less material would be used and the efficiency would rise.

The impacts of battery storage are also associated with uncertainty, especially because of their end-of-life modelling. Here, the recycling is not modelled with specific recovery processes for this technology owing to a lack of data.

Battery technologies are able to provide multiple services within island power systems not considered here. Some examples of such services are balancing of short-term fluctuations of PV units, power quality management and lowering the probability of loss of power supply. It must therefore be stressed that the results presented are applicable only to the provision of peak load.

Additionally, the need to upgrade electrical transmission and distribution (T&D) equipment is completely avoided through the use of stationary storage in scenario 2. Considering the environmental effort to produce and install T&D equipment in scenario 1 can potentially lead to more favourable results for the use of stationary battery storage in scenario 2.

In summary, the case study indicates that, although battery storage increases the share of renewables within the grid and therefore reduces the CO_2 emissions of the electricity consumed, it leads to increased impacts in some other categories.

7 Conclusion

The three case studies presented provide a comprehensive picture of the environmental impacts of selected energy storage technologies using the ILCD impact assessment method. All assessments rely on specific case studies, which imply specific operational conditions. This means that the conclusions also refer to these boundary conditions and cannot be generalized without restrictions. Under the given modelling assumptions, the use of battery storage results in increased environmental impacts in the majority of the assessed categories, both in comparison with pumped hydroelectric storage (case study 1) and in comparison with the standard small-scale energy system without battery storage (case studies 2 and 3). When looking at the breakdown of the environmental impacts on energy provision technology, it can be observed that the shift towards self-sufficiency in the battery-using micro grid is partially due to the increased share of electricity from PV, associated with higher environmental impacts than the assumed future electricity grid mix in 2035 (see Table 2) in several impact categories. In this respect, the desire for a self-sufficient energy system does not seem to be advantageous from an environmental point of view. However, the battery storage system is able to provide multiple services that are not considered within the presented use cases, *e.g.* short-term PV fluctuation balancing. Similar limitations apply to the results from case study 1, where fundamentally different storage technologies are compared using the same service (functional unit) as a basis for comparison, whereas in reality they would rather be complementary than competing technologies. However, the presented assessments provide interesting insights into the major strong and weak points of each technology and the potentially most appropriate fields of deployment for each of them. In general, there is a need for more research in this area to evaluate adequately different storage technologies providing valuable services for the grid. Developments in the field of battery production are currently moving at a fast pace and are certain to be considerable within the very long life cycle of 80 years considered in the comparison with pumped hydroelectric power. Such developments should be considered in the design and planning of future energy systems. Further research is also needed concerning disposal and recycling to assess the environmental impacts of battery storage systems. Recycling is of special interest because resource use is a critical point concerning battery use. Regarding heat storage, the source case study shows low relevance within the energy system. However, a general statement of environmental superiority seems inappropriate, taking into account that the case study presented assesses only one type of construction.

Abbreviations

CHP	combined heat and power
DoD	depth of discharge
ETP	energy-to-power
ILCD	International Reference Life Cycle Data System
LCA	life-cycle assessment
LCC	life-cycle cost
LCI	life-cycle inventory
LCIA	life-cycle impact assessment
LFP	lithium–iron phosphate with graphite anode
LIB	lithium-ion battery
LMO	lithium–manganese oxide with graphite anode
LTO	lithium–iron phosphate with lithium titanate anode
NCA	lithium–nickel–cobalt–aluminium oxide with graphite
NCM	lithium–nickel–cobalt–manganese oxide with graphite
PHES	pumped hydroelectric storage
PV	photovoltaic
SoC	state of charge
SOC	soil organic carbon
T&D	transmission and distribution

Acknowledgements

Parts of Section 3 are adapted from A. Immendoerfer, I. Tietze, H. Hottenroth and T. Viere, Life-cycle impacts of pumped hydropower storage and battery storage, *International Journal of Energy and Environmental Engineering*, 2017, **8**, 231–245, DOI: 10.1007/s40095-017-0237-5, which is distributed under the terms of the Creative Commons Attribution 4.0 International License (http://creativecommons.org/licenses/by/4.0/). The support from ENsource – Center for Applied Science, Urban ENergy Systems and Resource Efficiency (FEIH_ZAFH_562822), funded by the Ministry of Science, Research and the Arts of the State of Baden-Württemberg and the European Regional Development Fund (EFRE), is acknowledged.

References

1. European Commission-Joint Research Centre (EC-JRC), *ILCD Handbook: General Guide for Life Cycle Assessment – Provisions and action steps*, EC-JRC – Institute for Environment and Sustainability, Ispra, Italy, 2010.

2. J. Guinee, M. Gorrée, R. Heijungs, G. Huppes, R. Kleijn, A. de Koning, L. van Oers, A. Wegener Sleeswijk, S. Suh, H. A. U. de Haes, H. de Bruijn, R. van Duin and M. Huijbregts, *Life Cycle Assessment- An Operational Guide to the ISO Standards*, Leiden, The Netherlands, 2001.

3. ISO, *ISO 14040 – Environmental Management – Life Cycle Assessment – Principles and Framework*, International Organization for Standardization, Geneva, Switzerland, 2006.

4. ISO, *ISO 14044 – Environmental Management – Life Cycle Assessment – Requirements and Guidelines*, International Organization for Standardization, Geneva, Switzerland, 2006.

5. H.-J. Althaus, G. Doka, T. Heck, S. Hellweg, R. Hischier, T. Nemecek, G. Rebitzer, M. Spielmann and G. Wernet, in *Sachbilanzen von Energiesystemen: Grundlagen für den ökologischen Vergleich von Energiesystemen und den Einbezug von Energiesystemen in Ökobilanzen für die Schweiz*, ed. R. Frischknecht and N. Jungbluth, Swiss Centre for Life Cycle Inventories, Dübendorf, Switzerland, 2007.

6. M. Baumann, J. F. Peters, M. Weil and A. Grunwald, *Energy Technol.*, 2017, **5**(7), 1071.

7. European Commission-Joint Research Centre (EC-JRC), *Recommendations for Life Cycle Impact Assessment in the European context: Based on Existing Environmental Impact Assessment Models and Factors//ILCD Handbook. International Reference Life Cycle Data System*, Publications Office of the European Union, Luxembourg, 2011.

8. European Commission-Joint Research Centre (EC-JRC), Characterisation factors of the ILCD Recommended Life Cycle Impact Assessment methods//ILCD Handbook, *International Reference Life Cycle Data System: Database and Supporting Information*, Publications Office of the European Union, Luxembourg, 2013.

9. R. Frischknecht and S. Büsser Knöpfel, *Swiss Eco-Factors 2013 according to the Ecological Scarcity Method: Methodological Fundamentals and Their Application in Switzerland*, Bern, Switzerland, 2013.

10. http://ec.europa.eu/environment/eussd/smgp/communication/impact.htm# (last accessed January 2018).

11. G. Wernet, C. Bauer, B. Steubing, J. Reinhard, E. Moreno-Ruiz and B. Weidema, *Int. J. Life Cycle Assess.*, 2016, **21**(9), 1218.

12. J. C. Koj, P. Stenzel, A. Schreiber, P. Zapp, J. Fleer and I. Hahndorf, *Life Cycle Assessment of a Largescale Battery System for Primary Control Provision: (Poster)*, Muenster, Germany, 2014.

13. I. Tietze, A. Immendoerfer, T. Viere and H. Hottenroth, *Euro-Asian J. Sustain. Energy Develop. Policy*, 2016, **5**(2), 15.

14. Agora Energiewende, Stromspeicher in der Energiewende: Untersuchung zum Bedarf an neuen Stromspeichern in Deutschland für den Erzeugungsausgleich, Systemdienstleistungen und im Verteilnetz, *Studie*, Berlin, 2014.
15. F. Wandelt, D. Gamrad, W. Deis and J. Myrzik, *Power and Energy Student Summit (PESS) 2015*, 2015.
16. J. Völker, S. Peters and M. Teichmann, *Trendstudie Strom 2022: Metastudienanalyse und Handlungsempfehlungen*, 2013.
17. C. Bauer, R. Bollinger, M. Tuchschmid and M. Faist-Emmenegger, in *Sachbilanzen von Energiesystemen: Grundlagen für den ökologischen Vergleich von Energiesystemen und den Einbezug von Energiesystemen in Ökobilanzen für die Schweiz*, ed. R. Dones, *et al.*, Dübendorf, 2007.
18. VISPIRON, *Vergleich der Wirtschaftlichkeit von Pump- und Batteriespeicherkraftwerken*, München, 2015.
19. P. Stenzel, J. C. Koj, A. Schreiber, W. Hennings and P. Zapp, *Primary Control Provided by Large-scale Battery Energy Storage Systems or Fossil Power Plants in Germany and Related Environmental Impacts*, 2015.
20. http://www.younicos.com/de/projekte/07_Schwerin/ (last accessed July 2016).
21. P. Wolfs, in *2010 20th Australasian Universities Power Engineering Conference (AUPEC)*, ed. IEEE, Piscataway, NJ, 2010, p. 455.
22. T. Struck and J. Broichmann, *Batteriespeicherprojekte der WEMAG AG: (Presentation)*, 2015.
23. M. Hiremath, K. Derendorf and T. Vogt, *Environ. Sci. Technol.*, 2015, **49**(8), 4825.
24. D. A. Notter, M. Gauch, R. Widmer, P. Wäger, A. Stamp, R. Zah and H.-J. Althaus, *Environ. Sci. Technol.*, 2010, **44**(17), 6550.
25. Bundesministerium für Umwelt, Naturschutz, Bau und Reaktorsicherheit, *Projektionsbericht 2015 – gemäß Verordnung 525/2013/ EU. März 2015*, Berlin, Germany, 2015.
26. http://www.ingenieur.de/Themen/Energiespeicher/Ingenieure-entwickeln-Hochleistungs-Batterien-fuer-10000-Ladezyklen (last accessed July 2016).
27. https://www.energystorageexchange.org/projects (last accessed February 2018).
28. J. F. Peters, M. Baumann, B. Zimmermann, J. Braun and M. Weil, *Renewable Sustainable Energy Rev.*, 2017, **67**(Supplement C), 491.
29. C. Bauer, *Ökobilanz von Lithium-Ionen Batterien*, Villigen, Switzerland, 2010.
30. L. A.-W. Ellingsen, G. Majeau-Bettez, B. Singh, A. K. Srivastava, L. O. Valøen and A. H. Strømman, *J. Ind. Ecol.*, 2014, **18**(1), 113.
31. G. Majeau-Bettez, T. R. Hawkins and A. H. Strømman, *Environ. Sci. Technol.*, 2011, **45**(10), 4548.
32. M. Zackrisson, L. Avellán and J. Orlenius, *J. Cleaner Prod.*, 2010, **18**(15), 1519.
33. J. F. Peters and M. Weil, *J. Cleaner Prod.*, 2018, **171**, 704.

34. http://www.ensource.de/index.php/en/mainau (last accessed February 2018).
35. https://www.carmen-ev.de/files/Sonne_Wind_und_Co/Speicher/Markt%C3%BCbersicht-Batteriespeicher_2017.pdf (last accessed January 2017).
36. *A Comparative Probabilistic Economic Analysis of Selected Stationary Battery Systems for Grid Applications*, ed. M. J. Baumann, B. Zimmermann, H. Dura, B. Simon and M. Weil, 2013.
37. P. Stenzel, M. Baumann, J. Fleer, B. Zimmermann and M. Weil, in *IEEE International Energy Conference (ENERGYCON)*, 2014, *Dubrovnik, Croatia*, ed. I. Kuzle, IEEE, Piscataway, NJ, 2014, p. 1334.
38. *Evaluation of Calculation Methods, Models and Data Sources for Life Cycle Costing on the Example of Stationary Battery Systems*, ed. M. J. Baumann, D. Poncette and B. Zimmermann, 2014.
39. B. Battke, T. S. Schmidt, D. Grosspietsch and V. H. Hoffmann, *Renewable Sustainable Energy Rev.*, 2013, **25**, 240.
40. C. Weng, J. Sun and H. Peng, *J. Power Sources*, 2014, **258**, 228.
41. D. Yang, C. Lu and G. Qi, *Int. J. Comput. Electrical Eng.*, 2013, 330.
42. D. Rastler, *Electricity Energy Storage Technology Options: A Primer on Applications, Costs and Benefits*, Washington DC, United States, 2018, vol. 00, p. 40.
43. J. F. Peters and M. Weil, *J. Power Sources*, 2017, **364**(Supplement C), 258.
44. J.-B. Eggers and G. Stryi-Hipp in *Proceedings of the Sustainable Buildings - Construction Products & Technologies conference* 2013, ed. A. Passer, K. Höfler and P. Maydl, *Verlag der Technischen Universität Graz*, 2013, p. 580.
45. J.-B. Eggers, G. Stryi-Hipp and S. Herkel, in *Proceedings of Building Simulation 2015: 14th Int. Conference of IBPSA*, ed. J. Mathur and V. Garg, 2015, p. 2080.

Business Opportunities and the Regulatory Framework

REINHARD MADLENER* AND JAN MARTIN SPECHT

ABSTRACT

This chapter focuses on existing and emerging business opportunities and the role of regulatory frameworks for energy storage. First, an overview is provided of the technoeconomic characteristics of the most relevant energy storage options available in the market today, highlighting the complementarity and competition among these technologies and the increasingly important role of energy storage in the globally ongoing sustainable energy transition. Second, the manifold, but sometimes mutually exclusive, value-creating potentials of energy storage units are pointed out. Third, some of the major barriers to the wider adoption and diffusion of storage technologies that go beyond economic viability and cost competitiveness (both between the different energy storage alternatives and, one level above, between the various flexibility options competing with each other) are discussed. The numerous uncertainties related to the growth of variable renewable energy as part of the sustainable energy transition in this rapidly evolving field are also discussed. Decision makers in the energy storage business are well advised to study carefully the market and regulatory conditions and also underlying uncertainties, as business models may erode over time, thus jeopardizing the anticipated desired return on investment. In light of the potentially disruptive nature of some energy storage technologies, such as those installed in e-cars and homes, *i.e.* distributed ones where millions of units may be adopted quickly,

---

*Corresponding author.

Issues in Environmental Science and Technology No. 46
Energy Storage Options and Their Environmental Impact
Edited by R.E. Hester and R.M. Harrison
Published by the Royal Society of Chemistry, www.rsc.org

especially investors in large-scale and long-lived storage solutions must be cautious. Finally, the issue that energy storage is only one among a number of flexibility options to balance supply and demand in an intertemporal and/or spatial manner, and not necessarily the least-cost option, is discussed.

1 Introduction

The transformation of the energy system towards decarbonization and high shares of variable renewable energy sources (VRES), such as wind and solar energy, requires flexibility options to an increasing extent that facilitate the balancing of supply and demand. These include grid extension and modernization (more cable and more smartness/digitalization), demand response, more flexible conventional power generation, sector coupling and energy storage. In a liberalized energy market, competition is fostered on the production and retail side, whereas traditionally the grid is considered as a natural monopoly and regulated as such.

The current global energy storage market, estimated at 4.67 TWh (2017) and expected to grow rapidly to 11.89–15.72 TWh (2030),[1] is strongly dominated by hydro power storage (around 99%).[2] Owing to the shift towards high shares of VRES, the need for storage capacity is expected to grow exponentially in the years to come, and increasingly also to contain many other technologies, especially battery storage.[3,4] Although the discussion often centers around electricity storage, sector coupling (electric power, heat, transport, *etc.*) will also play an important role. Therefore, the economic valuation of thermal storage units[5] and power-to-X (P2X, *e.g.* power-to-heat, power-to-fuel)[6] options that allow sector coupling can be expected to become increasingly relevant and attractive. So far, the literature on the economic valuation of energy storage is still relatively sparse and scattered; for instance, the economic viability of using the natural gas grid as a storage option (so-called 'linepack storage') has been investigated by Arvesen *et al.*,[7] among others,[8–13] whereas the economics of pipe container and underground reservoir storage systems for power-to-gas (P2G) load balancing has been studied by Budny *et al.*[14]

Interest in the literature on energy storage has grown tremendously in recent years, especially with regard to battery storage (see, *e.g.*, Figure 2 in Luo *et al.*[15] for time trends in the Web of Science regarding keyword combinations with 'energy storage'). A number of books, review articles and reports have been published lately that cover not just scientific and engineering fundamentals but also technoeconomic aspects[4,16–18] and the value of energy storage.[3,19–24] Several studies[25–28] provide useful reviews of the different types of storage technologies that exist, their valuation and their developments in recent years.

Energy storage offers flexibility in the form of buffering short-term variations of supply and demand, ancillary services (voltage control, frequency control, black-start capability *etc.*) and compensation of seasonal variability. The increasing number of commercially available energy storage technologies not only differ regarding their technical and economic characteristics (see Figures 1 and 2), but also their relative economic merit and thus competitiveness depend significantly on their suitability and application,[29] and also whether regulatory conditions are prohibitive or favorable. In other words, there is no 'one size fits all', even though some technologies have the benefit of being modular and thus scalable[30] and the optimal size depends on the specific situation. The economic viability of the same geophysical storage unit (*e.g.* a salt cavern) can also be expected to depend strongly on the medium being stored, such as compressed air, natural gas, hydrogen, synthetic methane or carbon dioxide.[31]

The future need for storage capacity depends strongly on grid expansion (on all voltage levels, locally, nationally and internationally) and modernization (replacement, digitalization), but also on flexible generation (supply side) and flexible consumption (demand side). This creates considerable uncertainty regarding the profitability of storage options in the long run, and how profitability might change over time. Private households may rapidly adopt electric cars and stationary batteries (see Figure 3 for the case of household batteries in Germany) for a number of reasons beyond simple economic or technical considerations, such as self-supply (see Section 2) of homes or local communities (neighborhoods, city quarters, *etc.*), requiring a better understanding of 'prosumer' (producer–consumer) preferences with respect to the adoption of energy storage.[32]

Many storage options are found to be economically not (yet) viable without subsidies,[9,10,14,33–35] whereas in other applications economic viability depends considerably on the regulation in place (*cf.* Section 4).

Substantial research and development (R&D) efforts are currently ongoing for almost all types of energy storage technology. Mass adoption of lithium-ion batteries in vehicles and homes could be a game changer once these batteries have become reasonably powerful (see Section 2.3) and sufficiently inexpensive to make e-vehicles attractive compared with conventional internal combustion engine (ICE)-powered cars and home batteries a 'no-regret' option. Still, some experts argue that the long tail of the learning curve is flat (see Figure 4), which would necessitate new, and ideally more radical (rather than just incremental), technological development.

The remainder of this chapter is organized as follows. Section 2 provides an overview of the different factors that influence the economic value of storage units. Section 3 deals with the design of business models by accounting for subsidy/tariff schemes (promotion policies), value stacking (creation of multiple economic benefits by serving different markets) and portfolio combinations [cloud solutions, virtual power plants (VPPs), virtual storage swarms (VSSs)]. Section 4 provides an overview of regulatory aspects to be considered, and finally Section 5 provides concluding remarks.

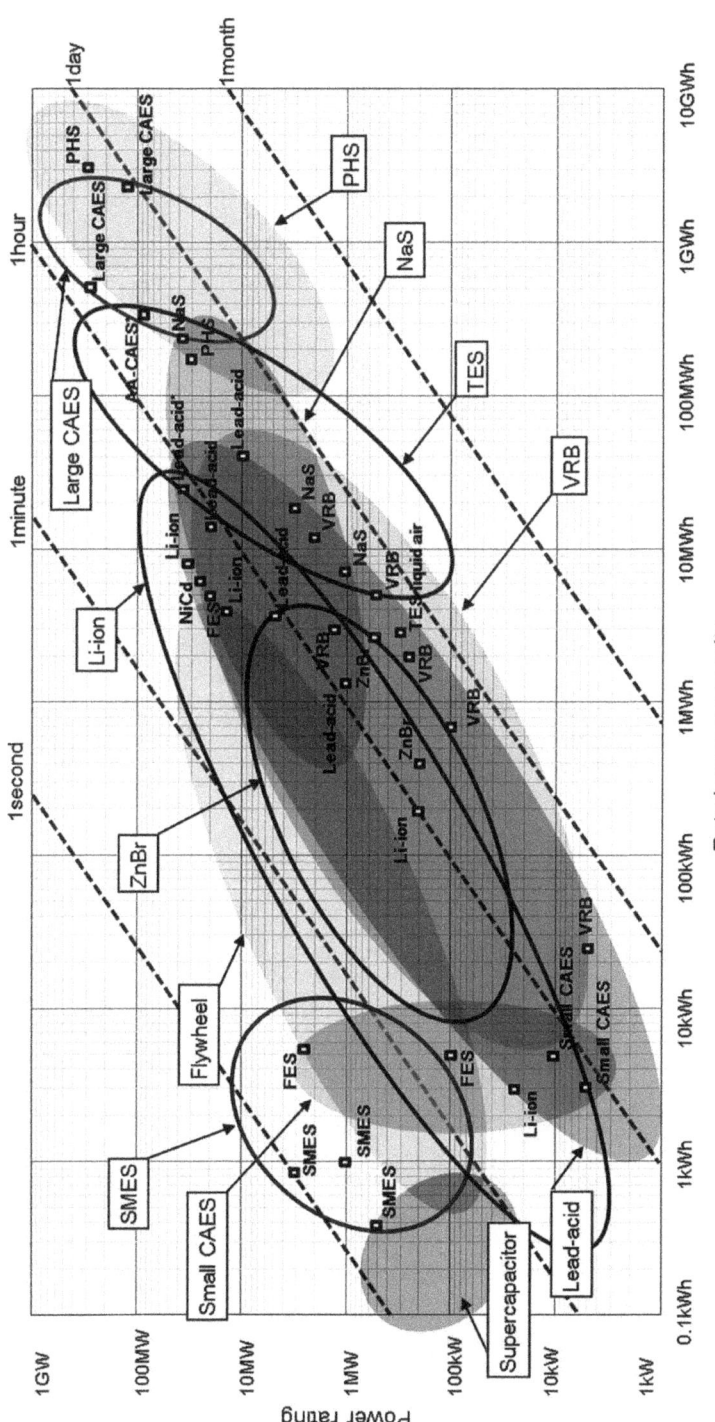

Figure 1 Technical characteristics of various energy storage technologies (power rating *versus* rated capacity). Reproduced from ref. 15, https://doi.org/10.1016/j.apenergy.2014.09.081, under the terms of the CC BY 3.0 license, https://creativecommons.org/licenses/by/3.0/.

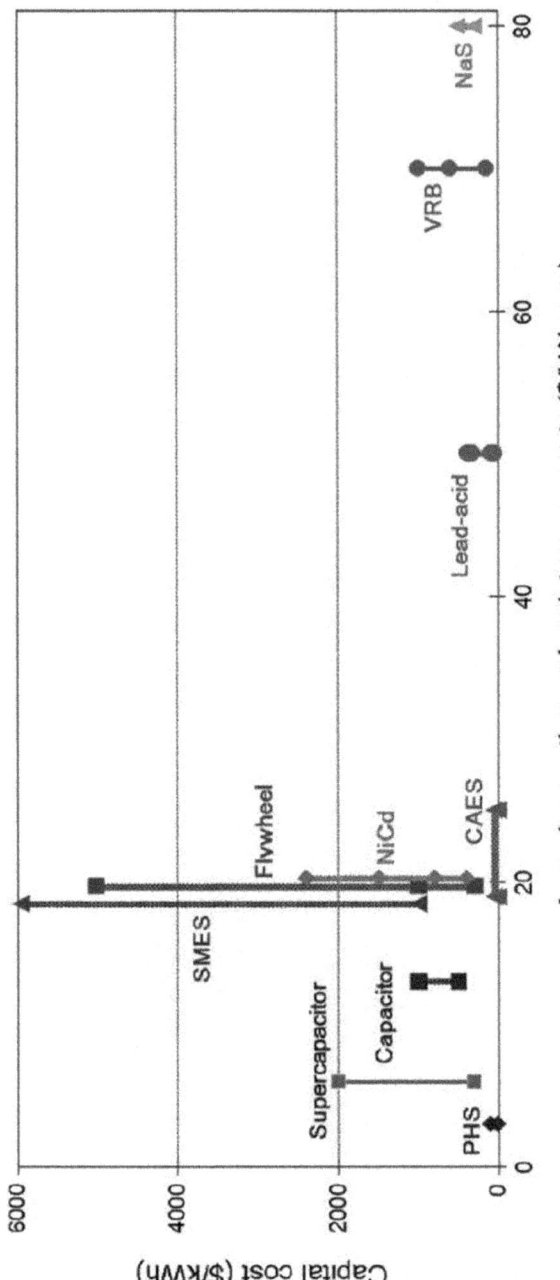

Figure 2 Economic characteristics of various energy storage technologies (capital costs *versus* O&M costs per year). Reproduced from ref. 15, https://doi.org/10.1016/j.apenergy.2014.09.081, under the terms of the CC BY 3.0 license, https://creativecommons.org/licenses/by/3.0/.

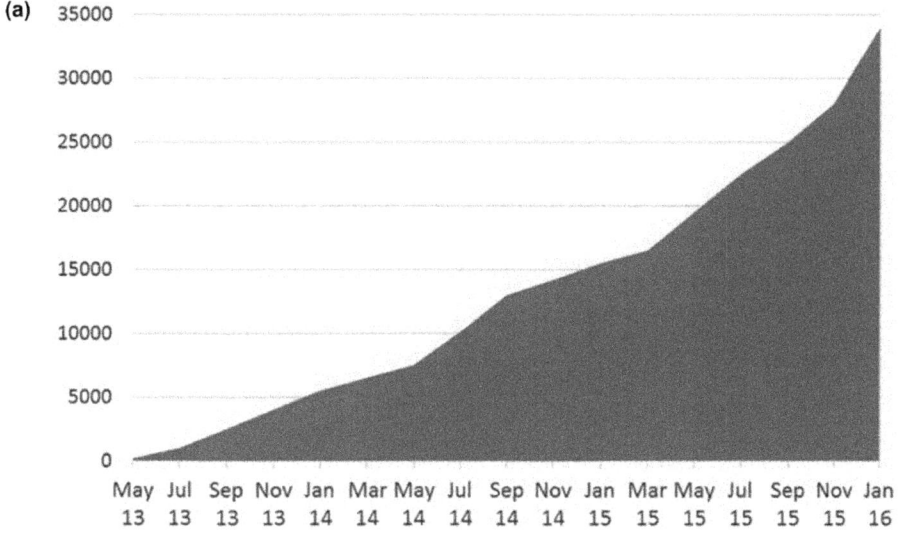

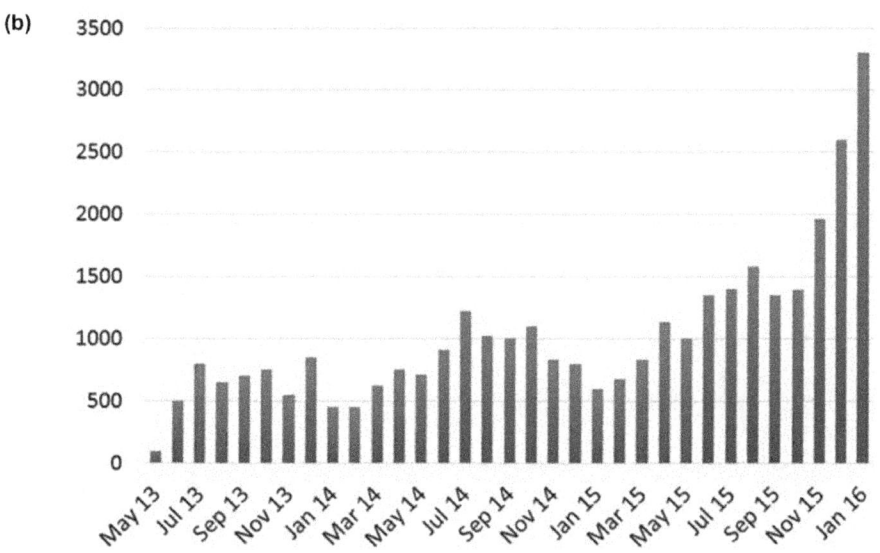

Figure 3 Adoption (a) and diffusion (b) of private household batteries in Germany, May 2013–January 2016.[9,10,36]

2   Economic Value of Storage

2.1   *Matching Technologies to Applications*

The full value of storage stems from the ability to provide a number of services across a variety of applications that often can be stacked (see Figure 5).

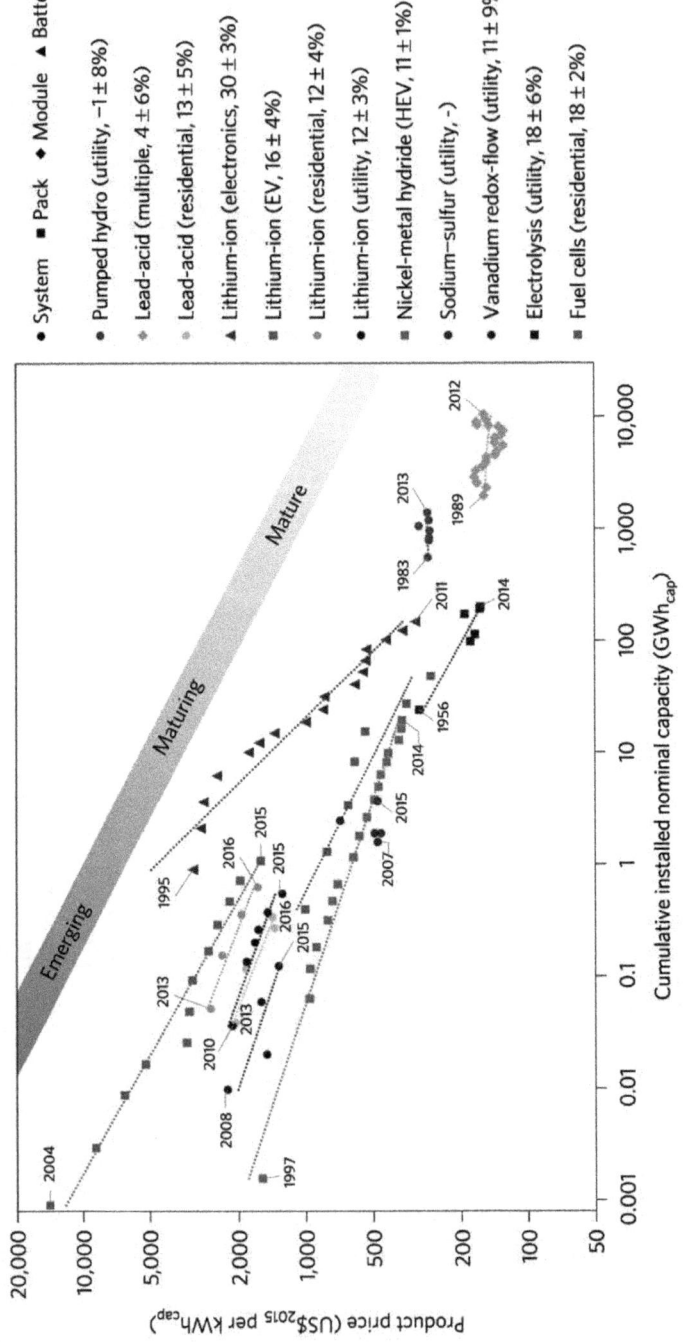

Figure 4 Learning curves for energy storage technologies regarding the maturity of each technology. Reproduced from ref. 37 with permission from Springer Nature, copyright 2017.

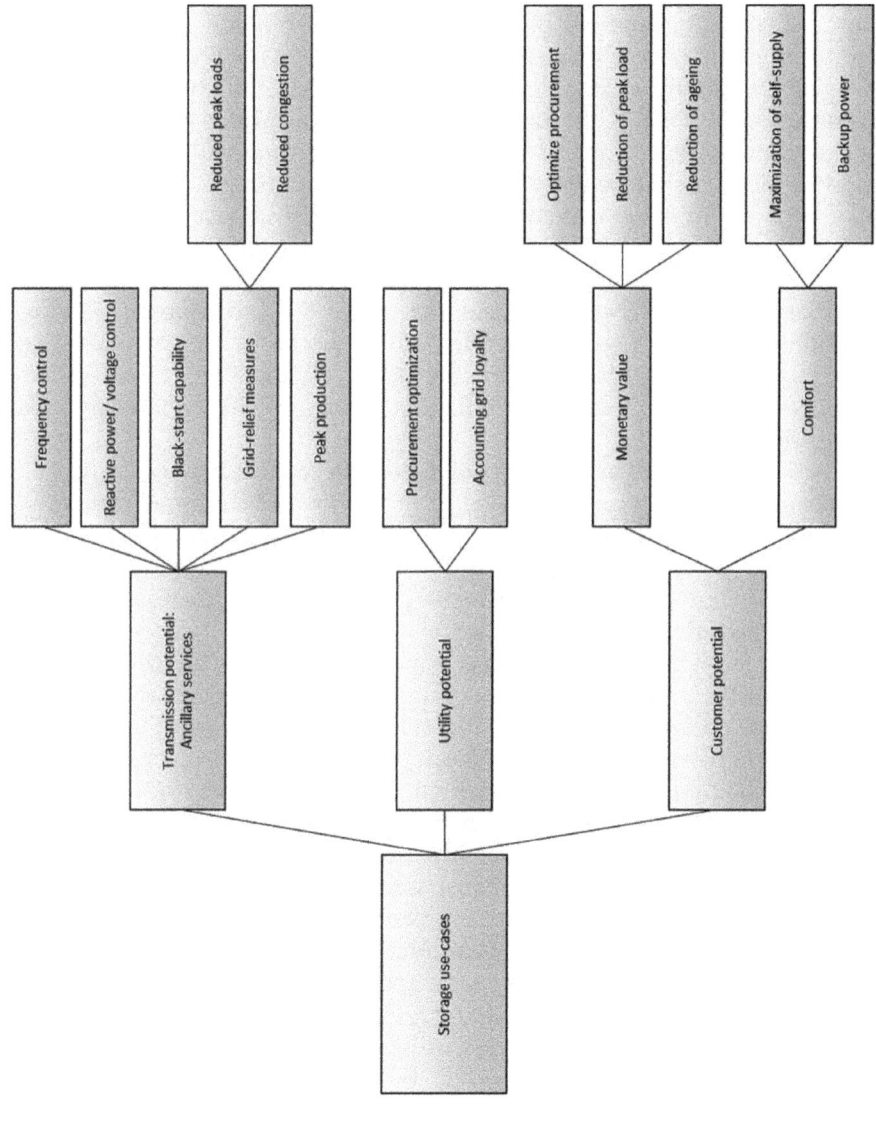

Figure 5 Alternative use-cases for energy storage devices.

The three main stakeholder categories are the grid operators (on the transmission and distribution levels), the utilities and the final customers. Whereas the grid operators benefit mainly in terms of the ancillary services shown in Figure 5, utilities benefit in terms of procurement and grid accounting benefits. Customers, in turn, are interested in either monetary or comfort gains, the latter of which can be divided into the provision of backup power and the enhancement of self-supply (100% self-sufficiency being referred to as 'autarky', *i.e.* when a grid connection would no longer be necessary all year, but where the costs might by far exceed the willingness-to-pay).

According to ACORE,[24] the most prominent applications can be grouped into generation services, grid services and behind-the-meter services. Generation services include energy time shifting/generation arbitrage, electric supply capacity and peak demand management and renewable capacity firming/smoothing. Grid services include transmission and distribution grid upgrade deferral, transmission grid congestion relief and the provision of ancillary services (frequency regulation, load following, voltage control *etc.*). Finally, behind-the-meter services include the provision of backup power, retail electricity time shifting and solar photovoltaic (PV) electricity self-consumption, demand charge management and the enhancement of micro grids.

The benefits provided by energy storage include economic, system and non-energy benefits.[38] The economic benefits stem from efficiency increases and reduced system costs, and the system benefits stem from increased reliability and flexibility. Finally, the non-energy benefits comprise, for instance, the increase in the amount of renewable energy (and thus a reduction in the use of fossil fuels or the compliance costs of generation unit owners or system operators[24]). For obvious reasons, the non-energy benefits are the most difficult to quantify (and monetize), but should not be forgotten when valuing energy storage and may provide justifications for granting subsidies.

Economic benefits may come in the form of additional revenues or cost savings (avoided costs). Revenues gained through the operation of storage devices may result from selling energy storage services or from participation in wholesale markets (*e.g.* bidding in capacity markets to provide reserve energy/capacity, or in ancillary services markets). Behind-the-meter energy storage can create value for both the customers and the grid operator, typically in the form of demand charge reductions, backup power or increased self-consumption from solar PV systems (see Fitzgerald *et al.*,[23] p. 7). Obviously, when trying to determine the net value of all use cases and services provided from energy storage devices utilized at different levels of the electricity system, valuation becomes very complex, involving a great many variables and interdependencies.

System benefits from storage units accrue when these offer greater flexibility to the system in responding to unanticipated changes, by enhancing the resilience of the system or by helping to reduce outage risks. Storage units support the grid through the provision of distributed capacity, frequency stabilization and voltage support (see ACORE,[24] p. 19).

The economic viability of storage units can be enhanced by so-called 'value stacking' (sometimes also referred to as 'benefits stacking', *e.g.* in ACORE,[24] p. 20). This means that if the value of single applications (*e.g.* self-consumption, backup power, reduction of peak load) can be offered by a single unit rather than separate units, the value created with the same storage unit can be increased by better exploiting its lifetime capacity. Note, however, that regulatory barriers (discussed in Section 4) can hamper the value-stacking opportunities. Also, the overall value that can be created depends on the specific situation and on various grid-specific factors (see, *e.g.*, Figure ES3 in Fitzgerald *et al.*,[23] showing four distinct examples of battery economics from value stacking in the USA).

Eventually, the viability and success of a business model depend on whether and how the value components created in a particular regulatory setting and a specific technology–application combination can be stacked and divided among the stakeholders, in such a way that they all benefit sufficiently to agree to participate in the business case in question. Also, the competition between business cases can jeopardize the profits expected, thus adding considerable uncertainty for decision makers due to the possibility of an unexpected erosion of their business case(s). The more dynamically evolving the energy markets concerned and regulatory frameworks are, the higher the uncertainty will be. Competition is desired by the regulator in the generation and retail parts of the value chain, whereas the transmission and distribution parts typically feature a regulated (natural) monopoly.

2.2 Merit Order of Alternative Storage Options

After discussing the applicability of the various storage technologies for certain energy system services, the next question is how to monetize the value of alternative energy storage options. This can help to compare different storage options and derive a merit order on which available technologies should be exploited from an economics perspective. Such a merit order could also be developed for all kinds of flexibility options in an analogous way.[19] As Figure 6 shows, this can be done in a two-dimensional way by scrutinizing the added value for the economy (social welfare perspective) on the one axis and the profitability for the storage device operator (business perspective). Cost–benefit analysis can be used to account systematically for the various benefits. However, the quantification of some of these benefits is often rather complex and market barriers and prevailing regulations will additionally limit the applicability of energy storage (see also Section 4). Overall, the net benefit to the individual stakeholders involved can be expected to be very case-specific.

In some cases, the direct benefits may be quantifiable, whereas in other cases, one might have to use proxies, such as the value of lost load (VOLL), which accounts for the value of avoiding costly service outages that energy

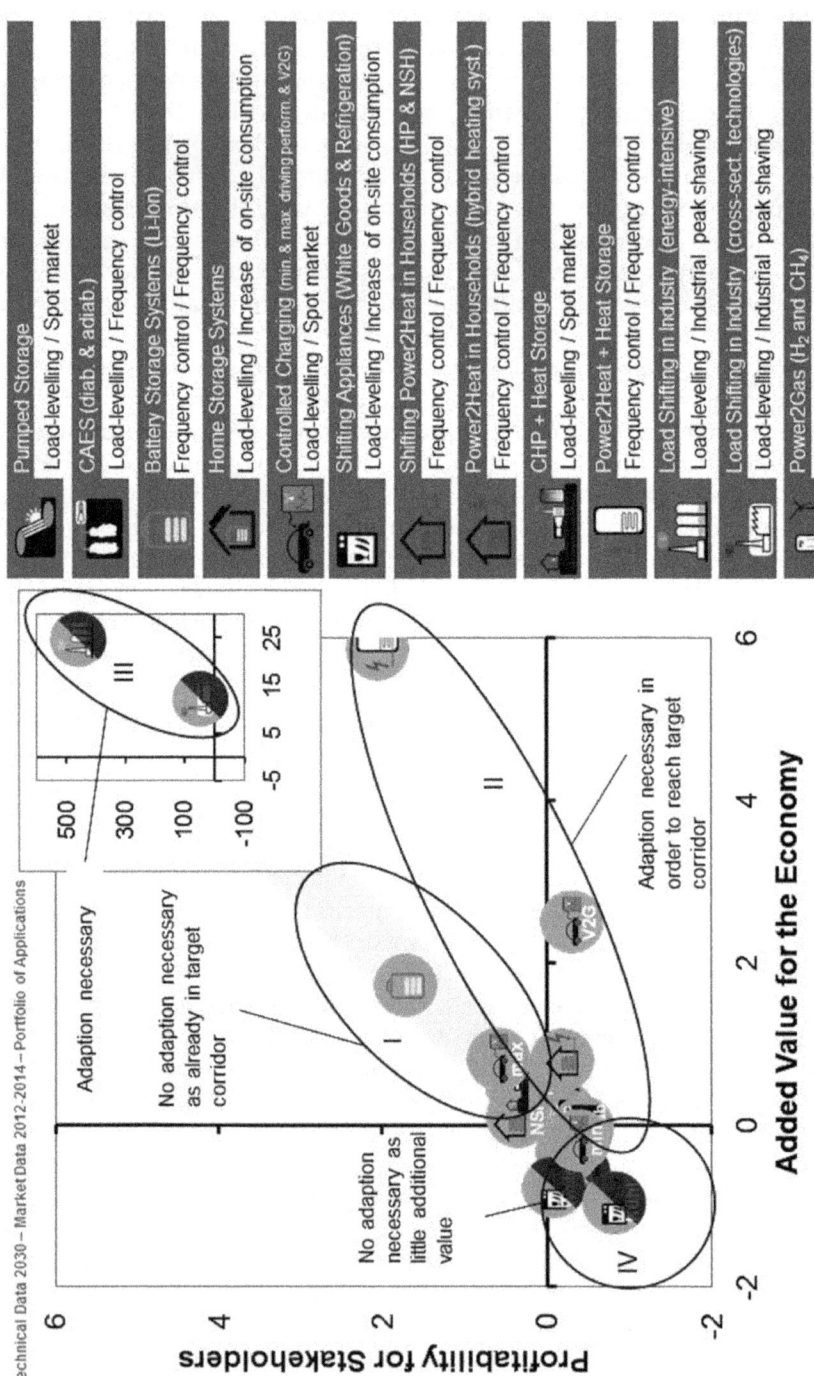

Figure 6 Merit order of various storage options,[39] © 2016 FfE.

storage provides due to enhancing the reliability of the system.[24] For the non-energy benefits, avoided external costs (*e.g.* from fossil fuel saving) can be used as a measure.

2.3 Location and Energy Density of Storage Units

The value of storage depends also on the location of the storage unit, which may be used in either a stationary or a mobile application. A frequent distinction for the connection of a storage device is between (1) transmission level, (2) distribution level and (3) 'behind the meter'. The further down the value chain the connection of a storage device to the system is, the more services it can potentially offer to the energy system at large.[23] Although a behind-the-meter energy storage device thus can provide the most services to the grid in principle, it might still not be the least-cost option from a system's perspective.

Madlener and Latz[41] studied the relative merits of centralized *versus* decentralized energy storage for the case of compressed air energy storage (CAES) units used for wind power integration. Using a profit-maximization algorithm and data for a 100 MW nearshore wind park, they found that a centralized CAES plant is relatively more attractive than wind turbines with integrated (distributed) CAES, and for the underlying market conditions diabatic CAES turns out to be more profitable than adiabatic CAES. Unless a minute reserve market exists where hourly contracts can be traded, none of the CAES power plants studied makes economic sense, and the economic viability depends considerably on how intensively the spot market and the market for minute reserve are used. CAES plants can be operated profitably only if combined trading in the spot market and minute reserve market is enabled and if some support scheme exists (such as the German Renewable Energy Sources Act).

A technical but also economic aspect of storage systems is the energy density. This aspect is of particular relevance in mobile applications (*e.g.* electric vehicles), since additional weight has a negative impact on the efficiency of the total system (*e.g.* the car). Berckmans *et al.*[40] identified three trends that might allow battery storage to catch up or even surpass fossil fuels in terms of energy density. The first is to use different electrodes with significantly higher capacities, such as sulfur (1672 mAh g^{-1}), silicon (4200 mAh g^{-1}) or lithium metal (3860 mAh g^{-1}), which would inherently increase the energy density of the cell (the electrode itself can store more energy). The second trend is to increase the voltage limit of a single cell to around 5 V, a common voltage level used in the field of electronics, which would also increase the cell's energy density (which can be simply defined as the integral of the actual capacity times the actual voltage). The third trend is to go towards solid-state electrolytes, because the use of liquid electrolytes can cause safety problems in the case of leakage.

2.4 Optimal Sizing of Storage Units

The economically optimal operation and size of a storage device depend not only on the technical characteristics but also on how it is used to create value. Different modeling approaches can be found in the literature, so the results obtained also depend considerably on the type of model being used. The literature addressing the optimal sizing of storage units is diverse and often focuses on this issue in a secondary manner only. In the following, we provide some exemplary studies illustrating the wide range of technologies, applications, valuation methods and business cases investigated.

Bradbury *et al.*[43] investigated the economic feasibility of 14 different energy storage systems used in seven real-time markets in the USA for arbitrage business, using a linear optimization model for finding the power and energy capacity per system that maximizes the internal rate of return (IRR). Their results show that although most storage systems have an optimal size of 1–4 h of energy storage, it is found to be 7–8 h of operation for the case of pumped storage hydro power and CAES.

McConnell *et al.*[42] found that the revenues from storage devices in the energy-only electricity market are similar to a peak generator and thus highly skewed to a few days per year. As a consequence, the economic value of the storage device is found to be fairly insensitive to the roundtrip efficiency, and little value can be gained from offering more than 6 h of storage. Similarly to conventional peak-power generators, storage units on energy-only electricity markets benefit from extreme prices, which in turn depend on capacity requirements, but in contrast to peaker plants can receive additional revenue streams from the provision of other energy services.

Weibel and Madlener[44] studied the economic viability and the optimal siting and sizing of a so-called ringwall hydro power plant (essentially a donut-shaped pumped storage hydro power plant made of concrete) that is combined with PV and wind, and where wind intensity, solar irradiation, electricity prices and investment costs were treated as uncertain. The results reveal not only the optimal size of the (scalable) storage device, but also refer to the market on which the electricity should be sold in order to maximize the profits, in addition to whether the investment should be made immediately or deferred.

Budny *et al.*,[14] based on an NPV (net present value) model and Monte Carlo simulation for electricity and gas prices and uncertainties, investigated the economic feasibility of P2G and gas storage options for hydrogen and methane from renewables. The three cases examined were the direct sale of gas in the market, temporal arbitrage between the electricity and gas markets and the use of storage for participating in the balancing markets (secondary reserve for electricity external balancing for natural gas). They found that the P2G system used for bridging the balancing markets is unprofitable. Pipe storage is the preferred option for both temporal arbitrage and balancing of energy. Still, the feed-in tariffs required for making pipe storage for P2G economically viable turn out to be relatively high (€100 per MW for H$_2$, €130 per MW for renewable methane), thus requiring support through the regulatory framework.

Kirmas and Madlener[47] studied the economic viability of second-life electric vehicle batteries for load shifting and peak shaving in residential applications, where there is an integrated PV system. Considering different scenarios for the case of Germany, they found that investment in second-hand lithium-ion batteries is profitable, and that for different electricity price paths assumed that the break-even battery price lies between €73 and 107 per kWh.

2.5 Economic Impact of Aging of Batteries

In comparison with other energy storage systems, chemical batteries have one major disadvantage: battery aging, which leads to an irreversible capacity decrease that still appears even when the battery is not in use.[45] There are different kinds of aging effects that cause an irreversible capacity decrease to which a battery is exposed during its lifetime: on the one hand, calendric aging, which occurs when the battery is being charged; and on the other, cycle aging, which is dependent on the use of the battery (see Vetter *et al.*,[46] p. 270). The economic impact of battery aging is important for at least two reasons: first, there is the possibility of using lithium-ion batteries from e-vehicles in stationary applications after their 'first life', and second, aging has to be considered in every storage cycle as additional wear costs. Owing to the high cost of such batteries and split benefits, *e.g.* between the grid operator and the private household, battery owners need to be willing to provide, and have to be adequately compensated for the value of providing, energy services (net of the value lost due to any increased devaluation caused by additional battery aging).

2.6 Prosumer Concept

The rapid diffusion of rooftop PV systems has been a striking feature of the sustainable energy transition process in many countries. The self-generation of electricity by private households turns them from passive consumers of electricity into active prosumers.[48,49] Such prosumers become interactive market participants – either independently as their own entities or represented by aggregators (*i.e.* specialized energy service providers). Such an aggregation of distributed energy resources is a process enabled by the liberalization of the electricity markets in combination with the progress towards smart grids. In the recent literature, concepts such as the Energy Internet, sustainable communities and local retail markets are investigated, covering the various degrees of prosumer participation in an unbundled market environment.[9,10,50]

2.7   Energy Cloud Concepts

Electricity storage services can be offered under the support of the technologies of the digitization and automation of storage operators.[9,10] The capacity of a distributed battery storage device can thus be shared in the sense of the 'sharing economy', as defined in Daunoriene *et al.*[13] Actors in the market are charged a fee for using the spare battery capacities of other actors. Note that this may include both mobile storage in batteries of (plug-in) electric cars and stationary storage systems. Note also that stationary storage systems in the sharing economy can be either centralized large-scale storage units or decentralized small storage units, the latter of which include home storage systems in residential households.

A sharing transaction is sequenced as follows.[9,10] First, the free storage capacities are aggregated. If a third party requests the use of free capacity, a schedule is generated for each individual storage device. Thereupon, the balancing group implements the bookings to illustrate the trading business. The booking is carried out in the balancing group of the storage cloud operator. Once the transaction has been executed, the real energy flow is recorded using the smart meter interfaces and this amount of electricity is billed to the participants of the transaction. The execution of a transaction within the storage cloud corresponds to a secondary usage of the storage. Potential users of spare storage capacities include network operators, energy suppliers, wind-farm operators and private households.

3   Value Creation for Business Models

3.1   Subsidies and Tariff Schemes

The economic value provided by storage depends on multiple factors. Popper and Hove identified the following eight categories as being relevant for the attractiveness of a region for storage systems: (1) storage experience, (2) storage subsidies and targets, (3) wind and solar penetration and growth, (4) distributed energy penetration and growth, (5) electricity price levels, (6) time-of-use (TOU) prices, (7) demand charges and (8) the structure and pricing of ancillary services markets.[51] They found that storage currently poses a positive value only in the USA, Japan and Australia. Among these countries, only Australia offers a positive value without funding, mainly owing to its low population density and thus long grid connections in combination with a high share of PV power. Even though many European countries have high shares of renewables in addition to substantial funding, a positive value of storage is in their view only expected to be reached between 2017 and 2020 (see Popper and Hove,[51] p. 11). This is mainly due to the small price differences on the European electricity markets, which are crucial for the profitability of bulk storage. Zafirakis *et al.*[52] found an average arbitrage value between €15 and 30 per MWh on the European Energy Exchange (EEX) market for 2011 (which, as we will elaborate on in Section 3.4, does not seem to have changed much in recent years). Several studies on the value of arbitrage from battery storage units can be found in the literature.[22,53,54]

3.2 Economic Value from Energy (Self-) Supply

In the European Union (EU), electricity prices for private households were, on average, around 20 €ct (€ct = euro cent) per kWh in 2017, with significantly higher prices in some EU member states (*e.g.* Denmark, 30.49 €ct per kWh; Germany, 30.48 €ct per kWh). Electricity prices rose significantly in recent years, by about 30% between 2008 and 2017 in the EU.[55]

The feed-in of renewable power generation into the grid, on the other hand, was heavily subsidized and thus lucrative in many countries in the past. However, the funding schemes have been steadily decreasing. For this reason, with the gap between increasing prices for electricity purchase and decreasing margins for electricity sale, self-consumption becomes more and more interesting for household customers. The direct economic value from energy (self-) supply mainly consists of the difference of avoided cost for the purchase of electricity and missed feed-in tariffs as opportunity cost.

However, there is more value in self-supply than just the direct cost savings. For a start, many customers favor the reliable costs of their own assets over the rather unpredictable volatility of (increasing) costs of electricity from the grid. Also, renewable self-produced energy is often valued higher than conventional electricity from the grid for ecological reasons. Further, a combination of local generation (*e.g.* by a PV system) and a battery storage system can also work as a backup power system and thus provide resilience against interruptions in the power supply. Several studies investigating the willingness to pay to avoid grid outages were able to verify this desire. However, the quantification in terms of a monetary value returned a rather broad span of results, ranging from €7.8[56] to €30 (US$37)[57] to avoid 1 h of outage, depending on the circumstances.

For large-scale industrial and commercial customers, the situation is somewhat different. They often pay significantly less for grid electricity (*e.g.* around 5 €ct per kWh in Germany, 3 €ct per kWh in France and 7 €ct per kWh in the UK for customers with an annual energy consumption between 20 and 500 MWh),[55] which is why the revenue gap is typically smaller than for households. However, there are other potential benefits of storage assets from which to profit. Under most regulations, these large consumers have to pay a peak demand fee (also referred to as a capacity charge), depending on their peak load within a given time span, in addition to the energy price.[58,59] To reduce the charges of the demand fee, it is important to shave load peaks, since a single power spike within a given year can add significant costs to the annual bill. In this context, energy storage assets can be a viable alternative to demand-side management or local fossil fuel-based generators. A local storage can release electricity stored earlier in the right quantity to reduce the overall load below a predefined target value. The trade-off is between reducing this target value to decrease the peak price gradually on the one hand, and not risking running out of power during one application and thus facing a completely untreated power spike on the other. This often constitutes a difficult optimization problem that also has to consider the degree of risk aversion of the individual customer. While many studies focused primarily on the conditions in Europe, other similar studies shed light on the conditions in the US markets.[60–62]

3.3 Economic Value from Ancillary Services

In the past, electricity was generated in large-scale centralized power plants and was typically transmitted unidirectionally through grids of decreasing voltage levels to the customers. Most ancillary services, such as reactive power provision or frequency control, were provided by the large power plants and various categories of small customers could be sufficiently described by a relatively small number of standard ('synthetic') load profiles. However, the transition towards decentralized renewable energy assets changes some fundamentals of this incumbent system:

1. Small-scale production assets have begun to replace the large conventional plants, but are only slowly starting to participate in the provision of ancillary services.[63] Some renewable assets, especially wind turbines, today begin with the implementation of predefined profiles for the provision of reactive power. The spinning and the supplementary reserve of the turbines in thermal power plants, however, cannot yet be directly replaced by wind or PV owing to a lack of inertia from rotating masses. At least regarding the spinning reserve, so-called synthetic inertia provided by wind turbines could be a future solution.[64]
2. The second issue is related to the production and consumption loads of renewable energy assets. Especially in rural regions, the local renewable electricity generation can significantly exceed the local consumption. In these cases, electricity has to be fed back into the next higher voltage level, an operation for which most electric systems were not originally designed.

In addition, in more and more cases, the maximum consumption becomes replaced by the maximum production as a key indicator for the grid dimensioning. In urban regions, in contrast, the significant charging capacity of battery electric vehicles may bring the grid capacity to its limits.[65,66]

At least two ways to deal with the problem of limited grid capacity exist: (1) nodal pricing (also referred to as local marginal pricing) and (2) zonal pricing. In a system with nodal pricing, the limited capacity in the grid is sold by means of congestion rights to the highest bidders, while actors with a lower value will have to waive their production or consumption. This model was implemented by most independent system operators (ISOs) in the USA and New Zealand.[67] In zonal pricing, the grid fees within one region are uniform. This again offers two options for how to deal with grid limitations. The short-term solutions are redispatching and/or counter-trading measures by the grid operator. This can quickly lead to massive costs, which then have to be allocated to the final customers (in Germany, redispatch costs increased 10-fold from 2011 to 2015[68] and currently exceed €1 billion per year). Therefore, in most countries, the long-term solution of grid expansion is intended in all cases where persistent shortages are to be expected. On the downside, this unconditional grid expansion can lead to high costs, especially when the full capacity is required for only a few hours per year. This was observed in particular for PV production, where the actual production reaches the installed capacity only occasionally. For these reasons, some countries introduced PV production peak-shaving mechanisms. In Germany, for example, small-scale PV systems must have either a ripple control receiver that allows the grid operator to limit the production or a pre-set production limit of 70% of the installed PV capacity (Art. 9 German Renewable Energy Sources Act 2017). This allows one to reduce the dimensioning of the grid and to keep the opportunity costs reasonably low (since most systems in Germany could produce with more than the 70% peak capacity constraint only during a few hours per year anyway).

On the consumption side, the forecasts regarding the impact of a broad diffusion of battery electric vehicles on grid expansion differ significantly. However, it requires little effort to imagine that a high penetration level of electric vehicles, each with a charging power of 20 kW as of today (compared with an average household consumption of roughly 0.5 kW), could quickly lead to distinct load peaks if a high proportion of the vehicle fleet is charged simultaneously.

However, this roll-out of renewable production, storage and consumption assets does not necessarily have to go along with either opportunity costs (when restricted congestion limits market activity) or high grid expansion costs. On the contrary, the diffusion of these technologies offers at least two types of solutions.

In solution 1, the grid operators could profit from falling costs for assets such as battery storage by purchasing and operating, for example, large-scale batteries in the grid. These batteries could theoretically be controlled by the grid operator, who would apply them both to even out the load and to provide ancillary services. In practice, sole operation for these purposes is not nearly profitable in most circumstances and other additional business models, such as using the storage for arbitrage, violate the strict unbundling restrictions in most countries (defined *inter alia* in the EU's 'Third Energy Package' of 2009[69]).

Solution 2 focuses more on utilizing the flexibility of distributed assets, typically located behind the meter of the customer. In particular, storage and consumption assets might potentially play an active part in safeguarding grid stability and actively avoiding peak loads by peak shaving. This seems possible for several reasons. Lithium batteries can adjust their power consumption or provision rapidly within (fractions of) a second, which potentially enables them at least to provide frequency reserve power or could in the future even support the frequency inertia. Moreover, the converters, which transform the dc power from the battery to ac power in the grid, could also provide reactive power precisely adjusted to the local needs of each individual connection node in the grid.[70] Also, a significant contribution to (both production and consumption) peak-load shaving could be realized with comparatively low opportunity costs. This is because for a home storage battery it makes only a minor difference if the battery is charged in the morning hours of a sunny day or during the midday hours. For the grid, such an operational change could mean that the battery is not just full when the highest production peak occurs in the midday hours but could instead relieve the grid when it is needed most. Similarly, on the consumer side, peaks may arise due to simultaneous car charging; yet it might be that many users of electric vehicles do not need an immediate full charging of their cars just when returning home from work. Instead, a large part of the charging process could be deferred into the night hours. However, again, this active involvement in electricity consumption would probably violate the rule of a grid operator in an unbundled market, which is probably part of the reason why this potential value of decentralized flexible assets is hardly used today: no business model has yet been able to allow the owners of the assets to participate in the value added, although some first attempts are currently being made, as described in Section 3.5. A further discussion on the potential of storage technologies to support the grid and provide ancillary services can be found in the literature.[71]

3.4   Economic Value from Arbitrage

The sustainable energy transition is also fundamentally changing the world for energy utilities. In the past, customers often had only a very limited choice in the selection of their electricity supplier. On the customer side, the companies therefore were able to rely on their supply monopoly and focused on maximizing the efficiency of their electricity production. In times of conventional thermal power plants, this often went along with an increasing scale of production. Only large players were able to finance large-scale, and thus efficient, coal-fired plants or stem the task of financing, constructing and operating nuclear power plants. However, the transition towards small-scale decentralized assets literally brings 'power to the people'. This is because for many of these assets both the electric and the economic efficiency scale less with the project size, so that even a private household can actively participate in electricity supply and storage in a profitable manner as a prosumer (see also Section 2.6). This development was accompanied by the liberalization of the energy market that allows customers to switch their supplier much more easily. These changes rocked the foundation of the incumbent utilities and resulted in significant falls in the stock market values of such enterprises (at least in Europe; see Figure 7).

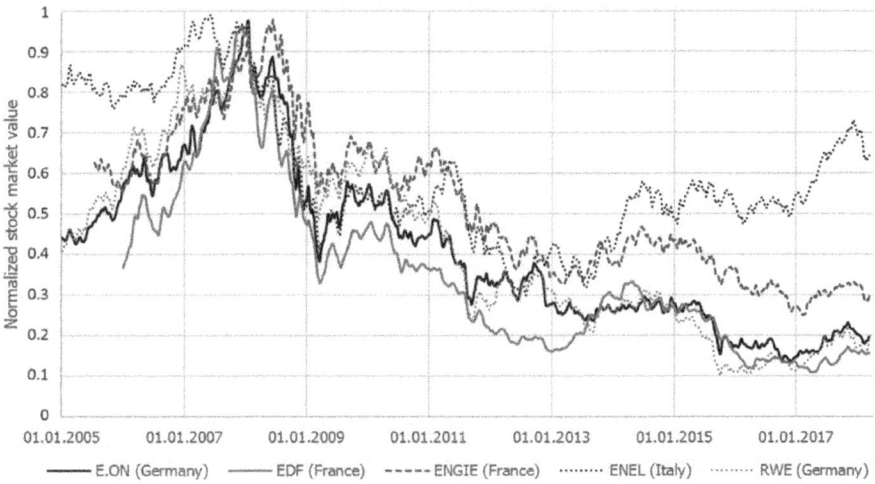

Figure 7 Development of the stock market valuation of the top five (according to power sales) utilities in Europe, normalized to the highest value since 2005.

Figure 7 shows that all five of the utilities considered reached their maximum market valuation between June 2007 (ENEL) and June 2008 (ENGIE). The graph also shows that some companies managed to recover better from the financial crash than others, even though none was able to reach their old all-time highs again. Hence, currently, utilities are rethinking their business model and switching their focus *inter alia* to renewables, customer proximity and competitive prices. One old issue regarding the prices from the energy supply perspective is that household customers in most countries pay a fixed price based on the energy rate, as introduced in Section 3.2. The supplier, on the other hand, has to purchase the electricity sold to the customers for volatile prices at the electricity exchange or in bilateral 'over-the-counter' trades. Therefore, the supplier can make a good profit at times when it can buy electricity for a low price (sometimes even a negative price), but has to spend from its own margin during times of short supply and thus high purchase costs. Consequently, it would be interesting for the energy supplier if its customers were to coordinate the use of their flexible assets to mitigate the procurement costs. A home battery storage therefore should potentially not be discharged at first chance in the early afternoon, but should keep its power to the evening hours when higher demand frequently causes higher wholesale market prices and thus procurement costs. Similarly, a shift of the charging of an electric vehicle from high-price hours in the evening to low-price hours at night would also save costs for the supplier. However, for this case, higher costs for the grid operator can be expected, which could arise from an increased simultaneity if, for example, all cars were to be charged at times of low electricity prices (see Section 3.3).

A second idea is to use storage for arbitrage. This could be conducted *via* either a large central storage or an aggregation of smaller devices within the households. The key figure here is the relationship of profit gained by means of arbitrage on the one hand and the costs for storage on the other. The profits from electricity arbitrage are calculated as the revenues from electricity sales minus the costs for electricity purchase. A special form of arbitrage that is discussed particularly in the USA is the use of storage units as so-called 'peaker plants'. Such storage units provide electricity during times of very high need and thus could potentially replace gas turbines, which were typically used in the past for this purpose. Since this scarcity of electricity at peak times should be included in the market price, we will consider this use-case of peak operation implicitly in the subsequent discussion of inter-temporal arbitrage.

To give an impression of the magnitude of the price spreads that can be realized, Figure 8 shows the hourly values in 2016 and 2017 at the Phelix day-ahead electricity stock price index for Germany, Austria and Luxembourg within an upper and a lower price limit (quantile). To provide examples, 50% of all hourly auctions ended with a price between 2.4 and 3.81 €ct per kWh, and 90% of all auctions resulted in prices between 1.02 and 5.49 €ct per kWh. Similarly, but seen from a different perspective, 0.1% of all results were below −7.6 €ct per kWh and 0.1% above 12.15 €ct per kWh. These numbers show that in the very best case every peak could be utilized; around 43 trades could realize a margin of more than $9.02 - (-2.69) = 11.71$ €ct per kWh, whereas only about eight trades could offer more than 19.75 €ct per kWh. This requires that the storage unit can completely charge and discharge within 1 h and that high and low prices occur in a perfectly alternating manner (which is not the case in reality), and without consideration of storage losses. In any case, the resulting margin has to be at least higher than the operational costs of any storage unit. For a large pumped storage hydro power plant with levelized costs of storage of about US$0.18–0.27 per kWh[73] (or about €0.15–0.22 per kWh), this price span is just at the border of economic feasibility.[52] The levelized costs for a large-scale lithium-ion battery storage device, however, are estimated to be between US$0.35 and 0.74[73] (or between €0.29 and 0.6 per kWh), which still seems far from what is needed for a profitable business model.

3.5   Virtual Power Plants (VPPs) with Storage

In the past, pumped storage hydro power plants were more or less the only relevant form of energy storage. These plants were large enough to justify a direct management on site. Upcoming technologies such as distributed battery storage, an adapted operation schedule of flexible renewable production assets and the use of flexible loads such as battery electric vehicles can be realized at a much smaller scale. Still, it is often not possible for single devices to participate in business models, either because they do not meet the minimum requirements to participate in the particular market, or

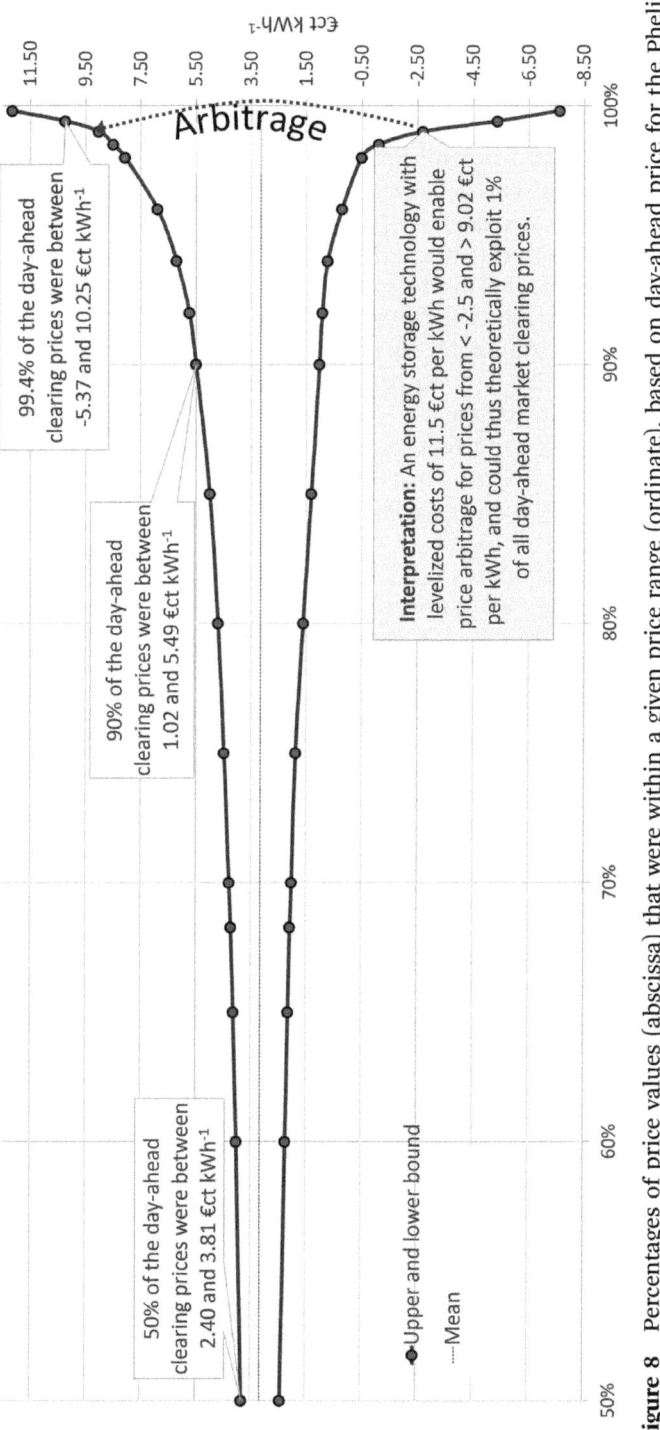

Figure 8 Percentages of price values (abscissa) that were within a given price range (ordinate), based on day-ahead price for the Phelix exchange market.[72]

the admission and/or operation are much too cumbersome to be in any relation to the expected profits. A promising solution to this is the virtual aggregation of large numbers of such (homogeneous or heterogeneous) assets into one pool, referred to as virtual power plants (VPPs) or virtual storage swarms (VSSs). This 'swarm' is then operated from a central location by means of information and communication technology (ICT). In this way, the assets can achieve the requirements for size in cooperation that is not possible for each device alone. Also, potential up-front costs (*e.g.* for prequalification measures) and the costs for operation and maintenance become reasonable when divided across the entire swarm. This concept is already widely applied in the marketing of renewable generation technologies such as wind power, PV or biomass, where it is also possible to achieve economies of scale by means of joint operation. For household-scale distributed storage technologies, however, this concept is rather new.[74] One of the first examples that this is possible was given by the German aggregator Caterva, which was prequalified for the provision of primary reserve (also referred to as frequency containment reserve).[75,76]

Primary reserve services are a first example for which the concept of aggregated assets shows advantages. Within the guidelines of the European Commission,[77] the framework of the European Network of Transmission System Operators for Electricity (ENTSO-E) and, *e.g.*, the local regulations of the German transmission system operators (TSOs), potential suppliers have to offer at least 1 MW of power to be able to participate in the provision of primary reserve.[12] Furthermore, the prequalification process itself is rather complex, so that both hurdles could only be overcome by very large battery storage projects on the megawatt scale that specialized in this single business case in the past. However, the aggregation of a huge number of small-scale batteries, which were originally used for maximization of self-consumption in households, were also able to meet the requirements as a swarm and were granted access to the market for primary reserve. Similar models could be imagined for pools of electric vehicles, where the operators utilize their storage potential and/or their flexibility in charging.

Another related topic is the aggregation of households' generation and storage assets in one large pool to maximize self-consumption. The idea is to integrate a large number of households virtually in one large swarm with one joint production and one joint storage capacity.[9,10] In this way, excess electricity from one region (*e.g.* with sunny weather at that time) can be stored in regions where the yield was low (*e.g.* due to unfavorable weather conditions) or the consumption was relatively high. The system also ensures that the potential of customers with an oversized storage is not wasted but used to store the excess electricity for other customers with insufficient storage capacity. This concept became especially popular in Germany due to legislative exceptions. A closely related yet different form of cloud storage is the concept of offering the capacity of large centralized storage systems to households, similar to the idea of data cloud storage.[78] The customers do not have to maintain a private system in this case, and only an accounting system keeps track of who stored or withdrew what amounts of power.

The optimization of the assets' operation schedule is a business case closely related to the energy cloud storage concept. Today, the control algorithms of decentralized assets, and especially home energy storage systems, are fairly simple and based on limited information. This is a problem especially for battery storage systems since some operation routines can lead to significant aging of the battery cells, which in turn leads to a mid-term decrease in storage capacity and an earlier need for replacement in the long run. Potent algorithms can include weather and load forecasts in their scheduling and can have a precise aging function, which allows operation on sub-optimal charge levels to be avoided. Although this sort of algorithm is typically too sophisticated to run at the individual location, a central aggregator can optimize the operation strategy efficiently in a central location for all its assets.

Finally, a virtual aggregator might solve some of the pressing issues for the grid operators and energy suppliers. While regulation might deny grid operators the operation of storage assets themselves, as discussed earlier, a contract with an external virtual aggregator, *e.g.* to operate its assets in a grid-friendly manner at peak times, could be a potential solution with some mutual benefit. Energy suppliers, in contrast, could either cooperate with aggregators to, for example, reduce procurement costs or extend their own role by entering the business of aggregating flexible distributed energy resources.

4 Regulatory Considerations

Important barriers for the deployment of storage technologies come from the current market design and regulation. Traditionally, energy storage often has not been treated as a category on its own in the regulatory framework, but considered either as generation or demand instead. Still, in some regulations, storage has benefited from privileged treatment, *e.g.* in the form of grid-use fee being waived, electricity tax credits or exemptions from levies (such as the one that is part of the Renewable Energy Sources Act in Germany).

Ideally, the real value of storage to the energy system ought to be reflected in the regulation, and regulation must both take a technology-neutral stance and take other flexible mechanisms (grid expansion, sector coupling, demand response, *etc.*) into account.

From an economics (welfare-maximizing) perspective, it makes sense to establish a technology-neutral regulatory framework that is geared towards economic efficiency enabled by well-functioning markets and thus the optimal allocation of scarce resources. At the same time, undesirable welfare distribution issues need to be taken care of by policy makers according to social preferences.

The system operator, in principle, is best suited to orchestrate the storage services to balance the system. In light of the trend towards more distributed generation (*e.g.* prosumer households) and distributed storage, the extent to which energy storage units should be directly controlled by the system operator is a critical question.[2] However, according to the EU regulation, TSOs are currently not allowed to own or operate any form of energy storage owing to unbundling regulations (*cf.* EC Directive 2009/72/EC). In contrast, distribution system operators (DSOs) are able to do so for certain applications, provided that the operational independence between their grid-related and other businesses is maintained (which is safeguarded by legal, accounting and functional unbundling requirements, if they serve less than 100 000 connected consumers[2]).

Market participation requirements, *e.g.* in terms of minimum capacity and minimum duration to be offered, impose limitations on what energy storage unit operators are allowed to offer. Market regulation needs to be developed in such a way that it enables energy storage system operators to participate in the market in a non-discriminatory manner, wherever these can provide benefits to the electricity system.[24]

5   Conclusion

In this chapter, we have shed some light on business opportunities and the role of the regulatory framework in the context of energy storage. To this end, we first provided an overview of the technoeconomic characteristics of the main energy storage options commercially available today and highlighted, on the one hand, the technological complementarity and competition among these technologies and, on the other, the key role of energy storage in the zero-carbon transition of energy systems worldwide. We pointed out further that the value-creating potential of energy storage units lies in the storage of energy and the provision of ancillary services to the system. Decision makers are well advised to study the market conditions carefully from a dynamic perspective in order to make sure that their business model does not erode over time to a point where the desired return on investment can no longer be achieved. Given the potentially disruptive nature of some (especially distributed) energy storage technologies, such as those installed in e-cars and homes, investors in large-scale storage solutions must be especially cautious. After all, energy storage is just one amongst a number of flexibility options to balance supply and demand, and not necessarily the lowest cost option.

References

1. IRENA, Electricity Storage and Renewables: Costs and Markets to 2030, *International Renewable Energy Agency*, 2017, Abu Dhabi, UAE.
2. G. C. Gissey, J. Radcliffe and P. E. Dodds, Regulatory barriers to energy storage deployment: the UK perspective, 2017, URL: http://www.restless.org.uk/documents/briefing-paper-1.pdf, p. 5.

3. J. Cho and A. N. Kleit, Energy storage systems in energy and ancillary markets: A backwards induction approach, *Appl. Energy*, 2015, **147**, 176.
4. H. C. Hesse, M. Schimpe, D. Kucevic and A. Jossen, Lithium-Ion Battery Storage for the Grid – A Review of Stationary Battery Storage System Design Tailored for Applications in Modern Power Grids, *Energies*, 2017, **10**, 42.
5. K. Risthaus and R. Madlener, Economic Analysis of Electricity Storage Based on Heat Pumps and Thermal Storage Units in Large-Scale Thermal Power Plants, *FCN Working Paper*, 2017, 3/2017; condensed version published in *Energy Procedia*, 2017, **142**, 2816.
6. P. D. Lund, J. Lindgren, J. Mikkola and J. Salpakari, Review of energy system flexibility measures to enable high levels of variable renewable electricity, *Renewable Sustainable Energy Rev.*, 2015, **45**, 785.
7. Ø. Arvesen, V. Medbø, S. E. Fleten, A. Tomasgard and S. Westgaard, Linepack storage valuation under price uncertainty, *Energy*, 2013, **52**, 155.
8. K. Bradbury, L. Pratson and D. Patiño-Echeverri, Economic viability of energy storage systems based on price arbitrage potential in real-time U.S. electricity markets, *Appl. Energy*, 2014, **114**, 512.
9. H. Broering and R. Madlener, Beyond Grid Parity: Conceptualization of the Merit of Cloud Energy Storage for Prosumers, FCN Working Paper, 2016, **13/2016**, (condensed version published as: Simulation and Evaluation of the Economic Merit of Cloud Energy Storage for Prosumers: The Case of Germany, *Energy Procedia*, **105**, 3507).
10. R. Madlener and H. Broering, Direct Marketing and Cloud Energy Storage in Liberalized Markets: The German Regulatory Framework as a Benchmark, *FCN Working Paper*, 2016, **14/2016**.
11. J. Speirs, P. Balcombe, E. Johnson, J. Martin, N. Brandon and A. Hawkes, A greener gas grid: What are the options, *Energy Policy*, 2018, **118**, 291–297.
12. Consentec GmbH, Description of load-frequency control concept and market for control reserves. Study commissioned by theGerman TSOs, Edited by 50Hertz Transmission GmbH, Consentec GmbH, 2014, https://www.regelleistung.net/ext/download/marktbeschreibungEn, retrieved October 26, 2017.
13. A. Daunoriene, A. Draksaite, V. Snieska and G. Valodkiene, Evaluating Sustainability of Sharing Economy Business Models, *Soc. Behav. Sci.*, 2015, **213**, 836.
14. C. Budny, R. Madlener and C. Hilgers, Economic Feasibility of Pipeline and Underground Reservoir Storage Options for Power-to-Gas Load Balancing, *Energy Convers. Manage.*, 2015, **102**, 258.
15. X. Luo, J. Wang, M. Dooner and J. Clarke, Overview of current development in electrical energy storage technologies and the application potential in power system operation, *Appl. Energy*, 2015, **137**, 511.
16. P. Komarnicki, P. Lombardi and Z. Styczynski, *Electric Energy Storage Systems*, Springer-Verlag, Berlin/Heidelberg/New York, 2017.

17. M. Sterner and I. Stadler, *Handbook of Energy Storage*, Springer-Verlag, Berlin/Heidelberg/New York, 2018.
18. M. Aneke and M. Wang, Energy storage technologies and real life applications – A state of the art review, *Appl. Energy*, 2016, **179**, 350.
19. G. Strbac and M. Black, Value of Bulk Energy Storage for Managing Wind Power Fluctuations, *IEEE Trans. Energy Convers.*, 2007, **22**(1), 197.
20. C. K. Ekman and S. H. Jensen, Prospects for large scale electricity storage in Denmark, *Energy Convers. Manage.*, 2010, **51**, 1140.
21. P. Denholm, J. Jorgenson, M. Hummon, T. Jenkin, D. Palchak, *et al.*, The Value of Energy Storage for Grid Applications, NREL Technical Report No. NREL/TP-6A20-58465, 2013.
22. A. Shcherbakova, A. Kleit and J. Cho, The value of energy storage in South Korea's electricity market: A Hotelling approach, *Appl. Energy*, 2014, **125**, 93.
23. G. Fitzgerald, J. Mandel, J. Morris and H. Touati, The Economics of Battery Energy Storage: How multi-use, customer-sited batteries deliver the most services and value to customers and the grid, *Rocky Mountain Institute*, 2015.
24. ACORE, Beyond Renewable Integration: The Energy Storage Value Proposition, *American Council On Renewable Energy*, 2016.
25. M. S. Guney and Y. Tepe, Classification and assessment of energy storage systems, *Renewable Sustainable Energy Rev.*, 2017, **75**, 1187.
26. S. Ould Amrouche, D. Rekioua, T. Rekioua and S. Bacha, Overview of energy storage in renewable energy systems, *Int. J. Hydrogen Energy*, 2016, **41**(45), 20914.
27. IEA, *Technology Roadmap: Energy storage*, OECD/IEA, Paris, 2014.
28. G. Strbac, M. Aunedi, D. Pudjianto, P. Djapic, F. Teng, A. Sturt, D. Jackravut, R. Sansom, V. Yufit and N. Brandon, *Strategic Assessment of the Role and Value of Energy Storage Systems in the UK Low Carbon Energy Future*, Report for Carbon Trust, 2012.
29. H. Krings and R. Madlener, Modeling the Economic Viability of Grid Expansion, Energy Storage, and Demand Side Management Using Real Options and Welfare Analysis, *FCN Working Paper*, 2015, 7/**2015**.
30. S. Franzen and R. Madlener, Optimal Expansion of a Hydrogen Storage System for Wind Power: A Real Options Analysis, *FCN Working Paper*, 5/**2016**.
31. X. De Graaf and R. Madlener, Optimal Time-Dependent Usage of Salt Cavern Storage Facilities for Alternative Media in Light of Intermittent Electricity Production and Carbon Sequestration, *FCN Working Paper*, 2016, 3/**2016**.
32. V. Galassi and R. Madlener, On the Prosumers' Side of the Electricity Markets: Preferences and Opportunities for Photovoltaic Systems with Storage, *FCN Working Paper*, 2014, 19/**2014**.
33. D. Kroniger and R. Madlener, Hydrogen Storage for Wind Parks: A Real Options Evaluation for an Optimal Investment in More Flexibility, *Appl. Energy*, 2014, **136**, 931.

34. G. Locatelli, E. Palerma and M. Mancini, Assessing the economics of large Energy Storage Plants with an optimisation methodology, *Energy*, 2015, **83**, 15.

35. L. Löbberding and R. Madlener, System Cost Uncertainty of Micro Fuel Cell Cogeneration and Storage, submitted to *Applied Energy*, 2017; pre-published as *FCN Working Paper*, 23/**2016**. (condensed version published in *Energy Procedia*, 2017, **142**, 2824.).

36. K. Kairies, D. Haberschusz, J. Van Ouwerkerk, J. Strebel, O. Wessels, D. Magnor, J. Badeda and D. U. Sauer, *Wissenschaftliches Mess- und Evaluierungsprogramm Solarstromspeicher, Speichermonitoring Jahresbericht 2016*, Institut für Stromrichtertechnik und Elektrische Antriebe der RWTH Aachen, 2016.

37. O. Schmidt, A. Hawkes, A. Gambhir and I. Staffell, The future cost of electrical energy storage based on experience rates, *Nat. Energy*, 2017, **2**(8), 17110.

38. WEC (World Energy Council) (2016). World Energy Resources, E-Storage: shifting from cost to value, www.energystorageexchange.org.

39. C. Pellinger, *Merit Order of Energy Storage in Germany by 2030*. Prague, 2016, Forschungsstelle für Energiewirtschaft e.V. (PowerPoint presentation held at the Distributed Energy Resources and Storage – Workshop at The Czech Academy of Sciences, November 29, 2016; https://www.ffe.de/publikationen/vortraege/784-merit-order-of-energy-storage-in-germany-by-2030).

40. G. Berckmans, M. Messagie, J. Smekens, N. Omar and L. Vanhaverbeke, and J. Van Mierlo, Cost Projection of State of the Art Lithium-Ion Batteries for Electric Vehicles Up to 2030, *Energies*, 2017, **10**, 1314.

41. R. Madlener and J. Latz, Economics of Centralized and Decentralized Compressed Air Energy Storage for Enhanced Grid Integration of Wind Power, *Appl. Energy*, 2013, **101**, 299.

42. D. McConnell, T. Forcey and M. Sandiford, Estimating the value of electricity storage in an energy-only wholesale market, *Appl. Energy*, 2015, **159**, 422–432.

43. K. Bradbury, L. Pratson and D. Patiño-Echeverri, Economic viability of energy storage systems based on price arbitrage potential in real-time U.S. electricity markets, *Appl. Energy*, 2014, **114**, 512.

44. S. Weibel and R. Madlener, Cost-Effective Design of Ringwall Storage Hybrid Power Plants: A Real Options Analysis, *Energy Convers. Manage.*, 2015, **103**, 871.

45. P. Keil, S. F. Schuster, J. Wilhelm, J. Travi, A. Hauser, R. C. Karl and A. Jossen, *Calendar Aging of Lithium-Ion Batteries, I. Impact of the Graphite Anode on Capacity Fade, J. Electrochem. Soc.*, 163(9), A1872.

46. J. Vetter, P. Novák, M. R. Wagner, C. Veit, K.-C. Möller, J. O. Besenhard, M. Winter, M. Wohlfahrt-Mehrens, C. Vogler and A. Hammouche, Ageing mechanisms in lithium-ion batteries, *J. Power Sources*, 2005, **147**, 269.

47. A. Kirmas and R. Madlener, Economic Viability of Second-Life Electric Vehicle Batteries for Energy Storage in Private Households, *FCN Working Paper*, 2016, 7/**2016**; a condensed version has been published in *Energy Procedia*, 2017, **105**, 3806.

48. A. Toffler, *The Third Wave: The Classic Study of Tomorrow*, Bantam, New York, NY, 1980.

49. M. Kubli, M. Loock and R. Wüstenhagen, The flexible prosumer: Measuring the willingness to co-create distributed flexibility, *Energy Policy*, 2018, **114**, 540.

50. C. Rosen and R. Madlener, Regulatory Options for Local Reserve Energy Markets: Implications for Prosumers, Utilities, and other Stakeholders, *Energy J.*, 2016, **37**(SI2), 39.

51. K. Popper and A. Hove, *Energy Storage World markets Report*, Edited by Energy StorageWorld Forum, 2017, https://energystorageforum.com/Energy_Storage_World_Markets_Report_2014-2020.compressed.pdf, retrieved March 29, 2018.

52. D. Zafirakis, K. J. Chalvatzis, G. Baiocchi and G. Daskalakis, The value of arbitrage for energy storage: Evidence from European electricity markets, *Appl. Energy*, 2016, **184**, 971.

53. I. Bakke and B. Norheim, Investment in Electric Storage Under Uncertainty: A Real Options Approach, Master's thesis, NTNU Trondheim, June 2015.

54. D. Metz and J. Tomé Saraiva, Use of battery storage systems for price arbitrage operations in the 15- and 60-min German intraday markets, *Electric Power Syst. Res.*, 2018, **160**, 27.

55. Eurostat (Ed.), Energy statistics. natural gas and electricity prices, 2017. http://ec.europa.eu/eurostat/web/energy/data/database, updated on 11/22/2017, retrieved on March 16, 2018.

56. S. Küfeoğlu, Doctoral Dissertation, *Economic impacts of electric power outages and evaluation of customer interruption costs*, 2015, Aalto University, School of Electrical Engineering, Helsinki.

57. D. A. Hensher, N. Shore and K. Train, Willingness to pay for residential electricity supply quality and reliability, *Appl. Energy*, 2014, **115**, 280.

58. J. McLaren, N. Laws, K. Anderson and S. Mullendore, *Identifying Potential Markets for Behind-the-Meter Battery Energy Storage: A Survey of U.S. Demand Charges*, Edited by NREL. National Renewable Energy Laboratory, 2017, https://www.nrel.gov/docs/fy17osti/68963.pdf, retrieved March 23, 2018.

59. K. Grave, B. Breitschopf, J. Ordonez and J. Wachsmuth, *et al.*, Prices and Costs of EU Energy - Final Report, ECOFYS/Fraunhofer ISI, 2016, 39.

60. R. Ferrera, T. Marvin, D. Larson, T. Lindsay and D. Falk, *Battery Energy Storage in Florida. Value, Challenges, andOpportunities*, Edited by University of California San Diego, School of Global Policy and Strategy, 2017.

61. *The Value of Energy Storage for Grid Applications*, ed. NREL, National Renewable Energy Laboratory, Boulder, CO, 2013.

62. P. Denholm, V. Diakov and R. Margolis, *The Relative Economic Merits of Storage and Combustion Turbines for Meeting Peak Capacity Requirements under Increased Penetration of Solar Photovoltaics*, National Renewable Energy Laboratory, Boulder, CO, 2015.

63. dena (Ed.), dena Ancillary Services Study 2030. Security and reliability of a power supply with a high percentage of renewable energy -Final report, Deutsche Energie-Agentur GmbH, Berlin, 2014.

64. ENSTO-E (Ed.), *Need for synthetic inertia (SI) for frequency regulation*, ENTSO-E guidance document for national implementation for network codes on grid connection. Brussels, 2017.

65. S. Nykamp, A. Molderink, V. Bakker, H. A. Toersche, J. L. Hurink and G. J. Smit, *Integration of heat pumps in distribution grids. Economic motivation for grid control*, In: Proceedings of the IEEE Power and Energy Society (PES) Innovative Smart Grid Technologies (ISGT) Europe Conference. Berlin, 2012.

66. J. Wood and S. Funk, Master's thesis, Can demand response help reduce future distribution grid investments? An economic study of peak shaving in the Norwegian distribution grid: SEMIAH pilot in Engene, Sørlandet (Southern Norway), Norwegian School of Economics, 2017.

67. L. Mautino, The EU electricity target model: the devil is in the details?, *Oxera*, Agenda, January 2013.

68. BNetzA (Ed.), *Monitoring Report 2016*. Bundesnetzagentur and Bundes-kartellamt, Berlin, 2017, https://www.bundesnetzagentur.de/DE/Sachgebiete/ElektrizitaetundGas/Unternehmen_Institutionen/DatenaustauschundMonitoring/Monitoring/Monitoringberichte/Monitoring_Berichte.html?nn=266276, retrieved on July 4, 2017.

69. EC, Evaluation Report covering the Evaluation of the EU's regulatory framework for electricity market design and consumer protection in the fields of electricity and gas, SWD, 412 final. European Commission, Brussels, 2016.

70. K. C. Divya and J. Østergaard, Battery energy storage technology for power systems—An overview, *Electric Power Syst. Res.*, 2009, **79**(4), 511–520.

71. ISEA, Technology Overview on Electricity Storage. Overview on the potential and on the deployment perspectives of electricity storage technologies, Smart Energy for Europe Platform GmbH, Institute for Power Electronics and Electrical Drives (ISEA) at RWTH Aachen University, 2012.

72. BNetzA (Ed.), Bundesnetzagentur, www.smard.de, updated March 23, 2018, retrieved March 27, 2018.

73. Lazard (Ed.), *Lazard's Levelized Cost of Storage – Version 2.0*, 2016, https://www.lazard.com/media/2391/lazards-levelized-cost-of-storage-analysis-10.pdf, retrieved March 21, 2018.

74. D. Westendorf and R. Madlener, Bundling of Distributed Battery Storage Units as a Virtual Storage Swarm, *FCN Working Paper*, 2017, **9/2017**.

75. Caterva GmbH and N-ERGIE AG, *For the first time: Frequency Containment Reserve by privately used swarm of Energy Storage Systems*, Nürnberg, Pullach im Isartal, 2015, https://www.caterva.de/pdf/Press_release_Caterva_N-ERGIE_Swarm-Prequalification.pdf, retrieved on February 1, 2017.

76. D. Steber, P. Bazan and R. German, SWARM - Increasing Households' Internal PV Consumption and Offering Primary Control Power with Distributed Batteries, in *Energy Informatics*, ed. S. Gottwalt, L. König and H. Schmeck, 2015.

77. EC, Commission Regulation (EU) 2017/1485 of 2 August 2017 establishing a guideline on electricity transmission system operation. In Official Journal of the European Union L 220, 2017, 90.

78. J. Liu, N. Zhang, C. Kang, D. Kirschen and Q. Xia, Cloud energy storage for residential and small commercial consumers: A business case study, *Appl. Energy*, 2017, **188**, 226.

Subject Index